Membrane Transport

PEOPLE AND IDEAS

.

Membrane Transport

PEOPLE AND IDEAS

Edited by Daniel C. Tosteson

American Physiological Society

BETHESDA, MARYLAND

© Copyright 1989, American Physiological Society
Library of Congress Catalog Card Number: 88-7502
International Standard Book Number: 0-19-520773-4
Printed in the United States of America by
    Waverly Press, Inc., Baltimore, Maryland 21202
Distributed for the American Physiological Society by
    Oxford University Press, New York, New York 10016

Table of Contents

———

I   *Introduction: Membrane Transport*
    in Physiology   1
    DANIEL C. TOSTESON

II *Biological Membranes as Selective Barriers*
    to Diffusion of Molecules   15
    HUGH DAVSON

III *Membranes, Molecules, Nerves, and People*   51
    J. DAVID ROBERTSON

IV *Transport Pathways: Water Movement*
    Across Cell Membranes   125
    ARTHUR K. SOLOMON

V   *Sodium-Potassium Pump*   155
    JENS C. SKOU

VI *From Frog Lung to Calcium Pump*   187
    WILHELM HASSELBACH

VII *Anion Exchanges and Band 3 Protein*   203
    ASER ROTHSTEIN

VIII *The Unfinished Story of Secondary*
     Active Transport   237
     ERICH HEINZ

IX *Electrodiffusion in Membranes*   251
    DAVID E. GOLDMAN

X *Reflections on Selectivity* 261
  CLAY M. ARMSTRONG

XI *Propagation of Electrical Impulses* 275
  LORIN J. MULLINS

XII *Membrane Transport in Excitation-*
  Contraction Coupling 291
  RICHARD J. PODOLSKY

XIII *From Cell Theory to Cell Connectivity:*
  Experiments in Cell-to-Cell Communication 303
  WERNER R. LOEWENSTEIN

XIV *Epithelial Transport: Frog Skin as a*
  Model System 337
  HANS H. USSING

XV *Flow and Diffusion Through*
  Biological Membranes 363
  JOHN R. PAPPENHEIMER

 Contributors 391

 Index 395

I

Introduction: Membrane Transport in Physiology

DANIEL C. TOSTESON

It is remarkable that the very existence of cell membranes was problematic when I and most of the other authors of this volume began our work. Permeability measurements had, of course, given rise to the hypothesis that membranes envelop cells (see chapter by Hugh Davson). However, it was not until the emergence of the electron microscope, and particularly, adequate techniques for the isolation, purification, and chemical characterization of membranes, that these structures were recognized as essential components of cells and organelles (see chapter by J. David Robertson). Moreover, the development of isotopic tracer and electrical methods during the 1940s and 1950s made possible a more quantitative description of the transport of substances across biological membranes (see chapters by Hans H. Ussing and Arthur K. Solomon). There ensued a period of intense activity in a broad range of subjects bearing on membrane transport in physiology. Rather than trace in detail the histories of development of these lines of research, many of which are addressed in subsequent chapters, I describe briefly my current picture of membrane transport in physiology drawn from the results of studies performed by many investigators over several decades.

A View of Biology From a Membrane Transport Perspective

I conjure a picture of a living organism as an extraordinarily complex, self-replicating, self-assembling, dynamic, open, physico-chemical system maintained in or near the steady state by the continual entry and exit of matter and energy. The system consists largely of small aqueous spaces ranging in thickness from 10^{-5} to

[1]

10^{-2} cm (intercellular and intracellular spaces in organelles such as nuclei, mitochondria, the endoplasmic reticulum, and various types of vesicles). The boundaries of these spaces are formed by thin membranes consisting of bimolecular sheets of phospholipids, cholesterol, and other amphipathic molecules perforated and decorated by proteins that are, in many cases, glycosylated. The thickness of the lipid bilayer is $\sim 4 \times 10^{-7}$ cm, and the total thickness of the membrane including the protein components is $\sim 10^{-6}$ cm.

Most of the chemical reactions that occur in this microhoneycomb or alveolar system take place in the small aqueous chambers or on the surfaces of their surrounding membranes. These reactions are catalyzed by enzymes that are either dissolved in the small aqueous chambers or adsorbed on or imbedded in membranes. In the latter case (e.g., enzymes involved in oxidative phosphorylation and photosynthesis), some reactions seem to require the relatively hydrophobic environment of the membrane interior. The transport of substrates and products between chambers is modulated by membranes. Uncharged, nonpolar molecules, like the respiratory gases or the undissociated forms of weak acids and bases, dissolve readily in the hydrocarbon chains in the interior of the bilayers and therefore diffuse rapidly between the aqueous chambers. The small size of the watery compartments assures that the dissolved solutes in these microphases are rapidly mixed by diffusion. Even for a relatively polar molecule like water itself, the solubility and mobility in the hydrocarbon chains of the bilayers yield a permeability coefficient $\sim 10^{-3}$ cm·s^{-1}, high enough to assure that water is almost always perfectly mixed by diffusion both between and within the aqueous phases.

By contrast, the permeability of bilayers to polar, and particularly, charged solutes is low. Movement of these substances between aqueous chambers takes place through proteins that span the bilayer. Many copies of each transport protein, many different kinds of transport proteins for each substrate (inferred from substrate and inhibitors specificity and transport mechanism), and many different proteins for different substrates offer polar molecules diverse parallel opportunities for movement across membranes between adjacent aqueous chambers. The rate of transport of a given polar substance (e.g., glucose) through a particular kind of membrane protein (e.g., the glucose transporter) is determined by the number of copies of the protein present in each unit area of membrane and the turnover number for transport through each copy. The direction and rate of movement of a particular polar substituent into one of the aqueous chambers are the vector sum of the rates of transport of that substance through all the transport proteins penetrating the bilayers surrounding that space. In the steady state, the sum of the net rate of transport and the net rate of production (or consumption) of the substance in the aqueous chamber equals zero.

[2]

The direction and rate of transport of a substance through a single particular transport protein depend on the rates of loading and unloading of a site on the transport protein at one membrane surface, and on the rate at which such a loaded site reaches and directly or indirectly unloads at the opposite membrane surface into the adjacent aqueous chamber. These rates depend, in turn, on the characteristics of the transport protein and the chemical reactions in which it is involved. In this sense, transport proteins differ from enzymes dissolved in the cytosol and free to rotate randomly in that the entrance of "substrate" to the active site is from one side while the exit of the "product" is to the other side of the bilayer.

In general, it seems useful to distinguish two different classes of transport proteins: those that form channels and those that form pumps. Channels can be open simultaneously to both sides of the membrane, and polar substances (including charged ions) move through them by some form of electrodiffusion depending on the detailed atomic anatomy of channel walls. These are dynamic structures that are frequently changing slightly from one shape to another. The transport properties of a particular channel are the result of an average of the several conformational states over time.

Transport through channels is always energetically downhill. By contrast, pumps are never open to both sides of the membrane simultaneously. They can move a solute energetically uphill either by coupling to the simultaneous downhill movement of a transport partner [secondary active transport by cotransport (symport) or countertransport (antiport)] or by coupling to a scalar chemical reaction like the hydrolysis of ATP (primary active transport). Most of the electrical currents between aqueous chambers are carried by ions moving through channels, but some are carried by ions moving through ATP and light-driven pumps.

The distribution of a particular transport protein in membranes is heterogeneous. Membranes surrounding different aqueous chambers contain different numbers of copies of the protein. These differences determine the relative permeability of membranes surrounding different aqueous spaces to the substrates transported by that protein. In some cases, a particular transport protein is not uniformly distributed in the bilayers surrounding a watery chamber. This arrangement is a necessary condition for directional movement of a solute through a chamber: for example, movement from one to the other side of a spatially oriented epithelial cell.

Like all proteins, the synthesis of channels and pumps is directed by the corresponding genes. Differences in the numbers of different types of transport proteins in different cells reflect differences in the expression of the various genes. Moreover, localization of transport proteins to the several kinds of bilayers within a particular cell (e.g., plasma membrane, mitochondria, nuclei) is directed by signal sequences that are genetically determined. Genes thus play a crucial

[3]

role in the regulation of membrane transport. Since such processes are involved in all responses of cells to changes in their environment, membrane transport plays an essential part in regulating gene expression. Genes and environment interact through molecules that move through membranes.

According to this picture, transport across membranes plays a central role in regulating the many chemical reactions that comprise a living system. Operating in concert with synthetic and degradative reactions, transport processes determine the concentrations of substrates and products in the numerous small aqueous reaction chambers. They permit such spaces to be close to one another [separated only by the thickness of the membrane (10^{-6} cm)] and yet have strikingly different concentrations of polar substituents. The proximity and small size of the reaction chambers allow for the rapid relaxation (or increase) of such concentration differences when transport pathways through the membranes become available.

FIVE WAYS OF THINKING ABOUT MEMBRANE TRANSPORT IN PHYSIOLOGY

Within this physicochemical vision of biology, there are several ways of thinking about transport through membranes. These ways include: *1*) the physiological role of the transport process; *2*) the anatomical location of the membrane; *3*) the chemical nature of the transported substance; *4*) the physicochemical mechanism of the transport process; and *5*) the primary, secondary, tertiary, and quaternary structure of the protein(s) that subserves transport. I comment briefly on each of these ways, emphasizing those ideas that seem to me most important.

Physiological Roles of Membrane Transport

It is satisfying to me to contemplate transport across membranes in the context of general physiology, that is, the functions common to all living systems. We now recognize that these functions depend on two general shapes of aggregates of macromolecules: membranes and rods. Rods subserve replication and movement. Membranes are essential for all other cellular functions, including secretion, signaling, metabolism, and energy transduction. In all of these cases, function depends not only on the shapes of the individual enzymes, receptors, and other macromolecules (tertiary structure) but also on the shapes that they form when they aggregate: rods or sheets, fibers or membranes (quaternary structure).

By *secretion*, I mean the net movement of solutes and water across membranes. When such movement occurs across a single membrane, it modifies the solute and water content and thus the volume of the space enclosed by the membrane, be it the cytosol of a cell, the inner compartment of a mitochondria, or of various secretory vesicles. The term *secretion* is usually used to describe the process by which net movement of solute and water occurs across two plasma membranes arranged in series and enclosing between them an aqueous space, for example, the apical and basal membranes of epithelial cells. Examples of secretion are seen in almost all epithelia, for example, gastrointestinal, renal, pancreatic, and other exocrine glands.

One important idea about secretion that has grown during my scientific lifetime is the primacy of solute (and particularly Na^+) transport in generating all forms of secretion. The movement of other solutes and water is usually a passive consequence of active Na^+ transport. An equally important notion to emerge from investigations of secretion across layers of epithelial cells such as the frog skin or the toad bladder is clear proof of the existence of "active" transport across biological membranes separating identical solutions. The chapters by Hans H. Ussing and by Erich Heinz present interesting perspectives on this discovery and make clear that such "active" transport can be mediated both by a pump driven by the hydrolysis of ATP (or some other energy-yielding scalar metabolic reaction) and by a coupled system that uses part of the energy dissipated in the downhill movement of one solute to move a second solute uphill.

Secretions often contain large molecules that do not penetrate bilayers. In most cases, such substances move into and out of cells by endo- and exocytosis. These processes involve the formation of vesicles and their subsequent dissolution by membrane fusion. In many instances, substances transported across membranes by this mechanism are dissolved in intravesicular water. In other instances, the transported molecule binds to a specific receptor protein on the bilayer surface.

By *signaling*, I mean the transfer of information within and between cells. Signals are carried electrically and chemically and often by both modes in series. Moreover, since electrical currents in biological systems are almost always carried by inorganic ions, both modes are essentially chemical in nature. In any case, both modes involve transport across membranes. Chemical messengers are accumulated and stored in aqueous spaces bounded by membranes and released either across membranes or by exocytosis. Electrically propagated signals involve the movement of ions through membrane channels that are opened and closed by electrical potential differences and/or by specific chemical compounds.

Some chemical signals, such as neurotransmitters, hormones, and mediators of cell growth, are released to the outside of the cells in which they are produced and act by combining with receptors on other cells. Other signals, such as Ca^{2+}, cAMP, cGMP, diacylglycerol, and inositol phosphates, are released into the cytosol and function by combining with binding sites on enzymes, nucleoproteins, and

[5]

other intracellular regulatory macromolecules. An example is the account in this volume by Richard J. Podolsky of the role of Ca^{2+} as the messenger regulating muscular contraction.

Perhaps the most impressive accomplishment of research on the role of membrane transport in signaling in recent decades has been the establishment of the connection between the movement of specific ions through voltage and chemically gated membrane channels, on the one hand, and electrical phenomena, on the other. In this volume David Goldman, Clay Armstrong, and Lorin Mullins describe some of the seminal discoveries in this field.

A third general function subserved by membrane transport processes is the *regulation* of metabolism. An example is the role of transport of glucose into cells (usually by cotransport with Na^+) in regulating the rate of glucose consumption. Glucose transport is, in turn, influenced by many chemical signals, including insulin. The role of membrane transport in regulating metabolism is evident in all types of cells and in all kinds of organisms, from prokaryotes to humans.

Life depends on the transformation of one form of energy to another, a process sometimes described by the term *energy transduction*. The primary reaction of present life on this planet is the conversion of electromagnetic radiation from the sun into the chemical energy contained in covalent bonds through the process of photosynthesis. The initial reaction in photosynthesis appears to be the conversion of light energy into a proton gradient, which then drives the synthesis of ATP. Both the formation of the proton gradient by a light-driven proton pump and the dissipation of the gradient coupled to ATP synthesis require specific membrane enzymes. An analogous process occurs in oxidative phosphorylation. In this case, the successive oxidation and reduction of cytochrome enzymes in mitochondrial membranes by electrons released from metabolites are converted into a proton gradient across mitochondrial membranes, which, in turn, also provides the direct source of energy for formation of ATP. Thus membrane transport of protons is essential for conserving chemical bond energy in the form of ATP both in photosynthesis, which converts CO_2 and H_2O to glucose and O_2, and in respiration, which carries out the reverse reaction.

Photosynthesis is not, of course, the only example of conversion of light to chemical energy in biology. Vision also requires transport across membranes. Rhodopsin is a membrane-bound enzyme found in light-sensitive cells. The absorption of light by retinal (the chromophore in rhodopsin) produces a shape change in the protein, which, in turn, alters the binding to rhodopsin of transducin (a guanine nucleotide–binding protein). These events set in motion a train of reactions that lead to a reduction in the concentration of cyclic guanine mononucleotide (cGMP). Because cGMP is an agonist for an ion-conducting channel in the retinal rod plasma membrane,

[6]

the reduction in cGMP concentration results in the closure of channels conducting Na^+ and Ca^{2+}, with the consequence that the electrical potential difference between the inside and outside of the retinal rods becomes larger (more inside negative). This change in membrane potential, in turn, alters the release of the neurotransmitter by the retinal rod or cone at its synapses with contiguous neurons. The neurotransmitters bind to receptors on the postsynaptic membranes, resulting in the opening of ion-conducting channels in the neuron. The currents flowing through the open channels at the synapse depolarize the neuron membrane, leading to the opening of voltage-sensitive channels and the initiation of an action potential that propagates along the axon membrane until it reaches the next synapse. Thus the detection of light by the nervous system involves not only the conversion of light energy to chemical bond energy but also a series of interconversions between chemical bond and electrical energy. An analogous series of energy transformations occurs in the detection of mechanical stimuli, including sound, chemical stimuli (taste and smell), and temperature. In all cases, transport across membranes is essential for many steps in the process.

From the point of view of general physiology, transport across membranes is essential for secretion, signaling, metabolism, and energy transduction. Membrane transport can also be conceived in anatomical terms and thus relate to branches of physiology that are defined by the biological structures they aim to understand, for example, cell physiology and organ physiology.

Transport Through Membranes in Different Locations

We now realize that all eukaryote and many prokaryote cells contain many different membranes. These intracellular membranes are arranged in several configurations. Some surround organelles, such as nuclei, and various kinds of cytosolic and endocytotic vesicles, such as synaptosomes. Others not only envelop but also are invaginated to form the lamellae of mitochondria and chloroplasts and the disks of retinal rods. Still others form complex reticular and tubular structures, such as the Golgi apparatus, and the rough and smooth endoplasmic reticulum. In all these cases, transport of molecules both along the surface of and through membranes is important for the economy of the cell. For example, the movement of mRNA from the site of transcription from nuclear DNA to the ribosomes and the endoplasmic reticulum in the cytosol is crucial for protein synthesis. To understand the function of each of the intracellular organelles as it relates to the function of the entire cell, it is necessary to characterize the distinctive transport properties of their several membranes.

Plasma membranes separate cells from their environment. They

[7]

regulate the exchange of matter between cytosol and extracellular fluid. The transport properties of plasma membranes generally differ from those of the various types of intracellular membranes. Both plasma membranes and intracellular membranes have different transport properties in different types of cells. Moreover, particular regions of plasma membranes of a given cell often differ in their transport properties, for example, synapses, cilia, and the apical and basolateral segments of epithelial cells.

Transport occurs not only through but between cells. The junctions between cells offer selective barriers to the movement of solutes and water. Although not strictly transport across membranes, these paracellular pathways are often critical in the operation of epithelial (as in the gastrointestinal tract, renal tubules, etc.) and endothelia (such as the capillary wall). The chapters by Ussing and John R. Pappenheimer address some characteristics of paracellular transport.

Transport also occurs directly between contiguous cells. This permits electrical and chemical coupling of large arrays of cells so that they can function in concert. Special structures called gap junctions, with transport properties different from other regions of the plasma membranes of connecting cells, subserve transport between such directly coupled cells. Werner R. Lowenstein's chapter describes such intercellular communication in some detail.

Within a given organism, different cells have membranes with different transport properties. Neurons, muscle cells, liver cells, and fat cells all express different relative numbers of channels and pumps for various physiologically important solutes. In blood cells and other types of cells that are not bound to their neighbors, transport proteins are presumably distributed relatively uniformly over the membrane area. In cells that adhere to their neighbors to form tissues, the distribution of transport proteins is often inhomogeneous to produce sidedness, as in polarized epithelia. Moreover, the same types of cells in different organisms have membranes with different transport properties. A much studied example is species differences in the cation-transport properties of mammalian erythrocytes. These comparatively simple, nonnucleated cells are surrounded by membranes that differ markedly in different organisms in their capacity to transport Na^+, K^+, and many other solutes.

Thus an anatomical, cytological classification of transport across biological membranes makes considerable sense. It complements a framework based on physiological function. Another way of thinking about kinds of transport is to consider the chemical nature of the transported substance.

Membrane Transport of Different Substances

The most ubiquitous chemical component of all living systems is water. Because membranes are very thin and because the ratio of area to volume of the enclosed intracellular or intraorganelle spaces is large, water equilibrates rapidly across the bilayer regions. Solomon's chapter summarizes work in his and other laboratories on the movement of water across cell membranes. Water-permeable paracellular channels are also sometimes present to assure sufficiently fast transport across such structures as the walls of the capillaries. The chapter by Pappenheimer describes his elegant approach toward a quantitative understanding of fluid movement through such pores.

By the same line of argument, specific transport processes may be discerned for a variety of physiologically important inorganic ions (Na^+, K^+, Ca^{2+}, Mg^{2+}, Cl^-, HPO_4^-, HCO_3^-, etc.), for organic metabolites (sugars, amino acids, and purines, etc.), and even for relatively nonpolar substances such as the respiratory gases O_2, CO_2, and N_2 that appear to traverse membranes mainly by diffusing among the aliphatic chains in bilayers.

Membrane Transport by Different Mechanisms

Another way of viewing transport across biological membranes is by the physicochemical mechanism of the process. This approach was more popular when I started work in this field than it is today. The most important reason for this change is that we now recognize more clearly the molecular entities, the transmembrane proteins, that subserve transport. In the 1950s and 1960s we thought more in terms of passive or dissipative versus active or conservative processes. It is still sometimes useful to distinguish between membrane transport processes that relax electrochemical potential differences between phases (dissipative) from those that maintain or increase such differences (conservative). Many (but not all) dissipative pathways involve some form of more or less restricted electrodiffusion. The chapters by Goldman, Armstrong, Loewenstein, and Pappenheimer deal mainly with dissipative transport processes.

Conservative transport processes generally require a chemical reaction of binding of the transported substance to a protein pump in the membrane. Binding allows for coupling of transport of more than one substance. Such coupling can occur between two or more transported substance moving in the same (cotransport, symport) or opposite (countertransport, antiport) direction as in secondary active transport. The chapters by Aser Rothstein and Heinz address this kind of conservative transport. Coupling can also arise between the transported substance and reactants that are not themselves transported, such as primary active transport driven by ATP hydrolysis. The chapters by Jens C. Skou, Wilhelm Hasselbach, and Ussing consider conservative transport systems of this type.

Transport Through Different Membrane Proteins

During this past decade, a number of membrane proteins that are known to promote the transport of specific inorganic ions or organic

[9]

metabolites, and in most cases the corresponding gene(s), have been isolated, sequenced, and partially characterized. Examples include the red cell membrane protein called band 3 or capnophorin that catalyzes Cl^--HCO_3^- exchange (see chapter by Rothstein), the Na^+-K^+ pump (see chapter by Skou), and the Ca^{2+} pump (see chapter by Hasselbach), the Na^+ channel (see chapter by Armstrong), and a rapidly increasing list of others. It therefore makes sense to organize thinking about membrane transport in physiology in terms of these proteins. Each provides a regulated pathway for movement of specific substances through the membrane. Each is made and regulated through the action of specific genes. Each may be viewed as an enzyme in which substances enter and products leave by different paths, on one side and on the other side of the membrane. One task of scientists interested in understanding transport across biological membranes is to describe in atomic detail the reactions catalyzed by these enzymes and their regulation by transcriptional, translational, and posttranslational mechanisms.

All of these ways of thinking about membrane transport inform a view of organs, organ systems, and organismic physiology. All organs, organ systems, and organisms depend on membrane transport in many ways. This dependence is particularly obvious for organs specialized for secretion, such as the gastrointestinal tract, the kidneys, and all exocrine glands, but it is equally evident in the neuromuscular and cardiovascular systems. The pervasive importance of membrane transport for all aspects of biological function makes it sensible to include this volume among the series celebrating the centennial of the American Physiological Society.

FUTURE DIRECTIONS OF RESEARCH ON MEMBRANE TRANSPORT IN PHYSIOLOGY

What of the future? This afternoon, gazing out over the Gulf of Maine, I see it in terms of exquisitely selective "chips" arranged in extraordinarily complex "circuits." Investigation of the molecular mechanism of transport involves the complete physicochemical description of each chip, each transport protein. Investigation of the cellular functions of membrane transport involves the integration of these chips into circuits that regulate the distribution of specific solutes between the countless tiny aqueous chambers of which all living organisms are made. There is reason for optimism about the prospects for progress at both ends of this spectrum. However, many formidable problems remain to be solved.

To make a complete picture in atomic detail of the path of a transported substance through its transport protein requires a knowledge of the dynamic, three-dimensional structure of the molecule. Moreover, because selectivity of binding of ions and other small hydrophilic molecules may depend on very small changes in interatomic distances (see the chapter by Armstrong), the structures must be determined at very high (10^{-10} m) resolution. At present, such resolution is barely achievable through the analysis of the pattern of diffraction of X rays by crystals of proteins. Making appropriate crystals is currently considerably more difficult with membrane as compared with water-soluble proteins. Moreover, it is uncertain whether the conformations assumed by membrane proteins crystallized from a single polar or nonpolar solvent will be identical to the conformations that they take when they are imbedded in a bilayer that contains both polar and nonpolar regions. Furthermore, the structure determined by X-ray crystallography is static, not a picture of atomic motion. Transport undoubtedly depends on the movement of atoms in the transport protein in relation to one another. Identifying such motion requires obtaining high-resolution structures of more than one functional state of the transport protein, for example, with and without binding of the transported substrate. Although not feasible now, it is possible that in the future high-resolution magnetic resonance and other spectroscopic techniques will allow for the development of a picture of the conformation of transport proteins while they are still imbedded in membranes and capable of the dynamic changes in shape that are presumably involved in transport. The agenda is clear, and it seems likely that progress in describing the characteristics of the chips will continue. Presently, we do not know how many different transport proteins will be discovered. I guess no more than a few hundred. Moreover, it is possible that there are considerable similarities in the structures of proteins that promote the transport of quite different solutes. No doubt these questions about the physicochemical properties of particular pumps and channels will preoccupy many scientists interested in membrane transport during the second century of the American Physiological Society.

Another face of the future of membrane transport in physiology is how individual transport protein are connected in series and in parallel to perform physiological functions. By connections in series, I mean situations where a transported substance moves through one bilayer membrane via one transport protein and thence through another bilayer membrane via another transport protein. An example is the movement of Na^+ from the outside to the inside of the frog skin first through an amiloride-sensitive system in the apical membrane of the epidermal cells and then through the Na^+-K^+ pump in the basolateral membrane. By parallel connections, I mean the movement of the transported substance through a single bilayer membrane via two or more different transport proteins that span the same membrane. An example is the movement of Na^+ into a neuron or muscle cell through the voltage-dependent Na^+ channel during an action potential and the subsequent outward movement of the accumulated Na^+ through the Na^+-K^+ pump. Even in relatively simple

[11]

cells like nonnucleated human erythrocytes, Na^+ is known to move through the plasma membrane via at least four different transport proteins. The Na^+ content of the cells is the resultant of the simultaneous operation of all of these pathways. In general, biological function that depends on transport across membranes always requires the activity of more than one transport protein connected in parallel and sometimes in series.

Several modes of regulation of membrane transport have been recognized. One is through the transcription and translation of the genes that code for transport proteins and for proteins that regulate transport proteins directly and indirectly. Once expressed, the activity of transport proteins is regulated by several mechanisms, including the concentrations of the transported substance on the two sides of the membrane. Increased movement of a substance through one protein produces accumulation on one side and depletion on the other side of the membrane. The resultant changes in concentrations affect the subsequent rates of transport not only through this protein but also all other proteins through which the substance can move. Thus the concentrations of transported substance in the aqueous chambers separated by bilayers have the effect of coupling the operation of several transport proteins arranged in parallel. Other chemical regulators, such as ions or compounds that may accumulate in an aqueous phase through transport or chemical synthesis, can also couple the operation of different transport proteins. Examples include Ca^{2+}- and cAMP-activated channels. The concentrations of protons in the aqueous phases on the two sides of the membranes are particularly ubiquitous modifiers of the activity of transport proteins. The availability of transport partners in co- and countertransport also regulate and couple movement through transport proteins. In addition to this myriad of chemical regulatory possibilities, the electrical potential differences across membranes are both the resultant and a driving force for transport through many different membrane proteins arranged in parallel.

A detailed description of how transport protein "chips" are arranged in series and in parallel in membranes and regulated to perform function will doubtless also fascinate physiologists for decades to come. How is genetic expression regulated to determine the number and kinds of different transport proteins in different membranes to accomplish the several essential biological functions that they subserve? What models are most useful for thinking about these circuits? What do recent observations about the importance of oscillatory and even chaotic behavior of biochemical cycles portend for our understanding of physiological cycles that include transport? These are some of the questions that now seem important to me.

During the four decades since I first began to enjoy the ideas of membrane transport in physiology, it has been my privilege to meet many scientists working in the field. Among them have been my teachers and most of my friends. It has been a special pleasure to work with some of them in producing this book. I have met and shared thoughts with them in many places, in their laboratories, in my laboratories, at small conferences in New Hampshire, Tokyo, Puschino, Granada, Stresa, Bratislava, Aarhus, Aquasparta, Woods Hole, and Crans-Sur-Sierre, to name but a few. The opportunity to know these people, to be a member of the community of scientists trying to understand membrane transport in physiology, has been as important to me as the ideas that we have discussed.

II

Biological Membranes as Selective Barriers to Diffusion of Molecules

HUGH DAVSON

THE initial concept of the membrane as a structure or interface that separates the cell from its environment—be it some sort of interstitium, as in the case of a tissue, or a suspension medium such as seawater or blood plasma—was, of necessity, a consequence of the primary generalization, usually attributed to Schwann, that the behavior of a tissue, be it plant or animal, is governed by the coordinated activities of the individual cells of which it is composed. Thus the cell, being the unit of tissue structure, must be limited by some layer that preserves its identity and prevents its fusion with adjacent cells. The history of the development of the concept of the cell membrane depended mostly on studies of the plant cell because it lent itself so much more readily to microscopical examination than the animal cell, especially under experimental conditions. The plant cell differs from the animal cell by the presence of a well-defined and microscopically resolvable cellulose wall that separates it from its neighbors in a tissue or from its fluid environment, as in single-celled organisms such as algae.

PLANT CELL

Figure 1 illustrates a typical plant cell characterized by the large central vacuole surrounded by a gelatinous, and often very thin, protoplasmic layer, or protoplast, and enclosed by the cellulose coat. The protoplasmic layer was described by Schleiden in 1842/43 as the plant slime (*Pflanzenschleim*) and by Nägeli (51) as the slimy layer (*Schleimschicht*). Both terms were displaced by von Mohl's *Primordialschlauch*, which is probably best translated as "primordial sac"; the contents of this sac he termed *protoplasm*. To quote him,

[15]

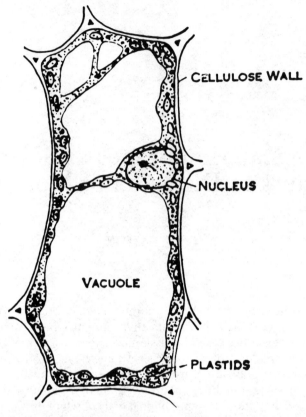

Figure 1. Typical plant cell with large central vacuole and thin protoplast attached to cellulose wall. [From Smith (63) by permission of the American Heart Association, Inc.]

"In full grown cells the protoplasm forms but a very subordinate part, as to mass, of the contents of the cell; while the watery cell-sap, which at first appeared only in isolated cavities, formed by degrees in the protoplasm, fills the whole cavity of the cell."

    According to Homer Smith (63), von Mohl's description of the protoplast as a *Schlauch* (sac) probably implied, in the light of his research on the changes in structure following gains or losses of water, that the protoplast was indeed enclosed in a membrane; thus the term *sac* came to mean the containing membrane without its contents. Smith (63) very acutely gives us a useful set of translations of the German words used in relation to the cell membrane. Because of the tendency to treat the protoplasmic layer as a membrane, the same German word may mean different things according to its author. Thus I quote: *"When von Mohl and Nägeli say* Zellenmembran, *we must read* the cellulose wall; *for* Schleim, Schleimschicht *and* Plasma *we must read* protoplasm; *for* Blase, *we must read* the central vacuole *or* smaller vacuoles; *and in Nägeli's paper the* Pri-mordialutrikel *midway becomes the* Plasmamembran *or* plasma membrane *in our present sense."*

Plasmolysis

The studies of Nägeli (51) gave us the definitive concept of a boundary layer that separates the protoplasm, or *Schleim*, from the vacuolar contents on the one hand and from the cellulose coat on the other. Thus, when a plant tissue is placed in what we would now call a hypertonic medium, for example, by addition of saccharose to the fluid in which it is bathed, the phenomenon of exosmosis (or *plasmolysis*, as it is now called) takes place as illustrated by Figure 2, *A* and *B*. The vacuole and protoplast shrink, and the protoplast separates from the cellulose coat; the process may be reversed (deplasmolysis) by placing the plant tissue in a more dilute medium. Nägeli's observations on the behavior of natural plant pigments such as anthocyanins (i.e., pigments dissolved in the watery medium of the protoplasm or vacuole) convinced him that there were barriers to the movement of these pigments at both the vacuolar and cellulose-coat sides of the protoplast. Nägeli, and later Pfeffer (58), gave the name *plasma membrane* (*plasmatische Membran*) to these suppositious layers bounding the protoplast. Nägeli emphasized the ability of protoplasm to repair itself after mechanical injury, provided the microscopically visible tear or defect is not too big, thus anticipating the more exact micrurgical studies of Robert Chambers. The repair depends on the capacity of the protoplasm to form new membrane.

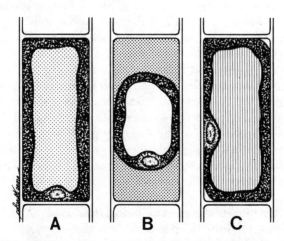

Figure 2. Typical plasmolysis (*A* → *B*) and deplasmolysis (*B* → *C*). [From Pfeffer (58).]

[17]

A rational concept of the cell membrane could not develop far, however, without an appropriate theoretical background for the behavior of solutions of crystalloids and colloids. It is interesting that the early study of plant material contributed to the formulation of some of our laws of the physical chemistry of solutions. Thus, with the concepts of osmosis, diffusion, and ionization so firmly embedded by our early education, it is difficult to put ourselves in the shoes of the biologists attempting to interpret changes in volume of cells at a time when these laws, with which we are so familiar, were being slowly deduced. Biological material, especially plant tissue, provided the intellectual stimulus and sometimes the experimental material for the enunciation of these laws of solutions (62). For example, Pfeffer's measurements of osmotic pressure were made with Traube's "precipitation membrane" produced in the convenient form of a membrane supported within a porous pot; however, these membranes were developed by Traube (65) during his attempts to simulate the cell growing under the influence of osmotic forces. He describes the formation of artificial cells by allowing a drop of, for example, gelatin solution into a medium containing tannic acid. The "precipitation membrane" forming at the interface allowed the endosmosis of water into the "cell," which grew in consequence. Addition of extra solute, such as lead acetate or sugar, increased the endosmosis. Traube quite clearly recognized the basis of osmosis in that it depended on the formation of a membrane of limited permeability to solute compared with that to water. By allowing a drop of copper choride to form in a solution of potassium ferricyanide he achieved a "cell" that was truly semipermeable in the sense that it was impermeable to inorganic salts but permeable to water. This precipitation membrane was utilized by Pfeffer in his porous pot experiments. Traube recognized that the permeability of his membranes was governed by the interstices between membrane-forming particles; expansion of the membrane during endosmosis tended to increase permeability, but this was counteracted by the intussusception of membrane particles formed by further precipitation of material. Growth of the true cell was probably, in Traube's view, conditioned by endosmosis together with intussusception of membrane-forming material into the expanding cell membrane.

Moreover, Pfeffer was primarily a botanist, as manifest by his classic book, *Pflanzenphysiologie*. Like Traube, he directed the whole weight of his investigations toward the interpretation of some physical phenomena in plant cells, especially as they related to transport of dissolved materials. Again, Van't Hoff (66) used not only Pfeffer's measurements on osmotic pressure but also de Vries' studies of isotonic coefficients for plant cells and Hamburger's osmotic measurements on blood cells to demonstrate the quantitative dependence of osmotic pressure on solute concentration and to add "osmotic pressure" to the tally of the colligative properties of a solute.

State of Knowledge at Turn of the Century

When Pfeffer published his *Pflanzenphysiologie* in 1890, a very clear concept of the physicochemical relations between the plant cell and its environment had emerged. The modern reader, who tends to date serious thinking on cell transport phenomena from a much later period, would find little to quarrel with in Pfeffer's chapter IV, "Mechanisms of Absorption and Translocation," some of its subtitles being: "Diosmotic properties of the cell"; "Plasmatic membranes"; "Ingestion and excretion of solid bodies"; "Diosmotic properties of cuticle and cork"; and "Quantitative selective power." Selective permeability of the cell membrane had been recognized by a number of workers using the deplasmolysis technique; thus the gradual return of the protoplasm and vacuole to their normal volumes after plasmolysis was correctly attributed by Pfeffer, among others, to the gradual penetration of the hyperosmolal solute into the protoplast and vacuole, so that the rate at which this happened was a measure of the protoplast's permeability to the solute. This plasmolysis technique had been applied to a variety of solutes by several workers, especially Overton, their results indicating a definite selectivity of the protoplast membranes to different solutes. Variations in this selectivity from plant to plant were emphasized by Pfeffer: "*As de Vries and Klebs have shown, glycerine and urea are substances which can penetrate all protoplasts, and often with marked rapidity. Here also are degrees of permeability exhibited, for the bud scales of* Begonia manicata *allow these substances to penetrate only with difficulty.*" Pfeffer describes the plasmatic membrane as a

living and dependent organ, by means of which the protoplast regulates the intercourse with the outside world, and hence, as might be expected, the plasmatic membranes of different plants exhibit certain specific differences and peculiarities. . . . The quality of the separating membranes, whether fluid or solid in nature, and the size of the molecules of the diosmosing substance, are of importance in determining the possibility of diosmosis as well as its character. Nevertheless, as the living organism shows, the relative size of the diosmosing molecules in relation to that of the micellar interstices through which they pass is by no means all-important. Thus the same protoplast, which is impermeable to NaCl *or* KNO_3, *may allow the larger molecules of methyl-blue, of albuminous substances, and perhaps also of other colloids to penetrate freely. This selective permeability, which is of the utmost importance to the plant, may be connected in some way with the fact, indicated by the ready penetration of solid particles,*

[19]

that the passage through the plasmatic membranes involved only a trifling expenditure of energy.

The word *diosmosis* is no longer current in transport literature, but its associated terms *endosmosis* and *exosmosis* are in use, with, however, different meanings from that implied by Pfeffer and his contemporaries. Diosmosis meant, in effect, diffusion, so that if solutes diosmosed out of the plant cell its turgor was reduced because of the decreased difference of osmotic pressure between the cell contents and its environment. Endosmosis and exosmosis simply connoted the directions of this transport, although today they refer to water movements only.

So far as the problem of active transport was concerned, Pfeffer was acutely aware of the accumulating power of plant cells, especially with respect to ions; he emphasized that this accumulation was selective, so that the salt content of the plant tissue (its ash) could be greatly different with respect to the concentrations of potassium, sodium, and calcium from that of the medium in which the plant was grown. He suggested various ways by which accumulation of a given solute might be brought about: "*Moreover, the nature of the plasma is such as to render it possible that a substance may combine with the plasmatic elements, thus being transmitted internally, and then set free again. When vacuoles are emptied, non-diosmosing substances do actually pass through the protoplasm.*" Clearly Pfeffer is outlining the possibilities of the carrier-mediated active transport so familiar to us today. His remark that substances not normally passing the plasmatic membrane may appear outside the cell through the emptying of a protoplasmic vacuole is certainly reminiscent of the "pinocytotic transport" so enthusiastically advertised by its main protagonist, George Palade.

UNIVERSALITY OF THE MEMBRANE

Besides recognizing that two membranes separated the plant protoplast from its environment, on the one hand, and from the central vacuole, on the other, Pfeffer generalized to the extent that:

Every mass of watery fluid present in protoplasm must be surrounded by a vacuolar membrane to form a larger or smaller vacuole, while masses of plasma which have escaped from the cell, also become clothed by a plasmatic membrane, and form large vacuolar bubbles in water, but not in plasmolysing solutions. . . . it follows that a plasmatic membrane must be immediately formed on the freshly exposed surface as it comes into contact, not with plasma, but with other media, and especially water.

In discussing the possible causes of the appearance of a new membrane on the surface of extruded protoplasm, such as that formed by

[20]

Figure 3. Formation of vacuoles in fragments of protoplasm immersed in water, which have been forced by pressure from a young root hair. [From Pfeffer (58).]

pressure from young root hairs as in Figure 3, Pfeffer has some very acute observations, not least of which was, *"the actual plasmatic membrane [i.e., the newly formed one] will in general be of such minimal size as to be incapable of measurement, and theoretically, a single or double molecular layer is sufficient for the maintenance of all the diosmotic properties possessed by the plasmatic membrane."*

PROTEIN MEMBRANE

Pfeffer rejected the notion of a lipoid membrane, preferring protein on the grounds that the precipitation phenomena on the surface of the exposed protoplasm were reminiscent of the coagulation that proteins undergo. *"Although the exact chemical constitution of the plasmatic membrane is not known, there can be no doubt that it is largely composed of proteid substances. This is indicated by the rigor caused by dilute acids."* Pfeffer rejected Quincke's suggestion that the membrane was a layer of oil covering the protoplast: *"that a slight solubility in oil should be shown by a single diosmosing substance has no general importance."* In the same footnote he does admit: *"E. Overton concludes that cholesterin or a cholesterin ether impregnates the plasmatic membrane and mainly determines its diosmotic properties and its permeability."*

MEMBRANE
TRANSPORT

So far as animal cells are concerned, we may turn to Hamburger's *Osmotischer Druck und Ionenlehre in den medizinischen Wissenschaften*, published in 1902 (36). While quoting the earlier work on plant cells, notably de Vries' studies on osmotic coefficients and those of Overton on plasmolysis, Hamburger applied the same principles to the erythrocyte, showing that the isotonic solution (e.g., for the horse) could be obtained by determining the concentration of NaCl that would give incipient hemolysis when the cells were suspended in a solution of this salt; this is 0.58% NaCl. He then found the degree of dilution of horse serum necessary to cause the same incipient hemolysis. From the two measurements he deduced that the isotonic solution for the horse erythrocyte is 0.92% NaCl. For the frog this is 0.62%. Hamburger's great step forward was the unequivocal demonstration that the erythrocyte, although "effectively" impermeable to NaCl and KCl, is nevertheless permeable to anions such as chloride and nitrate. He was led to this concept of separate ionic permeabilities by the phenomenon to which his name is irrevocably attached, namely, the Hamburger shift, which is the changes in chloride and bicarbonate concentrations that take place with altered CO_2 tension. Note that the concept of specific ionic permeability of membranes had been adumbrated by Ostwald (1890) and demonstrated somewhat equivocally, according to Hamburger, by Koeppe (48). Koeppe showed that ammonium salts penetrated the erythrocyte and were thus hemolytic; ammonium sulfate penetrated more slowly than ammonium chloride, presumably because the sulfate ion penetrated more slowly than the chloride ion. This finding formed the basis for later studies on the permeability of the erythrocyte to anions, for example, Höber (41).

STRUCTURAL STUDIES

While considering mainly the early morphological approaches to the establishment of the presence of a barrier separating the cell from its environment, we may briefly review the contribution of Robert Chambers to our knowledge of the outer structures of the animal cell, which he summarized in his *Harvey Lectures* (6) and later in the *American Naturalist* (6a). I met Chambers several times at Woods Hole and Cold Spring Harbor; he is best remembered, perhaps, for his absentmindedness. It is related of him that, in packing up to leave Woods Hole at the end of the summer he arrived at the railway station to find that, of his children, the baby was missing. He had put it in the bath to keep it out of the way during packing and left it there.

[22]

Chambers worked with animal cells, one of which was the egg of the sea urchin, *Arbacia punctulata*, a cell that was once the cell par excellence for kinetic studies; in fact its popularity as a subject for experimentation threatened it with extinction in the neighborhood of Woods Hole. His concept of animal cell structure is illustrated in Figure 4. As I summarized the situation in the first edition of my *Textbook of General Physiology*:

We may divide the outer layers of the cell into a group of extraneous coats and an inner protoplasmic layer; the former are not essential to the life or reproduction of the cell and consist, in the case of Arbacia, *of a secreted jelly which may be removed by shaking; a pellicle, called the vitelline membrane, which on fertilisation is lifted from its under- lying stratum and becomes a tough protective outer shell within which the embryo develops; and a hyaline plasma layer, secreted by the underlying cortex after the lifting of the fertilisation membrane. The two first mentioned layers are protective in function—the tougher the vitelline the more resistant is the cell to mechanical damage—whereas the hyaline plasma layer exerts its function as a cement when the egg divides, holding the blastomeres together in a definite pattern. The inner layer is made up of the plasma membrane and the cortex; damage to these by tearing or chemical agents, unless rapidly repaired, leads to the destruction of the cell; the plasma membrane probably repre- sents the true barrier which separates the interior of the cell from its external medium, one of the functions of the cortex being to act as a support for the membrane.*

At the highest magnification of the light microscope the plasma membrane could not be resolved as a separate structure.

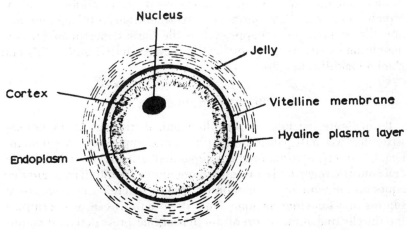

Figure 4. Typical animal cell. [Davson (21b, p. 331).]

[23]

This failure to resolve the plasma membrane meant that the investigator studying its structure, and even its existence, had to rely on indirect methods. The most rewarding of these methods was the study of the permeability of the cell, namely, the selectivity with respect to the solutes allowed to pass in and out, the hope being that the structure of the membrane would be revealed by this selectivity. The classic kinetic studies are those by Overton (56) on plant cells and by Gryns (35) and Hedin (38) on the erythrocyte, especially the studies by Hedin. These studies emphasized the selectivity of the barrier controlling penetration into the cell and the importance of lipid solubility, so that the partition coefficient of the solute when allowed to distribute itself between an oily and aqueous phase was a dominant factor (the higher the lipid solubility, or partition coefficient, the more rapidly the solute passed into the cell). Among other generalizations was that of Gryns that *"substances with the same chemical groupings behave similarly"* and the rule of Hedin that with polyhydric alcohols the rate of penetration decreases with the number of OH groups in the molecule. Hedin mentioned as exceptions to the lipid-solubility rule that urea and ethylene glycol penetrated rapidly into ox erythrocytes. Thus by the turn of the century the principle of the importance of specific chemical groupings, as well as the lipid solubility, of the penetrating solute had been established. Subsequent quantitative experimental studies, for example, that of Collander and Bärlund (9), many of which are reviewed in this volume, have confirmed the earlier generalizations, with an emphasis on the exceptions to the lipid-solubility rule. These exceptions are critical for deducing the nature of the physical barrier that retards the free diffusion of solutes from one side of the cell to the other. Among these exceptions is the phenomenon that has been described by Danielli (e.g., ref. 13) as "facilitated diffusion," by Reiner and me (26) as "enzyme-catalyzed permeability," and by Monod (e.g., ref. 7) as "permease-governed transport," a phenomenon that demanded a reappraisal of the basic concept of the cell membrane structure put forward by Gorter and Grendel (33) and Danielli and Davson (14).

Osmotic Equilibria

Before entering into the development of theories of membrane structure, we must pause to consider some simple osmotic phenomena, the interpretation of which depended critically on whether the cell could be regarded as an aqueous medium separated from another aqueous medium by a membrane that restricted the passage of solutes in a selective manner. Only if we do this can we recapture the intellectual atmosphere of the period that preceded and continued into the era of the unequivocal recognition of the importance of a cell membrane. Thus as late as 1921 M. H. Fischer (30) stated categorically, "*There are no membranes about cells. All the phenomena which are so difficult to explain when we assume a membrane to exist about cells are readily interpreted without recourse to such postulates on the basis of the colloid constitution of protoplasm.*"

The book that influenced me greatly in my early studies on permeability was Rudolf Höber's *Physikalische Chemie der Zelle und der Gewebe.* In those days we had to learn to read German fluently, so that I was able to read with fascination through the whole of the two volumes. I first met Höber when I was a research student at University College London; I had given the manuscript of the first paper I wrote to my professor and he had passed it on to Höber, who was one of several German physiologists who came to England at this time; I met him again some years later at Philadelphia, where he had been made a professor. He was a charming man, modest and unassuming and just the opposite of the "Herr Professor" that one associates with German academic life at that time.

Höber was one of the earliest to establish that the cell contents have the high electrical conductivity of the outside medium, a fact that demanded the existence of some restraint on the movement of ions into and out of the cell in view of the radically different ionic composition of the cell from that of its environment, especially the high concentration of K^+. Thus it had been argued that the contents of the cell could be regarded as a "colloid" so that a high internal concentration of an ion might be due to "binding" in a nonionic and therefore nonosmotic manner. This *kolloidchemische Standpunkt* that led people to regard the effects of changed ionic environment of the cell as the consequence of specific effects on its colloids (the lyotropic series is a favorite theme in so many of the papers quoted in Höber's book) was in obvious conflict with the interpretations of the same events in terms of exchanges of ions between two essentially aqueous media through a selectively permeable membrane. Thus, although we owe a great debt to Höber for, among other experimental studies, establishing the high internal conductivity of the cell contents, in retrospect I fear that his enthusiasm for the *kolloidchemische Standpunkt*, as evidenced by his book, tended to distract the attention of the younger school from the application of ionic equilibrium theory to the description of many cellular phenomena, applications that have resulted in the interpretation of the resting and action potentials of excitable cells, for example.

Osmotic Exchanges in the Erythrocyte

The unequivocal demonstration of the importance of a cell membrane in governing the exchanges of electrolytes and water between the erythrocyte and its plasma we owe, I think, to Merkel Jacobs of Philadelphia. He developed a simple technique for measuring what were, in effect, changes in volume of the cells as a result of ionic movements. Accurate measurements of volume of any cell were difficult in those days; for the erythrocyte, with its biconcave shape (Fig. 5A), this was especially difficult because swelling was unaccompanied by any measurable change in surface area (Fig. 5A–C). By maintaining a suspension of cells in a hypotonic environment that led to the hemolysis of some of the cells (the most fragile), small changes in osmotic pressure of the inside of the cell were reflected in changes in the degree of hemolysis, that is, the percentage of the cell population that could not tolerate the difference of osmolality across the membrane. With this technique Jacobs could measure minute changes in total osmolality within the cell and compare them with those predicted by application of simple equilibrium theory, on the assumption of free exchange of negative ions but effective impermeability of the cell membrane to positive ions such as K^+ and Na^+.

Jacobs was probably not the best choice as mentor for a young scientist visiting the United States for the first time. Scientifically he was brilliant, largely because he could handle his problems in a simple mathematical form; he combined this mathematical skill with the ability to develop very simple experiments requiring a minimum of apparatus (e.g., 46a). However, he was a recluse both physically and mentally; he lived in a remote part of the Philadelphia suburbs and rarely entertained, so that a visiting scientist had to make his own contacts with other scientists in Philadelphia or elsewhere. At Woods Hole he had a similar hideout. So far as discussion of ideas was concerned he was similarly withdrawn; whatever the proposition I put forward, it was invariably greeted with "very interesting," but that was all. In spite of these difficulties I learned a great deal from

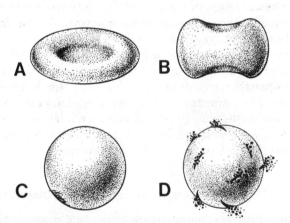

Figure 5. Changes in shape of erythrocyte on swelling until cortical hemolytic volume is exceeded. [Davson (21a, p. 1031) by permission of the American Heart Association, Inc.]

him about the necessity to prove things up to the hilt before rushing into print and also to experiment on more than one species.

In a series of papers, Jacobs, generally working with his colleague A. K. Parpart, showed that the swelling of the erythrocytes when the plasma was made acid and their shrinking when the plasma was made alkaline could be measured accurately by his hemolysis technique and accorded quantitatively with ionic equilibrium theory based on a Donnan equilibrium. Remarkably the erythrocyte often behaved as a perfect osmometer; there was no need to make allowance for the hypothetical escape of cations during the study. Thus the cells were effectively cation impermeable insofar as net exchanges were concerned. Thus the swelling and shrinking of the erythrocyte in response to changes in pH of the outside medium, so analogous with the swelling and shrinking of a fibrin gel under the same conditions, could be attributed to changes in distribution of ions across a cell membrane. Later I encountered a similar problem (21). When erythrocytes of most species are suspended in an isotonic KCl medium, that is, one in which NaCl is replaced by KCl, they show no change in volume, a phenomenon to be expected if the cell is effectively impermeable to Na^+ and K^+. However, dog erythrocytes hemolyze when suspended in isotonic KCl. The *kolloidchemische* explanation of this would have been that the KCl favored breakdown of the cell membrane or, if no membrane was allowed, increased the hydration of the colloidal protoplasmic mass to allow hemoglobin to escape from its interstices. In fact, the explanation turned out to be the consequence of the imperfection of the dog's erythrocyte as an osmometer. It exhibited quite a high permeability to K^+ and Na^+, so that in the KCl medium K^+ tended to penetrate and Na^+ to escape. Because K^+ tended to penetrate more rapidly than Na^+ tended to escape, the osmolality of the cell contents increased. The dog erythrocyte is a low-potassium cell, its main cation being Na^+; the cat erythrocyte also belongs to this class, but unlike that of the dog, when placed in isotonic KCl it shrinks: this is because permeability to Na^+ is much greater than that to K^+. Under conditions where permeability to Na^+ is inhibited, as with narcotics, KCl can become a hemolytic medium because under the same conditions permeability to K^+ is increased. Another example is the escape of cations from erythrocytes when placed in a nonelectrolyte medium. On the basis of the *kolloidchemische Standpunkt*, the effect of the absence of electrolytes on the cell surface would be attributed to the well-established effect of dialyzing colloidal solutions to remove their salts, which results in aggregation of their particles or micelles. In fact, however, the effect could be explained by simple equilibrium theory, the exchanges of anions creating large electrical potential gradients that would tend to accelerate positive ions out of the cell, breaking down its effective impermeability to these ions. Thus the highly permeable anions Cl^- and HCO_3^- tend to escape

[27]

from the cell; the only anion in a nonelectrolyte medium available to exchange with these anions is OH^-. The effect of this exchange is to make the outside of the cell acid and the inside alkaline, establishing a high internal electrical positivity (46a). I argued that this high potential gradient would drive K^+ out of the cell, as indeed happened; addition of small quantities of NaCl to the outside of the medium would reduce both the potential and the escape, as I later found (17). Finally we may take an example of the lyotropic series of ions, a series that played such an important part in the interpretation of cell phenomena in the heyday of the *kolloidchemische Standpunkt*. Thus, according to the theory, ions at the "hydrating end" of the series, such as CNS^- and I^-, would loosen the cell colloids so that if the membrane were colloidal, as the theory assumed, their permeability would increase. In a study of the permeability of the cat erythrocytes to Na^+ and K^+, a very definite lyotropic series was observed: penetration of K^+ into the cell, when placed in isotonic solutions of K^+ salts, was accelerated when CNS^- and I^- were substituted for Cl^- and decreased by substituting citrate or SO_4^{2-} at the "dehydrating end" of the series. However, when Na^+ permeability was measured under the same conditions, the reverse phenomenon occurred: thiocyanate tended to inhibit permeability (19). Thus the *kolloidchemische Standpunkt* could not have it both ways, and a different type of explanation had to be sought; the most likely explanation was that the ions were indeed having specific effects on the cell membrane but that these effects were directed toward the Na^+ in preference to the K^+. If this is true, then we are approaching a special aspect of this chapter, namely the inhomogeneity of the cell membrane with respect to its structure and hence with respect to its permeability to given solutes, an inhomogeneity that is revealed by its "unorthodox" behavior. Thus we may envisage an essentially lipid membrane and predict the relative permeabilities of a large number of different molecular and ionic types. When gross deviations from these predictions are found, for example, as in the relatively rapid penetration of some hexoses through cell membranes (a rapidity that bears no relation to the lipid insolubility of these highly polar compounds), we must invoke special "channels" or "active sites" that provide a way through the membrane that is forbidden to other molecules of comparable lipid insolubility. The same can be true of ions; on physical grounds we might expect K^+ to cross a cell membrane more rapidly than Na^+ because of its smaller hydrated radius; this is true of the diffusion of ions through water but by no means true of all cell membranes. Thus the study on cat erythrocytes just alluded to indicates a higher permeability to Na^+ than to K^+, suggestive of some special mechanism involved in Na^+ transport; the inhibition of this Na^+ permeability by thiocyanate, which accelerates K^+ transport, suggests that the thiocyanate interferes with the special mechanism, whatever this is, whereas its accelerating effect on K^+

might be due to its more rapid penetration into the cell than that of Cl^-, because penetration of the cation is associated with penetration of an anion too. The morphological basis for these inhomogeneities in the cell membrane is described elsewhere in this volume; my purpose here is to derive the origin of concepts of membrane structure, so I need make no apology for returning to those early days of permeability studies in which these concepts were first shadowed forth.

Membrane Structure

Because this chapter represents a personal account rather than a dry record of history derived from published papers, I must be forgiven if, before continuing to establish the background of facts and concepts that led to the promulgation of the bimolecular-leaflet concept, I mention the circumstances that attracted Jim Danielli and me to permeability problems. We were both postgraduates in chemistry at University College, London, in the early thirties. Almost by accident we discovered a course of lectures on advanced biochemistry given by Professor Jack Drummond, a course designed to attract the interest of pure chemists; from these students Drummond drew his own postgraduates. In dedicating our *Permeability of Natural Membranes* to Drummond, Danielli and I were doing a great deal more than making a formal acknowledgment of his encouragement, especially so far as I was concerned. Drummond gave me the chance to do research virtually on my own, rescuing me from the department of chemistry where research consisted of spending two years on a problem thought up by the professor, the answer to which one knew in advance; the experimental verification was tedious in the extreme.

Although a vitamin chemist, Drummond recognized that biochemistry did not consist exclusively of the application of organic chemistry to biological problems, so his course included a lecture on permeability in which he described the striking effects of Ca^{2+} on the osmotic responses of an estuarial worm (*Gundaulvae*) to changes in its external osmolality. We were extremely impressed with this effect of Ca^{2+} and, because the College had been fortunate to have brought into the chemistry department N. K. Adam, a surface chemist whose lectures on surface chemistry opened a new world to us, we saw an analogy with the effects of Ca^{2+} on surface films of fatty acids, the ion tending to compress them. Therefore we thought that Ca^{2+} would reduce permeability by causing a tighter packing of a surface film of lipid on the cell surface. This idea dominated our thinking, and when I persuaded Professor Drummond to take me into his department to work on a subject over which he could provide little or no specialized supervision (since the department was devoted to vitamin or hormone research), I decided to work on the permeability of red cells to K^+, having been influenced in my desultory reading prior to beginning this research by a paper by Kerr and Kirkorian (47a) on the escape of K^+ from erythrocytes.

Looking back, I can see why I did not usually find strong effects of Ca^{2+} on erythrocyte permeability; it was simply the failure of many of us to realize that washing a cell once with a Ca^{2+}-free solution does not remove all the ion from the cell membrane, and until ethylenediaminetetraacetic acid became available, the amounts remaining were adequate to maintain membrane function. Thus, rather unfortunately as it now appears, I tended to dismiss Ca^{2+} from my thinking. As we all now know, this ion plays an extremely important role in a wide variety of physiological phenomena, some of which depend on the transport of ions across membranes. An interesting example from my own work (20) involves the effects of F^- on the level of K^+ in the erythrocyte; this inhibitor of glycolysis causes a rapid escape of K^+, but this could not be due to the inhibition of a K^+ pump because the escape requires a long incubation period before beginning; thus I concluded that the effect of the inhibitor was to cause the accumulation within the erythrocyte of some product of metabolism that affected the permeability of the cell membrane to K^+. In fact, the effect is due to the inhibition of a Ca^{2+} pump that maintains a low internal concentration of this divalent cation, so that it is the accumulation of this within the cell that causes an increased permeability to K^+, not some metabolite (66a). Thus, insofar as internal concentrations are concerned, Ca^{2+} does not reduce permeability to the K^+, so that the original idea that led to the formulation of the bimolecular-leaflet hypothesis by Danielli and me was incorrect, but of course our instinct had been sound when we brought the relatively new branch of physical chemistry—surface chemistry—into the context of membrane structure and behavior.

While I was working in London on erythrocyte permeability, Danielli took up a Commonwealth Travelling Fellowship, during the tenure of which he spent two valuable years mainly in the laboratory of Newton Harvey at Princeton University and the Marine Biological Laboratory at Woods Hole. At Woods Hole he came in contact with a group of general physiologists that included W. J. V. Osterhout, M. H. Jacobs, K. S. Cole, and H. Fricke, who, with Newton Harvey, may be regarded as the leaders in the general physiological approach to cell biology. Harvey, whom I met a few years later at Princeton and Woods Hole, when I went to the United States on a Rockefeller Foundation Fellowship, was an ideal mentor to one so full of ideas as Danielli, and it was in his laboratory that the bimolecular-leaflet theory of membrane structure was formulated.

Current Thinking on Membrane Structure

Let us recapitulate the ideas current at the time. The lipoid nature of the membrane was postulated primarily on the basis of the early studies on permeability. The possible structures considered were either that the lipid was essentially a homogeneous layer or that it was an emulsion capable of changing from an oil-in-water to a water-in-oil system according to circumstances (6b). Standing in the wings was the theory, advocated by Pfeffer (58), that the membrane was primarily protein in nature since membranes derived from proteins could be made to exhibit selectivity based on molecular size—the sieve effect—that in some cells, such as those of the sulfur bacterium *Beggiatoa mirabilis*, seemed to predominate. Mosaic membranes involving regions of lipid and protein, spatially separate, would allow for the lipid solubility and sieve effects (41). As Danielli (11) emphasized, the decisive experimental requirement was a knowledge of the thickness of the cell membrane.

THICKNESS OF THE CELL MEMBRANE

The membrane was not resolvable in the light microscope, so this meant that it could not be much thicker than $0.12 \mu m$, that is, $\sim 1,200$ Å, but between this and the length of a likely fatty acid molecule, namely ~ 30 Å, there was a lot of room left for speculation. The classic contribution to this problem is that of Gorter and Grendel (33). They extracted the lipids from erythrocytes by means of large quantities of acetone; the evaporated residue was dissolved in a little benzene, and the lipid was spread on the surface of a Langmuir trough (Fig. 6). The area of the resulting film of lipid was slowly reduced by sliding a glass barrier across the surface of the water, forcing the lipid against an opposite retaining barrier; as soon as the film started to exert a measurable resistance to compression, the actual area occupied by it was measured. At this area the molecules of lipid were packed together to give a coherent film with the polar, or water-soluble, endings of the molecules attached to the water and the nonpolar, or hydrophobic, chains sticking out as illustrated in Figure 6. This measurement gives the total area that the lipids were capable of covering, supposing they were able to spread out in this

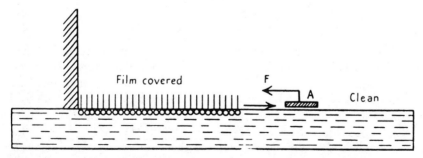

Figure 6. Typical arrangement for studying surface films of lipid. Film is compressed by flat glass bar (A) exerting a force (F). [Davson (21a, p. 1023) by permission of the American Heart Association, Inc.]

[31]

TABLE 1
Dimensions of Red Blood Cells

Species	No. of cells extracted $\times 10^6$	Area of single cell (μm^2)	Total surface of cells (m^2)	Spread surface of lipid (m^2)	Ratio of surfaces
Rabbit	6.6	74.4	0.49	0.96	2.0
Guinea pig	5.85	89.8	0.52	0.97	1.9
Human	4.74	99.4	0.47	0.92	2.0

From Gorter and Grendel (33).

manner to give a coherent film. Gorter and Grendel next measured the dimensions of the red blood cells; these are biconcave disks, so that the assessment of their area was not easy. By measuring the thickness and diameter and applying Knoll's (1923) formula, they were able to make a reasonably approximate estimate. Some results for three species are shown in Table 1, where the ratio for the area of lipid divided by the area of cells is ~2. Thus Gorter and Grendel concluded that the membrane was a bimolecular leaflet, as illustrated in Figure 7. Grendel (34) analyzed the total extracted lipids, and from the known surface areas per molecule occupied by the different constituents, he concluded that the relative contributions, in terms of spread area, were cholesterol, 36%; cephalin and lecithin, 50%; and sphingmyelin, 13%. He computed that the double layer of lipid would have an average thickness of 31 Å, but more recent studies, based on X-ray analysis of orientated lipids, suggested that the thickness of the orientated lipids in a double layer would be considerably higher, namely, 63 Å (29, 61). Thus Gorter and Grendel's study left us with a highly plausible basis for membrane structure. There were obvious uncertainties in their estimate of thickness, notably the total surface of the cells, the efficiency of the extraction of the lipid, and the question of whether the membrane might also have a protein constituent, which might act as a framework on which the lipoid film was deposited. Again, the protein, besides acting as a framework, might also cover the lipoid surfaces.

ELECTRICAL MEASUREMENTS

The classic studies that Höber (39, 40) did on the electrical conductivity of cell suspensions, already alluded to, showed that the cell could be considered electrically as a conducting medium (the cytoplasm) separated by a nonconducting dielectric (the membrane) from the outside medium, which was also conducting. Fricke (31) developed the theoretical treatment of Clark Maxwell and applied it to calculate the specific capacity of the cell membrane: this came out at 0.81 $\mu F/cm^2$. By assuming a value of 3 for the dielectric constant of the membrane, he arrived at a value of 33 Å for the thickness, or about 1 molecule thick. Danielli (11) questioned the

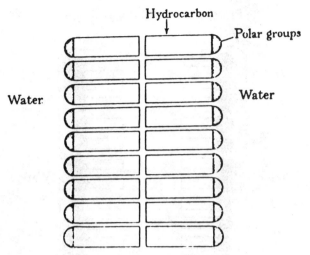

Hydrocarbon

Polar groups

Water

Water

Figure 7. Gorter-Grendel model of a bilayer.

use of the bulk solid value for the dielectric constant of material orientated in a thin film; however, examining the surface potentials of films and employing the simple relation between these and the dielectric constant gave estimates of 4 for the dielectric constant of the orientated film, not greatly different from Fricke's assumed value of 3. Thus, on the basis of the Gorter and Grendel study and that of Fricke, one had to decide between a mono- or bimolecular leaflet; on theoretical grounds the latter is the more stable configuration for lipid, if we consider the tendency for the lipid to take up another, highly stable form, namely, that of a micelle. Thus a single layer of lipid would show a strong tendency to transpose to a set of lipid micelles, whereas a bimolecular leaflet, anchored on each face to an aqueous phase, would have a greater chance of resisting this tendency to assume the micellar form (13).

ROLE OF PROTEIN

Protein clearly must play an important role in the structure of the cell; we have seen the importance of the cortex of the cell that presumably serves as a support for the plasma membrane. The cortex is doubtless made up of protein, exhibiting the power of reversibly liquefying during streaming movements of protoplasm and so on. Insofar as its incorporation into the structure of the plasma membrane was concerned, it was the X-ray and electron-microscopical studies of Schmitt (61), Robertson (59), Finean (29), and others that established a role for protein. Earlier, however, Cole (8) and Danielli and Harvey (15) demonstrated that the tensions at the surfaces of cells were very much lower than those expected of a simple lipoid layer, so that a coating of protein was incorporated into the model of Danielli and Davson (14) as shown in Figure 8. Later studies of

[33]

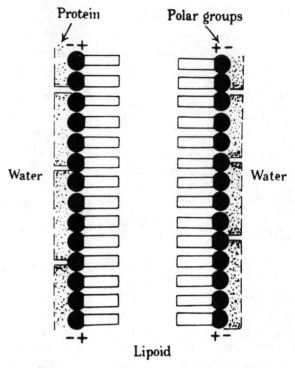

Figure 8. Danielli-Davson model of cell membrane. [From Danielli and Davson (14).]

the surface properties of proteins at an oil-water interface, carried out by Askew and Danielli (2), showed that a globular arrangement would be unlikely because proteins tended to unfold, or denature, at interfaces. Thus a more likely configuration for the lipid-protein complex than that suggested by Danielli and me was that shown in Figure 9.

X-ray Studies

So much for speculation. The high resolution provided by X-ray diffraction gave possibilities of defining more accurately the relation of lipid to protein. The difficulty with X-ray analysis, as with polarization optics, is that a layer only 60–100 Å thick would give such weak diffraction spots as to render them useless; only if a stack of membranes could be prepared would it be possible to obtain adequate X-ray diffraction patterns. Schmitt and Palmer (61), with what turned out to be brilliant prescience, thought that the myelin sheath of vertebrate nerve might be made up of layers of lipids similar in composition to those constituting the plasma membrane. They examined the X-ray diffraction patterns of individual lipids found in myelin (e.g., cholesterol, lecithin, etc.) and then examined the patterns of preparations of sciatic nerve. The comparison with the lipids

[34]

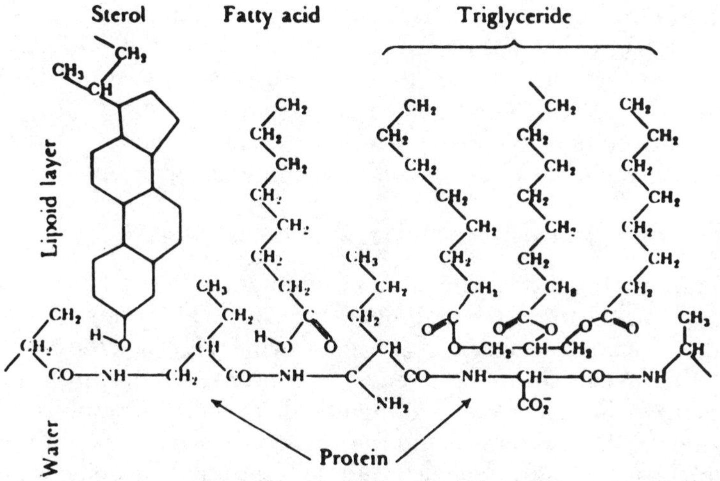

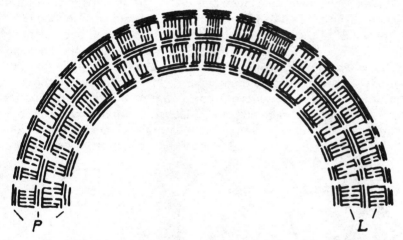

Sterol   Fatty acid   Triglyceride

Figure 9. Further development by Danielli of cell membrane model. [From Davson and Danielli (23).]

Figure 10. Molecular basis of the myelin sheath deduced from polarization optics and X-ray diffraction. [Davson (21b, p. 331).]

and the actual nerve root exhibited a difference in that the latter contained a spacing of ~160 Å, which was considerably larger than the corresponding spaces in the pure lipids, spacings that probably indicated the separation between successive laminae of the lipid or myelin, respectively. This larger spacing was interpreted as indicating the presence of protein interspersed between successive bimolecular leaflets of lipid, as indicated in Figure 10. In a more exhaustive investigation, Finean (29) confirmed this pioneering study of Schmitt and Palmer (61), concluding that the frog sciatic nerve myelin contains a low-angle repeat of 171 Å; an elaborate study of the intensities of the refractions indicated that large numbers of atoms, or groups of atoms, repeat at half of the unit-cell repeat, that

BIOLOGICAL
MEMBRANES

[35]

is, at 85.5 Å, so that the 171-Å repeating unit apparently contains a pair of almost, but not completely, identical units of 85.5 Å. This suggested that each half of the unit cell consists of a bimolecular leaflet of lipid plus one thin layer of protein, some slight variation in the constitution or symmetry within the plane of the layers necessitating the inclusion of two layers in the fundamental repeating unit.

EMBRYONIC DEVELOPMENT OF MYELIN

Schmitt's implicit assumption in attempting to extrapolate from the myelin sheath to the cell membrane was that the myelin sheath might be made up of layers of cell membrane; this posed the difficulty that the myelin sheath was readily permeable to ions, by contrast with the cell membrane, a difficulty that could be resolved if, in some way, these solutes could find their way around the individual lamellae. Geren (32) demonstrated that, during development of the nerve sheath in embryonic tissue, the myelin was indeed laid down by a spiral winding of cell membrane, the membrane material being derived from associated Schwann cells. The process is illustrated schematically in Figure 11. In more detailed studies, Robertson (59) confirmed this origin of myelin and, by introducing permanganate as a fixative, he was able to resolve the individual apposed membranes, which with osmium fixation appeared as single dense lines, into two dense lines 25 Å thick separated by a light zone of the same thickness. Thus there seemed to be a unit of 75 Å that might be the basis of cell membrane structure. The translation of the electron-microscopically observed tissue into naturally occurring dimensions in the living tissue is a difficult matter that need not concern us here;

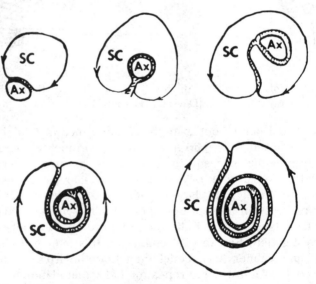

Figure 11. Origin of the myelin sheath from Schwann cell membrane. [Adapted from Geren (32).]

[36]

it is sufficient for our purposes that the X-ray diffraction and elec-tron-microscopical studies of myelin are valid studies of cell mem-brane material that has been "spun off" a living cell. The resolution of the units of 75 Å into three layers presumably has as its basis the bimolecular nature of the lipid leaflet and the presence of protein at each side. In the spinning process illustrated by Figure 11, each myelin layer will contain the protein that was attached to the outside and inside of the membrane respectively, and thus we can account for the finding of Schmitt and Bear that the main repeating unit of 171 Å can be resolved into two half-units that are slightly different.

SOME FURTHER KINETIC STUDIES

We have now reached the point at which structural studies have provided reasonable support for the general concept of a plasma membrane capable of exerting some selective control over the trans-port of solutes into and out of the cell. I return now to some of the experimental studies that, as far as I was concerned, were governed by the concept of the membrane whose permeability would be determined by the stability of a lipoprotein film.

NARCOTICS AND HEAVY METALS

Ørskov (52) showed that low concentrations of lead cause a striking escape of K^+ from erythrocytes. Danielli and I were seeking ways to alter the permeability of cells, considering, as outlined earlier, that alterations in the packing of the lipid film molecules might do this. We examined the effects of lead and other heavy metals in more detail (22); especially active was silver, the increased permeability caused by concentrations as low as 2.5×10^{-7} M being reversed by the addition of chloride to precipitate the silver. We considered that the heavy metals caused such tight packing of the lipid that they broke up the film reversibly into islands, leaving gaps through which ions could move freely. In the same work we studied a number of highly lipid-soluble substances that, like heavy metals, would cause hemolysis of the erythrocyte. Higher alcohols such as amyl alcohol, phenolic compounds such as guaiacol, and surface-active substances such as sodium oleate all brought about this breakdown in what we considered an "absolute impermeability" of the erythrocyte mem-brane to K^+. In those days, because we described the erythrocyte as "moribund," lacking a nucleus and subserving only the carriage of oxygen and acting as a buffer, we considered that there was an absolute normal impermeability to cations, the initial accumulation having occurred at an earlier stage in the cell's life. We remained blind to the possibility of an active accumulation of K^+ by the normal erythrocyte until J. E. Harris (37), working on stored blood, showed that the slow leakage of K^+ that occurs during storage could be reversed by warming the cells to body temperature. Actually Ørskov

(52) had shown that the escape of K^+ that occurs from erythrocytes in vivo after injection of lead was at least partially reversed. It therefore seemed that the actions of heavy metals, lipid solvents, and surface-active substances had the common feature of destabilizing a lipid film, thereby inducing a high permeability to K^+ and possibly other cations. This work led us to two concepts: *1*) the colloid-osmotic theory of cell destruction and *2*) the principle of facilitated transport.

Colloid Osmotic Hemolysis

The Gibbs-Donnan equilibrium, determined by the impermeability of a membrane to one of the ions in a system, was applied to biological systems most notably by Starling (63a), when he analyzed the forces governing the absorption and formation of extracellular fluid in relation to the protein-rich plasma, and by Henderson (38a) when he analyzed the distribution of anions across the red cell membrane, governed by the high concentration of the impermeant hemoglobin anion within the red cell. We were struck by the fact that Starling found a difference in osmotic pressure—the colloid osmotic pressure—between plasma and extracellular fluid that caused expansion of the volume of the extracellular fluid until opposed by a resisting tissue pressure. On the other hand, in erythrocyte the equilibrium could be described in terms of osmotic equality between the two media—plasma and intracellular fluid—a situation that was possible by virtue of the erythrocyte's apparent impermeability to cations, so that the equilibrium distributions applied only to the anions of the system, hemoglobin, chloride, and bicarbonate. It became clear to us that, in any cell with a significant internal concentration of impermeant colloidal ions, the loss of its apparent or virtual impermeability to cations would change the situation osmotically, permitting the development of a colloid osmotic pressure that draws fluid into the cell unless opposed by an internal pressure. In most animal cells, with their easily expansible membranes, the development of significant internal pressures was not feasible, so that the process of fluid intake lasted indefinitely until the cell burst. In our study of the effects of heavy metals and lipid solvents, we suggested that the permeability to cations induced by these agents could not only lead to the observed losses of K^+ but, if these losses were accompanied by increases in Na^+ because of breakdown of the effective impermeability of the membrane to this ion as well, a new form of Gibbs-Donnan equilibrium would be established, comparable with that described by Starling. This would lead to swelling of the cell often after shrinkage due to the initial escape of K^+ being greater than any gain in Na^+. This swelling would continue until the capacity of the cell to take up water was exceeded, in which case the membrane would break up and allow the escape of protein (see Fig. 5). Subsequent studies (24, 25) amply confirmed

[38]

the hypothesis. Thus when erythrocytes are treated with a photosensitizer, such as rose Bengal, and are subsequently exposed to light, they hemolyze (5). In 1940, Ponder and I (25) showed that, first, this hemolysis was preceded by an escape of K^+, thus bringing the phenomenon into the category of prohemolytic events. Second, we showed that if the light was turned off after a certain period of exposure, the escape of K^+ continued, proceeding to the point that hemolysis also continued. Thus, if we regard the suspension of erythrocytes as a population of cells of differing susceptibility to light, after a period of light exposure some will have become so highly permeable to cations that they will have reached the stage of hemolysis; others will not have been so severely damaged, so that the ionic movements—loss of K^+ and gain of Na^+—will not have progressed to the point that the swelling maximum has been exceeded. These latter cells, at the end of the exposure, will be physically intact, containing their hemoglobin. However, the increased permeability caused by the light exposure will continue to show its effects in the dark, with continued losses of K^+ and gains of Na^+ leading ultimately to further hemolysis. Thus the hemolysis that occurs after light exposure cannot be due to gross destruction of the cell membrane sufficient to allow hemoglobin to escape; instead, the effect must have been confined to an increased permeability to cations that gave expression to the colloid osmotic pressure that drew water into the cells.

I cherish fond memories of Eric Ponder, in whose laboratory I spent a month in 1936 and whole summers in 1937, 1938, and 1939. The "folding up" of the Long Island Biological Association that occurred in 1940 put an end to these visits. Ponder devoted himself almost entirely to the problems of erythrocyte hemolysis, publishing a monograph (*Haemolysis and Related Phenomena*) that was a classic. Many an evening did we spend at the "local" on the other side of the harbor over our respective Hankey Bannister (he was a Scotsman) and beer, quoting usually Gibbons *Decline and Fall* to each other.

Reversal of Hemolysis

In this context, we may say a few words about the extent of membrane damage that occurs during hemolysis. At the time we were working, the theory was generally held that hemolysis represented a severe rupturing, or dissolution, of the cell membrane. The demonstration that the sufficient cause need only be a membrane leakiness to cations to which we supposed the membrane was completely impermeable meant that the primary damage could be far less severe. This reminded us of a neglected phenomenon, the apparent reversal of hemolysis; thus when cells had been made to hemolyze (for example, by placing them in dilute saline) on addition of a strong salt solution and centrifugation, apparently intact eryth-

rocytes containing hemoglobin were found at the bottom of the tube
(4). The phenomenon was investigated in some detail (24); the
reversal consisted in the shrinking of the "ghost" by osmotic means,
which meant that the membrane must have retained some degree of
its semipermeability, that is, its effective impermeability to cations.
The hemoglobin remaining in the ghost was apparently sealed in by
a reassumption of hemoglobin impermeability. The relatively high
concentration of hemoglobin in the reconstituted ghosts was due to
the osmotic withdrawal of water during the reversal process. These
experiments showed that the formation of holes at least as large as
the diameter of the hemoglobin molecule, that is, in the region of
60 Å, was reversible. Moreover, Ponder and I (24) showed that the
extent of the damage to the membrane depends on the rapidity with
which the swelling is brought about. We measured the "fading times"
of erythrocytes undergoing hemolysis, that is, times required for a
cell to disappear from view in the microscope; the longer the time,
the smaller the hole sizes or the fewer the number, or both. By
allowing swelling to occur extremely slowly, when the difference in
osmotic pressure causing the lysis was minimal, very long fading
times of the order of 5–10 s could be achieved, compared with a
fraction of a second when lysis was induced by strongly hypotonic
solutions (Ponder and Marsland, 1935). Similarly long fading times
were obtained when cells were caused to lyse in the presence of
lipid solvents, such as amyl alcohol (24). Note that the cells were
made hypertonic to their medium by maintaining them in a saline
diluted to the point that some had already burst, so that there would
be a number with volumes just below the critical hemolytic volume;
by allowing the temperature of the microscope stage to fall slowly,
the changed dissociation of hemoglobin led to a small increase in
cell osmolality that was reflected in the bursting of these cells that
had been brought very close to their critical volumes. The long
fading times with the lipid solvents are to be expected, since the
development of the difference in osmolality between cell and its
environment depends on the movements of Na^+ and K^+ rather than
that of water.

Ponder and I found a relatively high permeability of the ghosts to
cations, suggesting that the holes in the membrane had not been
resealed completely; consequently the ghosts tended to swell and
burst when maintained in isotonic saline, with a second colloid
osmotic hemolysis taking place. Subsequent studies in other laboratories (42–44, 64) have shown that Mg^{2+} had to be present in the
medium if, during reconstitution and after, the ghosts were to reac-
quire and maintain a relative impermeability to cations. This probably explains why those prepared by Ponder and me were very leaky
in respect to cations, and it serves as a warning to all of us that
Ringer's solution is not necessarily an adequate substitute for a more
natural fluid during reconstitution of lysed cells.

[40]

CELL LOADING

The demonstration that the ghost may be reconstituted into a cell with low cation permeability has been of great value experimentally in the study of the active transport of ions across its membrane, because the reconstituted ghost can be loaded with an entirely different ionic composition from that of the original erythrocyte. In this way Hoffman (42–44) showed that Ca^{2+} inhibits ATPase inside the cell and as a result inactivates the Na^+-K^+ pump, although when presented to the outside of the cell it is ineffective. Again, Whittam (1962) demonstrated the different effects of Na^+ and K^+ on the cell's ATPase depending on whether the ions acted on the outside or inside of the cell.

FACILITATED TRANSPORT

Studies on the effects of heavy metals and narcotic substances (such as the alcohols) on permeability led me to investigate further the effects of narcotics on permeability in the hope of forming some sort of generalization as to their effects, for example, to explain the action as a result of "loosening" of the lipid film constituting the basis of the cell membrane. A review of the literature at the time (23) suggested that where accurate quantitative studies had been done, as in the study by Bärlund (3) on *Chara ceratophylla*, the only effect of ether, the narcotic substance he used, was to increase permeability reversibly. There were exceptions, however, and it is these exceptions that have turned out to be of the most interest. Thus Anselmino and Hoenig (1) found that the carbamates had a strong inhibitory effect on the permeability of human erythrocytes to glycerol. When working in Jacobs' laboratory, I tried very hard to find a narcotic or surface-active substance that would inhibit permeability of erythrocytes to nonelectrolytes such as glycerol but always I found an acceleration until, by accident, I used rabbit erythrocytes instead of ox erythrocytes and I was struck by the very strong inhibition that could be obtained with ethyl alcohol. Jacobs and Parpart (47) followed up this discovery and showed that the effect was indeed dependent on species, so that alcohols would inhibit glycerol permeability in erythrocytes of (besides the rabbit, as I had found, and humans, as Anselmino and Hoenig had found) rats, guinea pigs, groundhogs, and several birds. It is the erythrocytes of these species that show unusually high permeability to the highly lipid-insoluble glycerol molecule, so that the narcotics affect what is now recognized as a form of facilitated transport. Danielli (ref. 23, appendix A) showed that the Q_{10} for this type of transport suggests a nonuniform type of membrane, thus providing a further piece in the picture puzzle that was slowly emerging. Again, the exchanges of anions across the erythrocyte membrane are extremely rapid, so that special methods have had to be developed to give these exchanges quantitative expression (see, e.g., ref. 31). There is now no doubt that these exchanges involve a facilitated process, and it is interesting that here, too, the exchange is inhibited by narcotics (e.g., ref. 47).

Catalyzed Permeability

In my studies on the permeability to Na^+ and K^+ of cat erythrocytes, I observed that cells placed in an isotonic KCl medium lost Na^+ to the medium (high-Na^+ cells) and gained K^+. As Figure 12 shows, the rate of loss of Na^+ is very much greater than the rate of the concomitant penetration of K^+, despite the approximately equal gradients of concentration. This behavior may be described as unorthodox because, in any simple membrane system, we expect K^+ to have the higher permeability in view of its smaller hydrated radius. A further indication of the unorthodox behavior of Na^+ was given by the effects of temperature; as Figure 13 shows, raising the temperature does not necessarily increase the rate of escape of Na^+; in fact, on going from 35°C (close to the cat's body temperature) to 40°C there is a decrease in the rate of escape, and at 45°C escape is almost completely inhibited. The behavior of K^+, on the other hand, was orthodox: permeability increased monotonically with temperature.

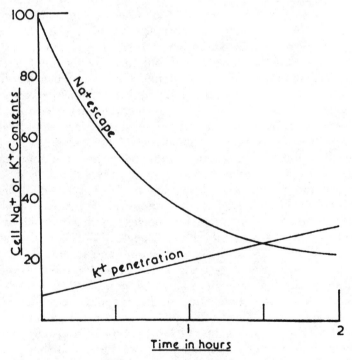

Figure 12. Penetration of K^+ and escape of Na^+ into and out of cat erythrocyte suspended in isotonic KCl. Concentration gradients of separate ions across the membrane are approximately equal.

When the pH of the medium was altered, the permeability to Na^+ showed striking variations (Fig. 14), rising from about zero at pH 6 to a maximum over the range pH 7.0–7.6, when it began to fall. Thus the optimum pH for Na^+ permeability was in the normal blood range; permeability to K^+ was barely affected by pH. Reiner and I (26) described the permeability to Na^+ as being governed by an "enzymelike factor"; I subsequently described it as catalyzed. The enzymelike factor was further characterized by the finding that narcotics and heavy metals inhibit Na^+ permeability but accelerate K^+ permeability. This enzymelike factor proved to be far more general than I had thought; thus Monod described "permeases" in bacteria that were either normally present in the membrane or that could be induced [for review see Cohen and Monod (7)]. These permeases enabled the bacterial membrane to transport specific

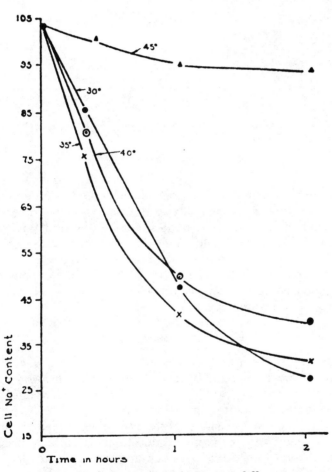

Figure 13. Escape of Na^+ from cat erythrocyte at different temperatures. Note that optimum permeability is near 35°C, suggesting an "enzymelike factor" governing Na^+ permeability.

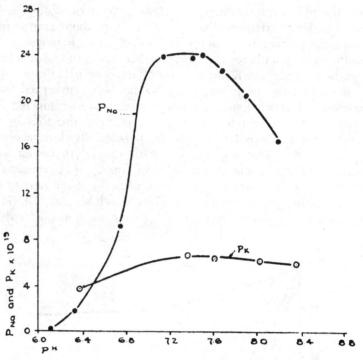

Figure 14. Permeability coefficients for Na⁺ and K⁺ as function of pH of suspension medium.

substrates into the bacterial cell. Finally, to complete the early history of the facilitated transport concept, the human erythrocyte has an unorthodox permeability to hexoses, as shown originally by Wilbrandt (68); subsequent studies, notably those of Widdas (67), LeFevre (49), and Wilbrandt et al. (70), showed that the kinetics of transport could be treated like the kinetics of an enzyme-catalyzed reaction, so that, employing Michaelis-Menten kinetics, affinity constants and transport maxima could be computed from plotting transport rate as a function of solute concentration.

MEMBRANE INHOMOGENEITY

I have traced the history of the development of the concept of the cell membrane to a period beyond the time when I devoted myself to specific membrane problems, having turned my attention to the elucidation of transport phenomena in the eye and brain, problems that permitted the application of the principles of membrane transport revealed by the experiments and thinking outlined previously in this chapter. Since this period, significant advances have been made in determining the fine structure of the membrane, which I shall not enumerate in detail. The introduction of the freeze-fracture technique for cell membranes has shown the presence of bodies

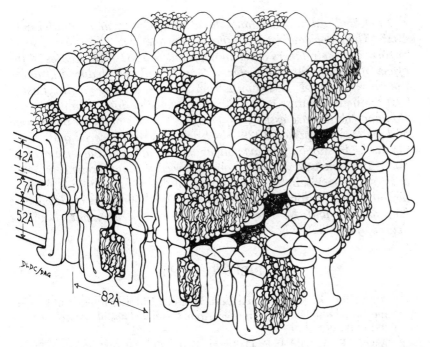

Figure 15. Model of gap-junctional region of contact of cells. [From Makowski et al. (50).]

within the bimolecular leaflet that, besides subserving specific functions such as the absorption of light, as in the disks of photoreceptors
and the grana of photosynthesis, may well be concerned in the
facilitated and active transport of solutes. These bodies may also play
a role in cell-to-cell transport as in the gap junction illustrated
schematically in Figure 15. Hand in hand with the morphological
attack has been the biochemists' attempts at isolating specific proteins from cell membranes that are apparently the morphological
basis for facilitated transport, as in the anion exchange across the
erythrocyte membrane (60).

James Danielli

In concluding this chapter, I quote my late colleague and friend,
James Danielli (13), speaking at a 1961 meeting of the American
Heart Association devoted to the plasma membrane and referring to
the fact that critical hypotheses or discoveries often arise independently more than once.

*Within my experience, there have been at least three such instances
in this field. First the concept of the lipoid layer of the plasma
membrane as a bimolecular leaflet was originally proposed by Gorter
and Grendel in a paper published in a Dutch journal in 1925. I was
quite unaware of this work when putting forward the same hypothesis*

[45]

in 1934, and did not encounter the work of Gorter and Grendel until 1939. Then the concept that the membrane was composed also of adsorbed protein layers arose in my mind in 1934 from studies of the effect of proteins upon interfacial tension and was deduced at about the same time by F. O. Schmitt from X-ray studies of myelin. Lastly, that the active centres in membrane responsible for facilitated diffusion and the like are essentially enzyme-like was deduced by Davson and me over the period 1939–1941 and was rediscovered quite independently by Monod and his colleagues in studies on bacteria around 1955. The conclusion to which one is forced is that there is commonly a degree of inevitability about scientific discovery; within a shortish period, if one person does not put forward a hypothesis or discover a fact, another will. No scientist can afford to be arrogant about the degree of originality he achieves.

BIBLIOGRAPHY

1. ANSELMINO, K. J., and E. HOENIG. Weitere Untersuchungen über Permeabilität und Narkose. *Pfuegers Arch. Gesamte Physiol. Menschen Tiere* 225: 56–68, 1930.
2. ASKEW, F. A., and J. F. DANIELLI. Discussion on surface phenomena. *Proc. R. Soc. Lond. A Math. Phys. Sci.* 155: 695–696, 1936.
3. BÄRLUND, H. Einfluss des Äthyläthers auf die Permeabilität der *Chara* Zellen. *Protoplasma* 30: 70–78, 1938.
4. BAYLISS, L. E. Reversible haemolysis. *J. Physiol. Lond.* 59: 48–60, 1924.
5. BLUM, H. F., N. PACE, and R. L. GARRETT. Photodynamic hemolysis. I. The effect of dye concentration and temperature. *J. Cell. Comp. Physiol.* 9: 217–228, 1937.
6. CHAMBERS, R. The nature of the living cell. *Harvey Lect.* 22: 41–58, 1926–1927.
6a. CHAMBERS, R. The physical state of protoplasm with special reference to its surface. *Am. Naturalist* 72: 141–159, 1938.
6b. CLOWES, G. H. A. On the role played by electrolytes in determining the permeability of protoplasm. *Proc. Am. Soc. Biol. Chem.* xiv, 1916.
7. COHEN, G. N., and J. MONOD. Bacterial permeases. *Bacteriol. Rev.* 21: 169–194, 1957.
8. COLE, K. S. Surface forces of the *Arbacia* egg. *J. Cell. Comp. Physiol.* 1: 1–9, 1932.
9. COLLANDER, R., and H. BÄRLUND. Permeabilitätsstudien an *Chara ceratophylla*. *Acta Bot. Fenn.* 11: 1–14, 1933.
10. CURTIS, H. J., and K. S. COLE. Membrane resting and action potentials of squid nerve. *J. Cell. Comp. Physiol.* 19: 135–144, 1942.
11. DANIELLI, J. F. The thickness of the wall of the red blood corpuscle. *J. Gen. Physiol.* 19: 19–22, 1935.
13. DANIELLI, J. F. Structure of the cell surface. *Circulation* 26: 1163–1166, 1962.
14. DANIELLI, J. F., and H. DAVSON. A contribution to the theory of permeability of thin films. *J. Cell. Comp. Physiol.* 5: 495–508, 1934.
15. DANIELLI, J. F., and E. N. HARVEY. The tension at the surface of mackerel

egg oil, with remarks on the nature of the cell surface. *J. Cell. Comp. Physiol.* 5: 483–494, 1935.

16. Davson, H. Studies on the permeability of erythrocytes. III. The cation content of erythrocytes of rabbit's blood in hyper- and hypo-tonic sera. *Biochem. J.* 30: 391–393, 1936.

17. Davson, H. Studies on the permeability of erythrocytes. VI. The effect of reducing the salt content of the medium surrounding the cell. *Biochem. J.* 33: 389–401, 1939.

18. Davson, H. Ionic permeability. The comparative effects of environmental changes on the permeability of the cat erythrocyte membrane to sodium and potassium. *J. Cell. Comp. Physiol.* 15: 317–330, 1940.

19. Davson, H. The influence of the lyotropic series of anions on cation permeability. *Biochem. J.* 34: 917–925, 1940.

20. Davson, H. The effect of some metabolic poisons on the permeability of the rabbit erythrocyte to potassium. *J. Cell. Comp. Physiol.* 18: 173–185, 1941.

21. Davson, H. The haemolytic action of potassium salts. *J. Physiol. Lond.* 101: 265–283, 1942.

21a. Davson, H. Growth of the concept of the paucimolecular membrane. *Circ. Res.* 26: 1023, 1962.

21b. Davson, H. *General Physiology* (3rd ed.). New York: Little, Brown, 1964.

22. Davson, H., and J. F. Danielli. Studies on the permeability of erythrocytes. V. Factors in cation permeability. *Biochem. J.* 32: 991–1001, 1938.

23. Davson, H., and J. F. Danielli. *The Permeability of Natural Membranes.* Cambridge, UK: Cambridge Univ. Press, 1943.

24. Davson, H., and E. Ponder. Studies on the permeability of erythrocytes. IV. The permeability of "ghosts" to cations. *Biochem. J.* 32: 756–762, 1938.

25. Davson, H., and E. Ponder. Photodynamically induced cation permeability and its relation to hemolysis. *J. Cell. Comp. Physiol.* 15: 67–74, 1940.

26. Davson, H., and J. M. Reiner. Ionic permeability: an enzyme-like factor concerned in the migration of sodium through the cat erythrocyte membrane. *J. Cell. Comp. Physiol.* 20: 325–342, 1942.

27. Dirken, M. N. J, and H. W. Mook. The rate of gas exchange between blood cells and serum. *J. Physiol. Lond.* 73: 349–360, 1931.

28. Dutrochet, R. J. H. Nouvelles observations sur l'endosmose et l'exosmose, et sur la cause de ce double phénomène. *Ann. Chim. Phys.* 35: 393, 1827. [Quoted by Smith (63).]

29. Finean, J. B. The nature and stability of nerve myelin. *Int. Rev. Cytol.* 12: 303–361, 1961.

30. Fischer, M. H. *Oedema and Nephritis.* New York: Wiley, 1921.

31. Fricke, H. The electric impedance of suspensions of biological cells. *Cold Spring Harbor Symp. Quant. Biol.* 1: 117–124, 1933.

32. Geren, B. B. The formation from the Schwann cell surface of myelin in the peripheral nerves of chick embryos. *Exp. Cell Res.* 7: 558–562, 1954.

33. Gorter, E., and F. Grendel. On bimolecular layers of lipoids on the chromocytes of blood. *J. Exp. Med.* 41: 439–443, 1925.

BIOLOGICAL
MEMBRANES

[47]

34. GRENDEL, F. Über die Lipoidschicht der Chromocyten beim Schaf. *Biochem. Z.* 214: 231–241, 1929.

35. GRYNS, G. Über den Einfluss gelöster Stoffe auf die rothen Blutzellen mit den Erscheinungen der Osmose und Diffusion. *Pfluegers Arch. Gesamte Physiol. Menschen Tiere* 63: 86–119, 1896.

36. HAMBURGER, H. J. *Osmotischer Druck und Ionenlehre in den medizinischen Wissenschaften.* Wiesbaden, Germany: Bergmann, 1902, vol. 1.

37. HARRIS, J. E. The reversible nature of the potassium loss from erythrocytes during storage of blood at 2–5°C. *Biol. Bull. Woods Hole* 79: 373, 1940.

38. HEDIN, S. G. Über die Permeabilität der Blutkörperchen. *Pfluegers Arch. Gesamte Physiol. Menschen Tiere* 68: 229–338, 1897.

38a. HENDERSON, L. J. *Blood: A Study in General Physiology.* New Haven, CT: Yale Univ. Press, 1928, p. xix.

39. HÖBER, R. Eine Methode, die elektrische Leitfähigkeit im Innern von Zellen zu messen. *Pfluegers Arch. Gesamte Physiol. Menschen Tiere* 133: 237–253, 1910.

40. HÖBER, R. Ein zweites Verfahren die Leitfähigkeit im Innern von Zellen zu messen. *Pfluegers Arch. Gesamte Physiol. Menschen Tiere* 148: 189–221, 1912.

41. HÖBER, R. The permeability of red blood corpuscles to organic anions. *J. Cell. Comp. Physiol.* 7: 367–391, 1936.

42. HOFFMAN, J. F. The link between metabolism and active transport of Na in human red cell ghosts. *Federation Proc.* 19: 127, 1960.

43. HOFFMAN, J. F. The active transport of sodium of ghosts of human red blood cells. *J. Gen. Physiol.* 45: 837–859, 1962.

44. HOFFMAN, J. F., and F. M. KREGENOW. The characterization of new energy dependent cation transport processes in red blood cells. *Ann. NY Acad. Sci.* 137: 566–576, 1966.

45. JACOBS, M. H. Early osmotic history of the plasma membrane. *Circulation* 26: 1013–1021, 1962.

46. JACOBS, M. H., and A. K. PARPART. Osmotic properties of the erythrocyte. II. The influence of pH, temperature and oxygen tension on hemolysis by hypotonic solutions. *Biol. Bull. Woods Hole* 60: 95–119, 1932.

46a. JACOBS, M. H., and A. K. PARPART. Is erythrocyte permeable to hydrogen? *Biol. Bull. Woods Hole* 62: 63–76, 1933.

47. JACOBS, M. H., and A. K. PARPART. The influence of certain alcohols on the permeability of the erythrocyte. *Biol. Bull. Woods Hole* 73: 380, 1937.

47a. KERR, S. E., and V. H. KIRKORIAN. Effect of insulin on distribution of nonprotein nitrogen in blood. *J. Biol. Chem.* 81: 421–424, 1929.

48. KOEPPE, H. Der osmotische Druck als ursache des Stoffaustausches zwischen rothen Blutkörperchen und Salzlosungen. *Pfluegers Arch. Gesamte Physiol. Menschen Tiere* 67: 189–206, 1897.

49. LeFEVRE, P. G. Evidence of active transfer of certain non-electrolytes across the human red cell membrane. *J. Gen. Physiol.* 31: 505–527, 1948.

50. MAKOWSKI, L., W. L. D. CASPAR, W. C. PHILLIPS, and D. A. GOODENOUGH. Analysis of the X-ray diffraction data. *J. Cell Biol.* 74: 629–645, 1977.

51. NÄGELI, C., and C. CRAMER. *Pflanzenphysiologische Untersuchungen.* Zurich, Switzerland: Schultess, 1855, pt. 1. [Quoted by Smith (63).]

52. Ørskov, S. L. Untersuchungen über den Einfluss von Kohlensäure und Blei auf die Permeabilität der Blutkörperchen für Kalium und Rubidium. *Biochem. Z.* 279: 250–261, 1935.

53. Osterhout, W. J. V. Permeability in large plant cells and in models. *Ergeb. Physiol. Biol. Chem. Exp. Pharmakol.* 35: 967–1021, 1933.

54. Osterhout, W. J. V., and S. E. Hill. Salt bridges and negative variations. *J. Gen. Physiol.* 13: 547–552, 1930.

55. Ostwald, W. *Z. Phys. Chem.* 17: 189, 1895. [Quoted by Hamburger (36).]

56. Overton, E. Über die allgemeinen osmotischen Eigenschaften der Zelle, ihre vermutlichen Ursachen und ihre Bedeutung fur die Physiologie. *Vierteljahrsschr. Naturforsch. Ges. Zuer.* 44: 88, 1899. [Quoted by Jacobs (45).]

57. Overton, E. Studien über die Aufnahme der Anilinfarben durch die lebende Zelle. *Jahro. Wiss. Botanik* 34: 669–701, 1900.

58. Pfeffer, W. *Pflanzenphysiologie.* Leipzig, Germany: Engelmann, 1890.

59. Robertson, J. D. The ultrastructure of adult vertebrate myelinated fibres in relation to myelinogenesis. *J. Biophys. Biochem. Cytol.* 1: 275–278, 1955.

60. Rothstein, A., and M. Ramjessingh. The functional arrangement of the anion channel of red blood cells. *Ann. NY Acad. Sci.* 358: 1–12, 1980.

60a. Schmitt, F. O., R. S. Bear, and K. J. Palmer. X-ray diffraction studies on structures of nerve myelin sheath. *J. Cell. Comp. Physiol.* 18: 31–42, 1941.

61. Schmitt, F. O., and K. J. Palmer. X-ray diffraction studies of lipide and lipide-protein systems. *Cold Spring Harbor Symp. Quant. Biol.* 8: 94–101, 1940.

62. Smith, H. W. A knowledge of the laws of solution. *Circulation* 21: 808–817, 1960.

63. Smith, H. W. The plasma membrane, with notes on the history of botany. *Circulation* 26: 987–1012, 1962.

63a. Starling, E. H. On the absorption of fluids from the connective tissue spaces. *J. Physiol. Lond.* 19: 312–326, 1895–96.

64. Teorell, T. Permeability properties of erythrocyte ghosts. *J. Gen. Physiol.* 35: 669–701, 1952.

65. Traube, M. Experimente zur Theorie der Zellenbildung und Endosmose. *Arch. Anat. Physiol. Physiol. Abt.* p. 87–165, 1867.

66. Van't Hoff, J. H. Die Rolle des osmotischen Druckes in der Analogie zwischen Losungen und Gases. *Z. Phys. Chem.* 1: 481–508, 1877.

66a. Whittam, X. X. *Nature Lond.* 219: 610, 1968.

67. Widdas, W. F. Facilitated transfer of hexoses across the human erythrocyte membrane. *J. Physiol. Lond.* 127: 318–327, 1954.

68. Wilbrandt, W. Die Permeabilität der roten Blutkörperchen für einfacher Zucker. *Pfluegers Arch. Gesamte Physiol. Menschen Tiere* 241: 302–309, 1938.

69. Wilbrandt, W. Die Permeabilität der Zelle. *Ergeb. Physiol. Biol. Chem. Exp. Pharmakol.* 40: 204–291, 1938.

70. Wilbrandt, W., S. Frei, and T. Rosenberg. The kinetics of glucose transport through the human red cell. *Exp. Cell Res.* 11: 59–66, 1956.

III

Membranes, Molecules,
Nerves, and People

J. DAVID ROBERTSON

THIS paper is, as requested, an autobiographical account of my involvement with the evolution of ideas about the molecular structure of biological membranes and nerve fibers since the electron microscope began to be used productively in this field in the late 1940s. I was asked to write about people as well as science and to include anecdotes and snapshots that I might have whenever appropriate. This chapter is an abridged version of a paper published in the *International Review of Cytology* (138) for a different audience and with a different emphasis.

One purpose of this chapter is to help young people in the early stages of a research career to see how science evolves and new ideas are generated and disseminated. I shall thus be quite personal and candid about my own career, including some of the mistakes I have made that might be instructive.

BACKGROUND AND TRAINING

I was born in Tuscaloosa, Alabama, in 1922, the son of a city policeman and a county elementary school teacher. I graduated from Tuscaloosa High School and entered the University of Alabama in 1939 at age 16, having decided that I wanted to be a physician and do medical research. As a premed student in biology, I attracted the attention of the chairman of the department, Dr. Charles Pomerat, who offered me a job as a teaching assistant in biology for the following year. I accepted with alacrity.

Charlie Pomerat (Fig. S1) was in his first year at Alabama. He had finished his Ph.D. in biology at Harvard a few years earlier, where (I learned much later) he was a classmate of Keith Porter, who was later to play a part in my scientific life. Pomerat, who had just returned from a postdoctoral training period with Houssay in South

Figure S1. Charles M. Pomerat in his office at the University of Alabama in the early 1940s.

America, was a very charismatic and fascinating teacher. He was also a talented and ambidextrous artist. His lectures in biology were beautifully organized, and he always filled the blackboard with marvelously detailed colored drawings constructed with both hands while he talked. He gave me a sense of excitement about biology that I have never lost. He was just starting his research in tissue culture, a field in which he was later to achieve considerable stature. He had a great deal of influence on me and a small group of students, who were also teaching assistants. I remember that he took a group of us, including John and Jimmy Gregg (now Professors of Zoology, Emeriti, at Duke and the University of Florida, respectively), to a marine biology lab on the Gulf of Mexico one summer. I became fascinated then by invertebrate marine animals and still like marine biology and working with invertebrates.

Toward the end of my college career, Pearl Harbor was bombed and I joined the navy. I was given a commission as an ensign HV(P) early in 1942 and told to continue in school. That fall I entered the two-year University of Alabama Medical School, from which I transferred to Harvard in 1943. The navy soon put me on active duty as an apprentice seaman.

I received the M.D. degree in 1945. After graduation, I interned at Boston City Hospital (BCH) in internal medicine on Harvard's famed IVth medical service. At that time, in places like BCH, interns were paid only room, board, and laundry, so I was poor again. However, I was so busy I didn't notice it until near the end of my internship when I began dating my future wife. This I managed by selling my microscope, an old brass monocular Spencer, which was the only possession I had. Because of the war, internships had been restricted to nine months, but because World War II had just ended as our internship began, we were allowed an additional three months. Initially I had wanted to intern in pathology and, while in medical school, had actually accepted an internship in pathology at Columbia

Presbyterian Hospital. However, the navy objected and encouraged me to intern in surgery or medicine. I withdrew from the pathology internship and, after a competitive examination, began one in internal medicine at BCH.

When we were given the additional three months as interns (before two years of obligated military service), I naturally chose to work at the Mallory Institute of Pathology at BCH, named for the late, great pathologist F. B. Mallory. I had completed a fourth-year elective there under Fred Parker, the Chief of the laboratory. Dr. Parker was a reclusive, but respected and stimulating, teacher. One of the sons of F. B. Mallory, Kenneth, was Parker's second-in-command, and I was also acquainted as a student with Mallory's other son, Tracy, who was the Pathologist-in-Chief at Massachusetts General Hospital. I found him an inspiring person: he helped sustain my interest in pathology. Although my experience at the Mallory Institute was very good, I became increasingly frustrated by the limitations of light microscopy.

In 1948, upon discharge from active naval duty, taking advantage of the GI bill, and using some savings from my practice and what I could earn as a staff physician in the Student Health Department (five dollars an hour, I think), I enrolled as a graduate student at MIT, where I expected to learn about the electron microscope (EM) and how it could be applied to biological problems. After the first year, I was awarded an American Cancer Society Fellowship ($3,500 per year) and my wife enrolled at Radcliffe College as a transfer student in English Literature from Huntington College in Montgomery, which she had attended after we were married. She was also able to take advantage of the GI Bill because she had been a first lieutenant in the Army Nurse Corps overseas during the war, so we were well supported through the next three years of graduate school.

Crayfish Giant Fiber Synapses

In 1949 I knew of nobody who had managed to get any meaningful pictures of synapses with the EM, so I decided that this would be a good topic for my Ph.D. thesis. My first inclination was to try to get some sections of rat or mouse stellate ganglion. However, Professor F. O. Schmitt, my thesis advisor, discouraged me because he thought that I would not be able to identify synapses with certainty. He had insisted that I should find a well-defined system in which I would be able to know exactly where I was in sections for the EM. I finally settled on the giant nerve fiber median-to-motor synapse in the crayfish. The light microscopy on this synapse had been done in 1924 by Johnson (79). It was known from the work of Wiersma (171–173) to be a one-way conducting synapse, and I felt sure I could locate and identify the pre- and postsynaptic fibers in electron micrographs.

The crayfish has a ladderlike central nervous system that runs the

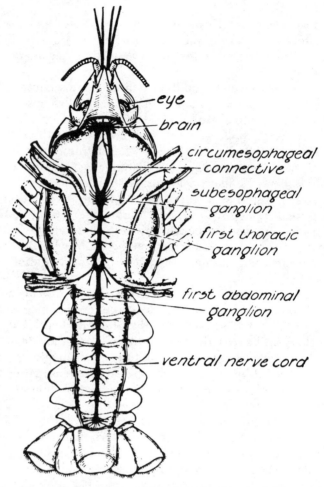

Figure 1. Diagram of crayfish with blackened nerve cord. [From Curtis and Guthrie (31a).]

whole length of the animal and consists of a bilaterally symmetrical, almost entirely ventral, nerve cord (Fig. 1). In common with other invertebrates, it is composed of unipolar neurons, in which the single process of the cell serves both axon and dendritic functions, and glia cells, commonly called Schwann cells. Most of the nerve cells are aggregated into the ventral parts of bilaterally symmetrical ganglia, with one ganglion for each body segment. The most cranial ganglion, dorsal to the others, is called the brain. It is connected by two circumesophageal commissures; the first subesophageal ganglion and all the other ganglia are ventral to the gut. The nerve cord contains four dorsally located giant nerve fibers, each about 100 μm in diameter, that run the entire length of the animal; all receive sensory impulses from the periphery. Figure 2A is a diagram redrawn from Johnson (79) of one of the abdominal ganglia, viewed from the dorsal surface, showing the four giant nerve fibers (two median and two

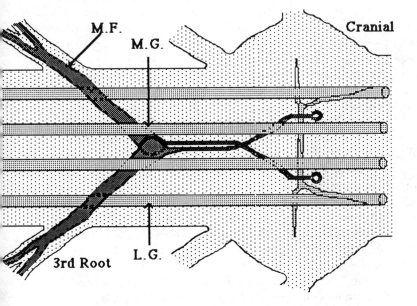

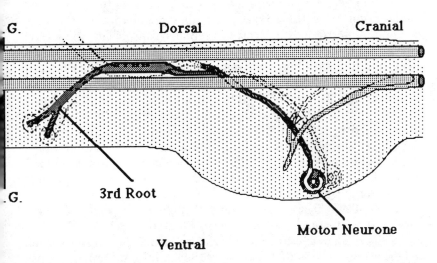

Figure 2. A: dorsal view of abdominal ganglion of crayfish nerve cord with two median giant (M.G.) and lateral giant (L.G.) fibers. Cells of origin of giant motor fibers are shown in ventral part of ganglion; axoplasm is darkly stippled. They decussate at midline as they course caudally and dorsally toward contralateral sides, from which they leave ganglion in third roots. They expand and make synapses, with median and lateral giant fibers contacting median giants ventrally and lateral giants dorsally. B: cells of origin of motor fibers are shown in side view. Cells of origin of lateral giant segmental fibers are not shown, but Johnson (79) shows branches of these axons contacting at midline, with another branch running ipsilaterally and expanding cranially as it forms septal synapse with terminating lateral giant from next caudal segment. [Adapted from Johnson (79).]

lateral) and three nerve roots leaving it on each side. Diagonally disposed septae interrupt the lateral giants in the cranial part of each ganglion; these are indicated at the top. The lateral giant fibers are spoken of as if they were single, continuous giant fibers, and indeed they appear so structurally and functionally. However, the septae that interrupt them really are the boundaries of separate giant fibers, one for each segment. The septae are very fast acting two-way–conducting electrical synapses that unite these fibers functionally so that they act together as if they were each individual giant fibers.

MEMBRANE
TRANSPORT

The cells of origin of the motor giant fibers are shown in Figure 2A at the level of the first nerve root in the ventral part of the ganglion. These are illustrated better in Figure 2B. Their axons run dorsally and decussate near the level of the second nerve roots. They then expand to become giant fibers as they each synapse respectively, first with the median giant fibers and then with the lateral giants, before leaving the ganglion in its third root to provide motor innervation to the body muscles. Johnson does not show the cells of origin of the lateral giant fibers but illustrates two axons that come up from the ventral side in each half of the ganglion and branch once. One branch contacts the corresponding branch of the other in the midline, and the other branch expands laterally to become the cranial segment of the lateral giant fiber proceeding to the next ganglion. As it expands it forms the septal synapse. Thus the lateral giant, despite being referred to as a single nerve fiber, is really a succession of giant fibers, one for each segment, in contact in each segmental ganglion in the septal synapse. The median giant fibers, in contrast to this, are believed to originate as axons of at least two separate cells in the brain that expand and run the whole length of the nerve cord without interruption. Figure 3, *top*, is a simpler diagram of an abdominal ganglion, showing how the motor fibers synapse with the afferent giant fibers in side view; Figure 3, *bottom*, is a similar diagram from a side view, indicating how the septum is formed in the lateral giant fibers.

Figures 4 and 5 are transections of abdominal ganglia showing some of these features [taken from my Ph.D. thesis (105)]. Figure 4 shows the median giant-to-motor synapses and Figure 5 is a section at a more cranial level that also shows the lateral giant septal synapse on the left side. Wiersma (171–173) showed that, although the afferent giant-to-motor and the septal synapses look the same morphologically, they are quite different physiologically. The median-to-motor and lateral-to-motor synapses are functionally polarized, conducting impulses only in the afferent-to-efferent direction. There is an appreciable synaptic delay of 0.3–0.5 ms. The septal synapses conduct equally well in either direction, and the synaptic delay is very short (~0.1 ms) for the whole length of the animal. Because of this, the latter were regarded as electrical synapses rather like the artificial "ephapses" that Arvanitaki (1) obtained by simply putting carefully dissected giant nerve fibers into apposition.

I soon found something by light microscopy that had been missed in previous work. I discovered that the postsynaptic giant fiber extended numerous small fingerlike synaptic processes only a few micrometers in diameter through the combined sheaths of the two giant fibers that made intimate contact with the median giant axon. Figure 6 is a light micrograph of a transection through an abdominal ganglion showing these synaptic processes. I thought this was an original discovery until I read a 1939 paper by J. Z. Young (175) that described essentially the same thing in the squid giant fiber

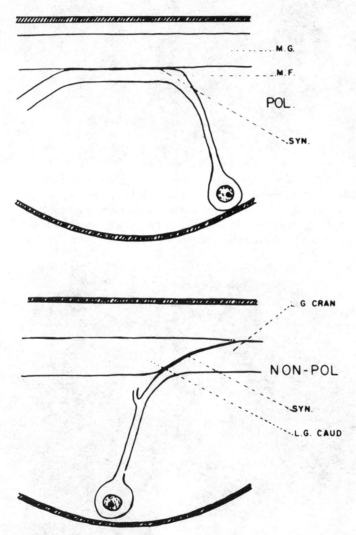

Figure 3. *Top*: simplified diagram showing how motor fibers (M.F.) make polarized (POL.) synapse with afferent giant (M.G.) fibers. *Bottom*: similar diagram showing how nonpolarized (NON-POL) lateral giant septal synapses (SYN.) are formed by apposition of caudal segment of lateral giant (L.G. CAUD.) with its cranial segment (L.G. CRAN.).

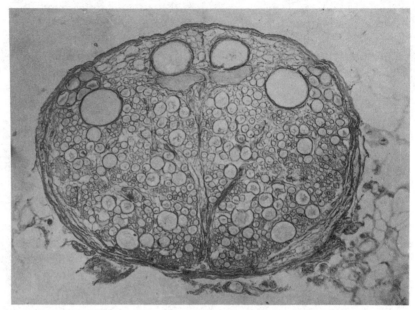

Figure 4. Transection of crayfish nerve cord showing all four giant fibers dorsally and two motor fibers synapsing with medial giant fibers. Note increased density of axoplasm of motor fibers. Fixed by vom Rath method (see ref. 107), embedded in paraffin, sectioned with steel knife at ~10 μm, and stained with hematoxylin and eosin. ×165.

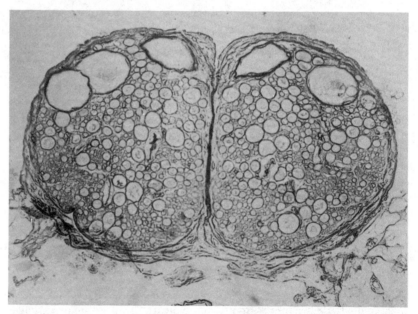

Figure 5. Same preparation as for Fig. 4 but sectioned more cranially to level of motor giant fiber. Lateral giant septal synapse appears on *left* but is not present on *right* at this level of sectioning. ×165.

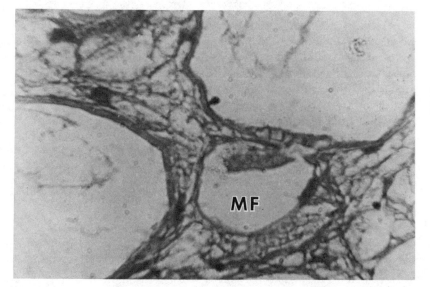

Figure 6. Same preparation as for Figs. 4 and 5 but at higher magnification
and showing fingerlike synaptic processes of motor giant fiber (MF). Axoplasm of motor fiber is greatly shrunken in this section. ×925.

synapse in the stellate ganglion of *Loligo*. The nice thing was that
these synaptic processes were so much smaller than the presynaptic
giant fiber that there was no doubt that I could unambiguously
identify the pre- and postsynaptic structures with the EM. It was
quite clear, in my first electron micrographs, that the pre- and
postsynaptic membranes were intimately apposed but continuous
structures. I think it is fair to say that this was the first confirmation
by electron microscopy of Ramón y Cajal's neuron doctrine. I first
presented this work at Cold Spring Harbor in 1952 (106). Alan
Hodgkin, Bernhard Katz, Andrew Huxley, K. S. Cole, H. J. Curtis,
and many other people whose work I knew about, were in the
audience. I gave a ten-minute presentation and remember that I got
a standing ovation, the first and only one of my career. This first
presentation at a scientific meeting had a profound effect on my
subsequent career because it led Katz to invite me in 1954 to come
to London to set up a laboratory at University College. I published
these findings in abstract form in the *Cold Spring Harbor Symposium
on Quantitative Biology* (106), in my thesis (107), and in several
papers from 1952 to 1955 (107, 108, 111), as well as more information in a review article in 1961 (124). Incidentally, in the review
article I included a diagram showing the two giant motor fibers
forming synapses with one another. Later I decided this was wrong
but have not yet finished the serial-section study that led me to this
conclusion and thus have not published it. I still intend to do this
unless someone else takes it up.

[59]

THE KANSAS PERIOD

In 1952 I received a Ph.D. degree in biochemistry from MIT. I considered staying on at MIT for a year or so as a postdoctoral fellow under Schmitt, but I had three offers of positions as an Assistant Professor of Pathology at the University of Pennsylvania in Philadelphia, the University of Washington in Seattle, and the University of Kansas in Kansas City, the latter of which I accepted.

At that time I intended to pursue my interest in pathology, so I worked essentially as a resident in pathology for three years. However, it soon became apparent to me that I could not do an adequate job of routine pathology and do the kind of research I wanted to do. I tried to do the clinical work, but my heart was not in it; I spent most of my time on my research efforts and ultimately abandoned pathology.

The Motor End Plate

Besides continuing my previous work, I thought it would be a good idea to try to get some electron micrographs of motor end plates because nobody had done that. I settled on the lizard *Anolis carolinensis* as a suitable substitute for the European adder that my reading suggested would be the best animal. I soon obtained some sections of lizard motor end plates and sent off an abstract to the American Association of Anatomists (AAA) for presentation at the annual meeting in Galveston, Texas, in April, 1954 (109). Later that year I presented essentially the same material at the annual meeting of the Electron Microscopy Society of America (EMSA) (110). When I wrote the abstract for the AAA meeting, I had not seen the synaptic vesicles in the endings, but by the time I got to the meeting, I had quite clear pictures of them. George Palade (95) and James Reger (102) also reported on EM studies of motor end plates at that meeting, and Palade and I had similar pictures. Reger was still working with thick sections with the plastic removed, so his pictures, though very clearly displaying important new features of motor end plates, did not show such important details as the vesicles. Palade had seen the vesicles before me because he had described them in his abstract. However, neither he nor Palay, who assisted him, published the pictures. Reger published his findings in 1955 (103), but because he had used the old thick-sectioning methods, his paper did not reveal the vesicles or other fine features of the endings. When my micrographs were published, I believe they were the first ones recorded showing synaptic vesicles and similar fine structural features in motor end plates. To be sure, Palay (96, 97), De Robertis and Bennett (42), Fernandez-Moran (50), and Sjoestrand (153) published pictures of synaptic vesicles in central nervous system synapses at about that time, and their probable function as carriers of chemical transmitters was immediately recognized.

[60]

Spiral Myelin

Two other discoveries worth describing were made during the Kansas City period while working on lizard muscle fibers. Perhaps I should begin with some background. In the late 1940s, Herbert Gasser, then Director of the Rockefeller Institute, began EM work on peripheral nerves. He was specifically interested in the structures of "C" fibers, and came to MIT to ask for some technical advice from Schmitt, who had been one of his colleagues at Washington University in St. Louis. Schmitt asked Betty Geren to show him how we did things, and Betty worked with him to some extent, although not to the point of being a coauthor. I was not involved in any of this directly. Gasser presented a preliminary paper at the previously mentioned Cold Spring Harbor Symposium in which he gave a new concept about how C fibers were constructed (59, 60). It had been thought previously that C fibers were bundles of axons completely embedded in syncytial Schwann cells. Gasser realized this was not so because he saw membranous structures leading from each axon to the surface of the Schwann cell. He thought the axon was simply embedded in the Schwann cell and was related to it rather like the gut in the body cavity of a vertebrate is related to the mesentery. So he coined the term *mesaxon* for each of the paired membrane structures connecting each axon to the surface of the Schwann cell. Figure 7 is a micrograph of a small unmyelinated nerve showing these features. It would be hard to overestimate the importance of this concept because it was the fundamental key to understanding the relationships between peripheral and central nerve fibers and the satellite cells. This was a most impressive accomplishment, which excited my respect and admiration. Here was a person who had spent his life as a neurophysiologist (earning a Nobel prize in 1934 with J. Erlanger), who in his later years had the administrative responsibility for a major research institute, and who nevertheless ventured into a field like electron microscopy and succeeded in making one of the most important advances of the time. I had the privilege of getting to know this remarkable and inspiring man during this period. The last time I saw him, not long before he died, we spent the afternoon at the Rockefeller Institute poring over my micrographs of nerve fibers; I was subjected to the most penetrating and insightful questions I had ever encountered. When we parted, he invited me to stay to have dinner with him, but I regret very much that I declined because I wanted to get back to Boston that night.

Betty Geren, my former colleague at MIT, began studying developing nerve fibers in 1953 in her new laboratory in the pathology department at Children's Hospital in Boston. She observed that before myelin appeared there were nonmyelinated nerve fibers having a one-to-one relationship with Schwann cells; they had long mesaxons wrapped in a spiral around the axon. This led her to suggest

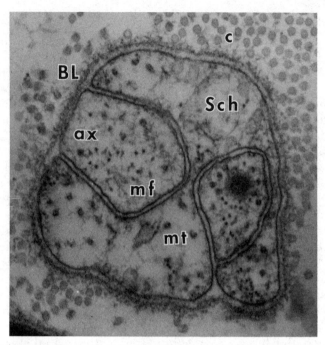

Figure 7. Cross section (ca. 1963) of unmyelinated C nerve fiber fixed in
OsO₄, embedded in Araldite, sectioned with diamond knife, and stained
with lead salts. Note very clear basement lamina (BL), distinct collagen fibers
(c), and clarity of microfilaments (mf) and microtubules (mt) in axoplasm
(ax) and Schwann cell (Sch). ×65,000.

(61, 63) that the myelin sheath might be simply a spirally wound,
tightly compacted mesaxon. She told me about this privately, but I
was quite skeptical. I knew of no other EM evidence for this, and I
think it is fair to say that neither of us thought of the spiral apparatus
of Golgi-Rezzonico in this connection as we should have. However,
while I was looking for motor end plates in lizard muscle, I naturally
saw many transections of small myelinated nerve fibers near the
nerve endings, as in Figure 8. I shall never forget the elation I felt
when I examined this micrograph with a hand lens and realized what
it meant. It showed exactly what would be predicted by Betty's
theory—an outer (*arrow 1*) and inner (*arrow 2*) mesaxon running
into the outer and inner surfaces, respectively, of the myelin sheath
in opposite directions—just as they should if myelin were simply a
spirally wound, tightly compacted mesaxon. I immediately realized
that she was right and telephoned her to tell her. I stopped in Boston
on my way to London in November 1954 to show the micrograph
to her and Schmitt. It was published in 1955 (112).

It seemed that Geren was right, but my micrograph did not prove
the main point because, although it seemed clear that myelin must
form by some mechanism involving spiral evolution of the mesaxon
around the axons, the micrograph was of such low resolution that

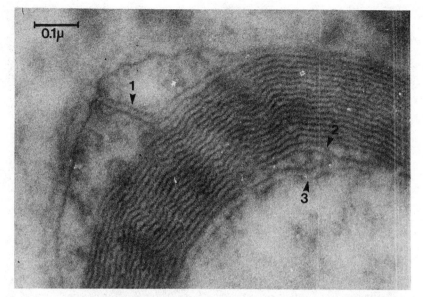

Figure 8. High-magnification micrograph of portion of myelin sheath in lizard nerve fiber, indicating outer mesaxon (*arrow 1*), inner mesaxon (*arrow 2*), and paired axon–Schwann cell membranes (*arrow 3*). This is original micrograph in which these relationships were first seen (published in 1955). ×126,000. [From Robertson (112).]

one could not be sure that there were not some other components added between the spiral loops of the mesaxon to make myelin. The essential details had to await elucidation by more technical advances, as described subsequently. I describe first another discovery made with lizard muscle while I was looking for motor end plates because I think there is a lesson here for young people.

The Transverse Tubule System of Muscle

While looking at longitudinal sections of lizard skeletal muscle fibers, I observed the relationships described in Figures 9 and 10. These are the original micrographs that I published in 1956 (113), and they show the presence of membraneous elements located between the myofibrils regularly at the A-I junctions. I recall vividly that I was working late at night in my laboratory in Kansas City sometime during the winter of 1953 to 1954, when I first saw these structures. Everything suddenly clicked in my mind as I realized what they meant, and I got that intense frisson that comes so rarely to a scientist when he makes a truly new and important discovery. I conceived of these tubules as parts of a continuous system of transverse membrane sheets running all the way across the muscle fiber from one surface of the plasma membrane to the other, enveloping the myofibrils. I realized that this explained an anomaly that A. V. Hill had pointed out earlier (75); that is, it is quite impossible for

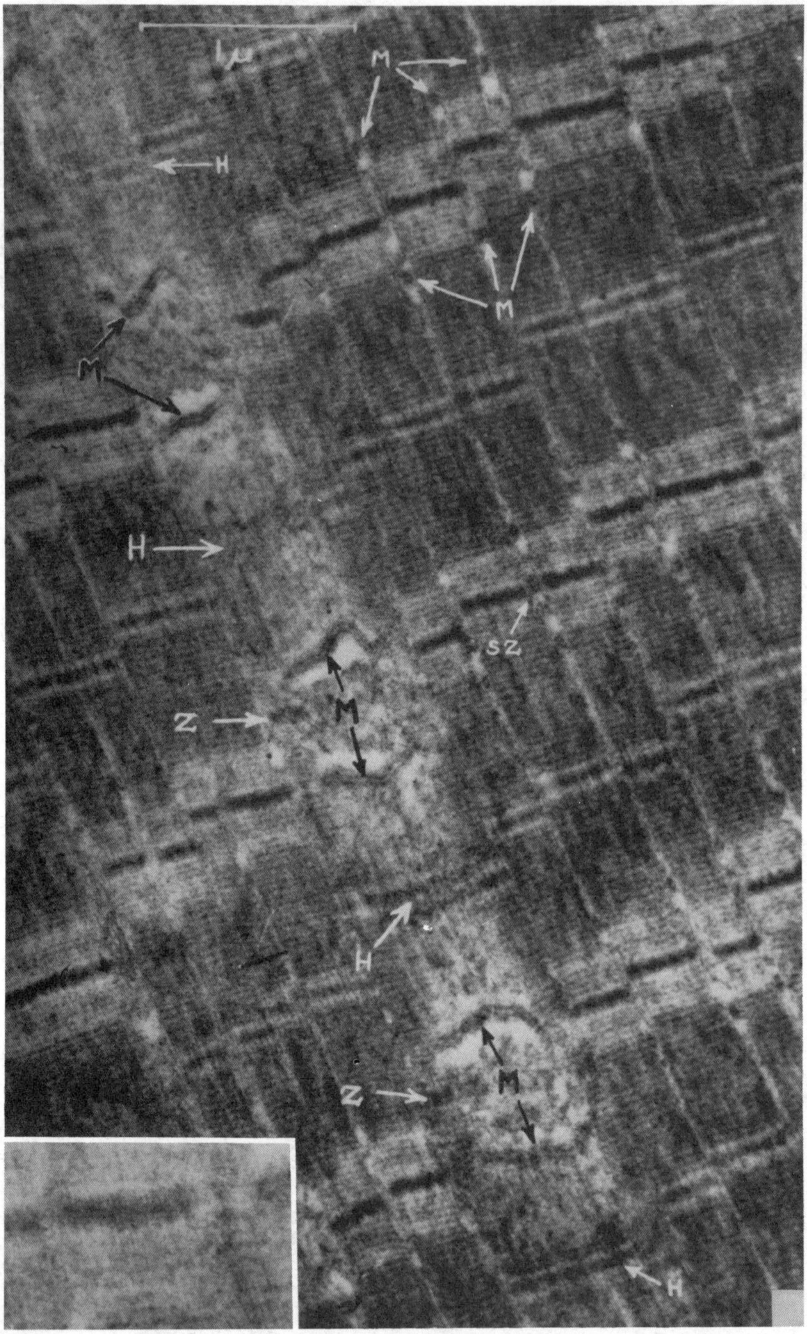

Figure 9. Longitudinal section of lizard muscle fiber showing A and I bands with clear H and Z bands. Note tubular paired membranes (M) between myofibrils cut in cross section (*upper right*). They are located regularly at A-I junctions. Plane of section (*left center*) passes between myofibrils; tubules are seen clearly in thin layer of sarcoplasm between myofibrils. Section was stained with phosphotungstic acid; ~40-nm axial periodicity is very clearly shown. ×28,000. *Inset (lower left)* shows Z band at higher power. ×59,000. [From Robertson (113).]

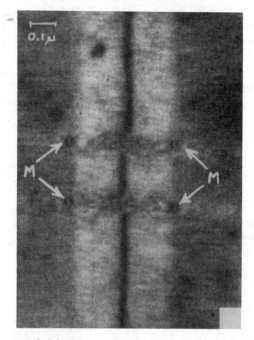

Figure 10. Longitudinal section similar to Fig. 9 showing paired membrane structures (M) in register at A-I junctions between myofibrils. ×71,000. [From Robertson (113).]

any diffusible agent released by the action potential at the surface of a muscle fiber to diffuse fast enough to be responsible for the uniform synchronous contraction of all the myofibrils. He suggested that there must be some structure running across the muscle fiber that was responsible for this. As soon as I saw these transverse membranes, I concluded that they were the structure Hill had predicted and that they probably were continuous with the surface membrane and conducted an electrical disturbance from the surface membrane across the fiber. Of course, this was jumping far ahead of the evidence, but the clues were all there and I was right. This idea, coupled with the hard evidence I had, constituted a major advance in thinking, but I dropped the ball and herein lies the lesson.

I was invited to give a lecture in a summer course in June 1954 that Schmitt had organized at MIT; I presented these micrographs and my interpretation of them to the group. Nobody claimed to have thought or heard of it before. I recall in particular that Geren was very skeptical. Shortly afterwards I was invited to give a paper at the annual meeting of the International Association of Medical Museums (later the International Academy of Pathology) early in 1955 in Houston, I believe. I wrote an abstract giving the essential findings and presented the material. Again nobody claimed to have had this idea before. Unfortunately, these proceedings were not published and I have not even been able to find a record of the meeting, so my

[65]

first two public presentations of what seemed to be a new way of looking at muscle fibers were not recorded, and this proved to be important. I did, however, send off a manuscript describing my findings and interpretations that was received by the *Journal of Biophysical and Biochemical Cytology* on May 9, 1955 (113). This paper originally contained the diagram in Figure 11.

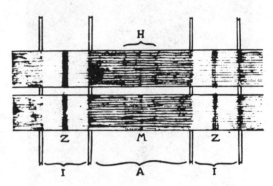

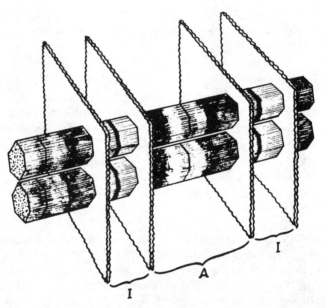

Figure 11. Diagram from original manuscript submitted to *J. Biophys. Biochem. Cytol.* in May 1955, intending to convey interpretation of paired membrane structures arranged regularly across muscle fibers at junctions of A and I bands. Drawing has not been altered and reveals lack of understanding of membrane compartmentalization at that time. Text described membranes as sometimes appearing tubular, but they were not diagrammed this way because unit membrane structure had not yet been seen and fundamental compartmentalization provided by internal membranes of cells was not at that time understood by author. [From Robertson (113).]

THE LONDON PERIOD

During the winter of 1953–1954, I received a letter from Katz asking if I would like to come to University College London to set up an EM laboratory jointly in his Department of Biophysics and in Professor J. Z. Young's Department of Anatomy. I was invited to come to London to talk about this, and in November of 1954 I flew there with my micrographs. When I showed the pictures of motor end plates to Katz he exclaimed that I was showing him something, the vesicles, that he would have had to invent to explain the miniature end plate potentials that Fatt (48, 49) and he and del Castillo (40) had just described. Katz asked me to let him include one of the micrographs showing the vesicles in a review article he and del Castillo were then preparing (41). I did this, and even though the same micrograph was published the same year (1956) in my paper on motor end plates (114), the first published picture showing synaptic vesicles in motor end plates was my picture published in the article by del Castillo and Katz (41).

Katz and Young were together getting electron microscopy started at University College, and although Katz had made the initial contact with me, it was Young who spent the most time with me on that visit. He wanted me to set up my lab in his department, though my office would be in Katz's. Figure S2 is a snapshot of John Young with his daughter Kate taken just after my period in London.

The Transverse Tubule System of Muscle, Continued

In 1955 I resigned from my job in Kansas and formally accepted the job in London. I soon had a laboratory operating in the anatomy

Figure S2. John Young holding his daughter Kate in the late 1950s.

department with my office in the biophysics department. I bought a new Siemens Elmiskop 1b, bringing the third of these new instruments into Great Britain. The lab was in operation by the end of 1955.

By this time I had learned how important the intellectual atmosphere of a school can be, and at University College I found one that exceeded my expectations by far and was most satisfying. We lived in a roomy but (to us) rather primitive flat in Earl's Court, belonging to the College. The flat did not have central heating at first, and we had no car. However, we could afford for my wife to go to the West End theatres frequently. There the best orchestra seats could be had for twelve shillings and sixpence. We soon developed a pattern of living in which she was at the theatre several nights a week with neighbors and I was in the lab. We would often meet in a pub at about 10:30 P.M. and then have a late dinner at home. Other nights I often worked until the small hours of the morning and usually slept late, rarely seeing the children at all. Sometimes before going off to school, they would come into the bedroom and say *"Hello and goodbye, Daddy,"* but I saw little of them during the week.

Of course, when I settled in London, I had brought the micrographs of all the things I had seen in Kansas City, and in January or February of 1956, I took some of them to Cambridge when I went there to give a lecture at the invitation of Alan Hodgkin.

I discussed the pictures of the transverse membranes in lizard muscle with Andrew Huxley and Hodgkin after my lecture. Huxley had recently done a classical piece of work with Taylor on frog skeletal muscle fibers (78), showing that if a micropipette was placed on the surface of a muscle fiber just over the Z line and a stimulating current was passed through its tip, a sharply localized contraction occurred involving only a few myofibrils and only the sarcomeres immediately under the microelectrode tip. He was receptive to the idea that transverse membranes might be responsible for this wave of contraction, but he tried to convince me that the Z line structure itself in my micrographs looked membranous and might be the membranous structure responsible. I didn't agree because I didn't believe that the Z line was a membrane. Some months later he came by my laboratory in London to tell me that he had been working on an invertebrate muscle with a 10-μm sarcomere length and that it showed contraction when the micropipette was placed on the surface opposite the A-I junctions. He suggested, therefore, that my interpretations of the lizard muscle might be right and that the differences between my findings and his with frogs might simply represent a species difference; later on this was found to be true, and indeed the T tubules in mammalian skeletal muscle are located at the A-I junctions.

By that time I had received the editorial comments on my paper from Keith Porter, acting for the journal. Two anonymous editors gave the manuscript very bad reviews. They said my pictures were overinterpreted and suggested that I remove the diagram of the transverse membranes and tone down the discussion. They were particularly concerned with my terminology regarding the sarcolemma, stemming from my failure to designate the thinnest dense line next to the cytoplasm as the cell membrane and to call it the sarcolemma. At that time I had not yet seen the unit membrane structure and was uncertain exactly where the cell membrane was located. The editors also clearly did not like my suggestion that the transverse membranes were responsible for electromotor coordination of contraction across the muscle fiber. I was completely intimidated, and despite receiving a telegram from Palade (soon after I got the reviews in the fall of 1955) saying that I should send the paper for publication, presumably as it stood, I proceeded to remove the diagram and much of the discussion. The original manuscript (113) contained the following sentence in reference to transverse membranes as the last sentence in the discussion section: "*It seems appropriate to indicate nevertheless that such a membrane system could obviously play an electromotor role in the synchronization of muscular contraction (22).*" Reference 22 was to A. V. Hill (75). In the final version this sentence was eliminated and I put a reference in the results section to the work of Huxley and Taylor (78) saying "*which have revived interest in an old idea that some kind of physical element probably exists to convey the influence of surface membrane depolarization across the myofibrils (22) (49).*" Evidently I put this in at the galley proof stage because it is not in my copies of the original manuscript. I must have done this after my discussion in Cambridge with Huxley. This, coupled with removal of the sentence about the electromotor role of the transverse membranes, completely distorted the meaning I originally wished to convey. About all that was left when this paper finally appeared in 1956 (113) were the pictures and the reference to A. V. Hill (75) to indicate that I had understood the findings. Partly as a result of this experience, I stopped working on muscle and concentrated on nerve structure.

The only person who has ever credited me with having anything to do with this major step in thinking about how muscle fibers work is Huxley, who mentioned in his Croonian lecture in 1973 (77) that I had shown him my micrographs and discussed their significance during the previously mentioned period. In fact there is no reason why I should have been given any credit for this idea. I was simply naive in having talked about it in public without having published it.

I have told this story for the benefit of the young and innocent who have not yet learned how science evolves; no doubt similar stories could be told by others. Good original ideas may give personal satisfaction, but if they are to be credited to an individual, they must be recorded in the literature in a timely fashion. Other kinds of communication are simply too easily forgotten. Another moral of this story is that a young person must have the strength to stand on convictions and not let journal editors deflect him or her beyond legitimate editorial matters. Certainly one should not be deflected from working on a topic. This experience should be taken as a basis for being cautious about talking about one's work before it is published, and this is my main reason for discussing it. However, I hope it will not be taken as a justification for the kind of excessively secretive behavior that I sometimes see today among some of my young colleagues. This experience obviously is a basis for some caution, but it represents an extreme case and should be viewed in this light. I still find myself talking freely about my latest work, even when I know very well that the person I am talking to is working on the same things and may well inadvertently pick up some ideas from what I say that might not be so obvious and forget from where they have originated. I don't really know why, but I simply prefer to work in the open; I feel compelled to share any new thinking I have with others. I know this is not always a good idea, but I still do it anyway. This could be classified as arrogance, humility, naivete, or simply stupidity, depending on how one looks at it! Whatever it is, I am always repelled by those who guard their ideas and data as if they were gold nuggets ready to be picked up by some thief. I don't advise my assistant professors to behave the way I do, but I must confess that it pleases me if they do.

The Unit Membrane

In 1956 John Luft (88) rediscovered $KMnO_4$ as a fixative for cytology, and almost simultaneously Glauert, Rogers, and Glauert (64, 65) introduced the epoxy resin Araldite as an embedding medium. At the urging of Audrey Glauret regarding Araldite, I tried both these methods together on peripheral nerve fibers; the first set of specimens were an abysmal failure. However, I decided to try it again, and this time it worked. Immediately I began to see membranes in a new way. In the past cell membranes had always appeared as a single, rather hazy and indefinite dense line (see Figs. 7 and 8). Now they appeared in my micrographs as sharp, triple-layered structures ~7.5 nm thick and composed of pairs of dense strata, each about 2.5 nm wide, separated by a light zone ~2.5 nm thick. They looked like they appear today in material accepted as superior (116, 117, 119). They gave the kind of triple-layered appearance seen today as the human erythrocyte membrane in Figure 12.

This preparation was fixed in glutaraldehyde and embedded in polyglutaraldehyde by the glutaraldehyde carbohydrazide (GACH) method (72); and sections were stained with uranyl and lead salts. The only essential difference is that the $KMnO_4$–fixed material was somewhat less thick and more uniform, probably due to shrinkage.

All the plasma membranes in the sciatic nerve fibers appeared

[70]

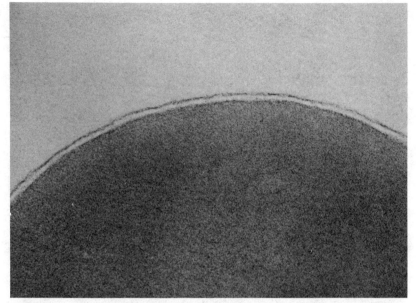

Figure 12. Section of human erythrocyte membrane fixed with glutaraldehyde, embedded in polyglutaradehyde, and sectioned with diamond knife.
Triple-layered pattern is similar to that shown by membranes fixed with
$KMnO_4$ during 1950s. ×158,000.

triple layered, as did the membranes of the membranous organelles.
I then applied this technique to sciatic nerves from newborn and
somewhat older rats and soon discovered what the structure meant.
I saw all the stages of myelination that Geren had observed (Figs.
13–15), but now the triple-layered structure of the membranes was
much clearer. There was a gap of ~15 nm between the two triple-
layered membranes of the axon and the Schwann cell, but there was
no such gap between the two apposed Schwann cell membranes
making the mesaxon (Figs. 14–15). Here the two triple-layered
structures in close contact made a pentalaminar structure ~15 nm
thick. The mesaxons were seen making less than one and up to three
or more loops around the axon in a simple spiral. In some places one
could see the mesaxon loops in contact along their cytoplasmic
surfaces making two lamellae of compact myelin. Finally, there were
many fibers showing several layers of compact myelin, with the outer
and inner mesaxons related to the myelin just as I had observed
earlier in lizard nerve fibers, except that now I could resolve the
substructure of the membranes. Figure 15 is one of the original
micrographs showing the triple-layered structure of the Schwann
cell membranes of the mesaxon and how the outer and inner mesaxons are related to compact myelin. The findings are summarized
in Figure 16, which shows in the diagrammatic cross sections how
myelin originates.

Most importantly, I could now say with certainty that, regarding

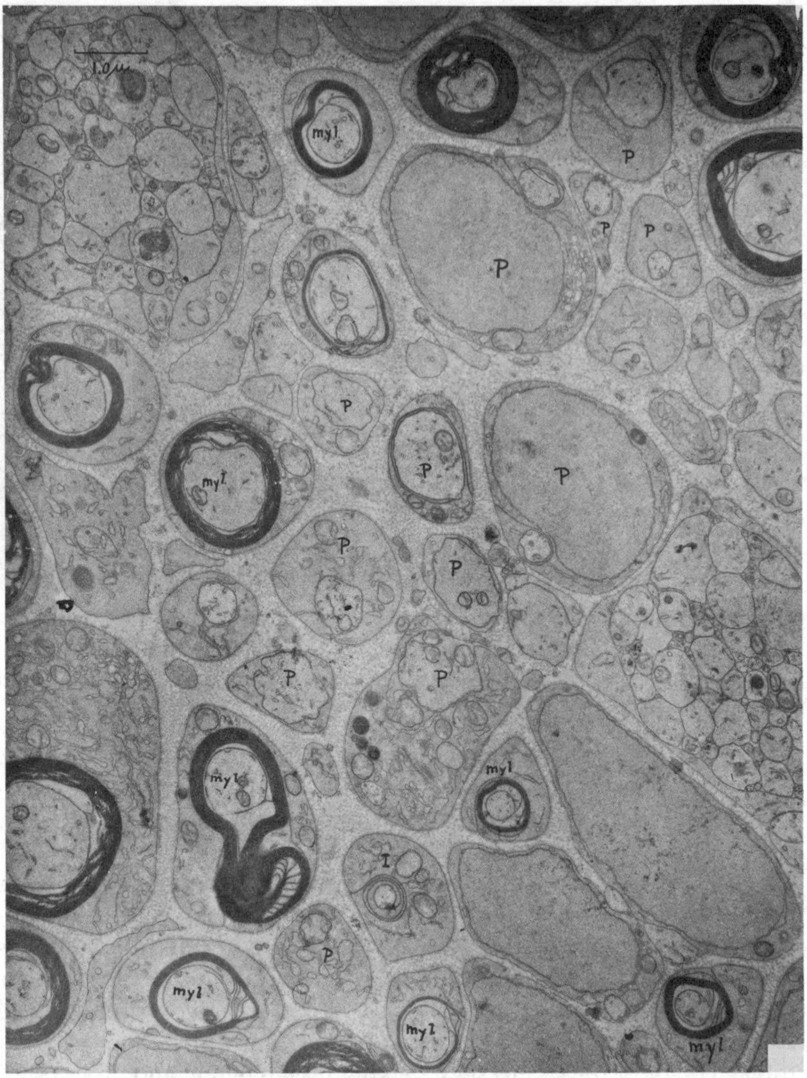

Figure 13. Low-power survey micrograph of transection of developing sciatic nerve of neonatal mouse showing myelin protofibers (P), intermediate fibers (I), and early myelinated fibers (myl). KMnO₄ fixation, epoxy embedding. ×10,250.

the resolution of the electron micrographs (~2.5–3.0 nm), there was nothing added to the membrane components as compact myelin was formed. The repeat period in the compact myelin of permanganate-fixed nerve fibers was ~16 nm. This was closer to the repeat period of 17.1–18.6 nm measured by X-ray diffraction in fresh intact nerve fibers of amphibians and mammals (148) than to the period of ~12 nm that Sjoestrand had observed in OsO₄-fixed myelin (152), so I believed that the triple-layered structure was a better representation of the real membrane structure than the fuzzy dense line seen in OsO$_4$–fixed material.

Having established these morphological relationships, I was now in a position to make a deduction about the molecular structure of the triple-layered structure. As previously mentioned, Schmidt studied myelinated nerve fibers by polarization-optical methods in the 1930s (144) and found that fresh fibers showed radially positive intrinsic birefringence in myelin but that the positivity reversed to negativity after lipid was extracted. This led him to propose that myelin consisted of layers of lipid molecules in a smectic or nematic fluid crystalline state, with the long axes of the lipid molecules oriented radially, alternating with layers of elongated protein molecules with their long axes predominantly oriented tangentially, as in Figure 17. Schmitt, Bear, and Clark (145) established the dimensions of the radially repeating unit by X-ray diffraction studies of fresh myelin. Because the repeat period of smectic fluid crystals of the kinds of lipids in myelin was ~4 nm, it was clear that the repeating unit of myelin, if it also incorporated protein, could have no more then two bimolecular layers of lipid and only a few monolayers of protein. There were several ways that the lipid and protein molecules

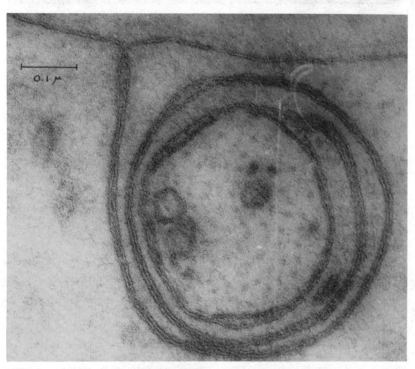

Figure 14. Cross section of intermediate myelinating fiber in same preparation as for Fig. 13 showing two unit membranes in close apposition along their outside surfaces to make future intraperiod line of compact myelin. ×148,000.

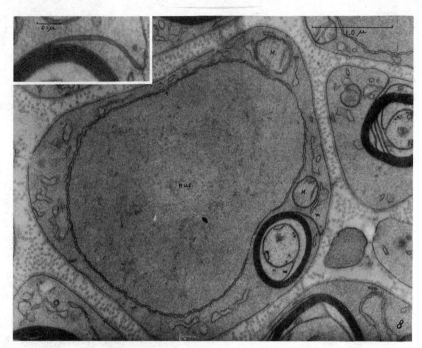

Figure 15. Same preparation as for Fig. 13 with section of early myelinated fiber including Schwann cell nucleus (nuc.). Outer and inner mesaxons (m) and two mitochondria (M.) are indicated; outer mesaxon is enlarged (*upper left*) to show two unit membranes coming together to make major dense line of compact myelin. ×22,000; *inset,* ×64,000.

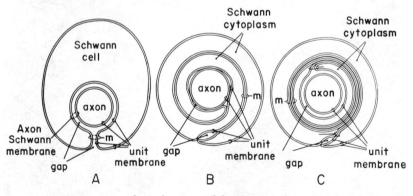

Figure 16. Diagrams of mechanism of formation of myelin sheath, as described in text. Myelin (m) is present after first junction of cytoplasmic surfaces of mesaxon (m) to make one major dense line. First intraperiod line begins in mesaxon with apposition of two external membrane surfaces.

could be arranged, as in Figure 18 (148). Brian Finean, a British X-ray diffractionist and postdoctoral fellow in Schmitt's laboratory at MIT when I was there, had made an educated guess about the most likely repeating structure and published the diagram shown in Figure

[74]

19 (52). It was apparent to me that, if Finean's diagram was correct, the molecular structure of the triple-layered structure of the Schwann cell membrane could be deduced by comparing the molecular diagram with the EM picture and establishing identities between

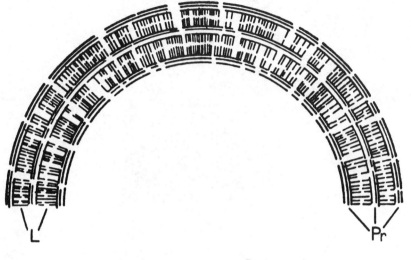

L: lipid   Pr: protein

Figure 17. Diagrammatic conception of molecular organization of myelin sheath as alternating layers of lipid and protein, with long axes of protein molecules primarily tangentially disposed and lipid molecules primarily radially disposed. [From Schmidt (144).]

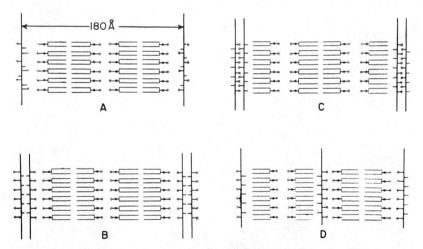

Figure 18. Four molecular diagrams (*A–D*) indicating four possible arrangements of lipid and protein molecules in radial repeating unit detected by X-ray diffraction in myelin. Lipid molecules are represented by tuning fork symbol, protein molecules by line for backbone chain, and side chains by perpendicular short lines. [From Schmitt et al. (148).]

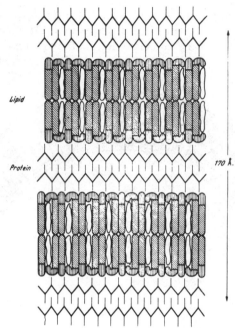

Figure 19. Conception of radially repeating unit of myelin. [From Finean (52).]

the various layers of the two repeating units, neglecting the small differences between the dimensions. The slightly smaller dimensions in the EM sections were assumably due to shrinkage. I did this, as indicated in Figure 20, and concluded that the triple-layered structure of the Schwann cell membrane represented the underlying molecular pattern in Figure 21. The core of the membrane was depicted as a single bilayer of lipid with the polar heads directed outward and covered by monolayers of protein. It was obvious from the electron microscopy that there was a chemical difference between the protein on the outside surface and that on the inside because the reactivity to fixatives and stains was clearly very different. This was also apparent from the X-ray diffraction work of Finean because the repeat period would have had to have been half the value obtained if there hadn't been a significant chemical difference between the two surfaces of the bilayers. Finean referred to this as the "difference factor," which I guessed was the presence of carbohydrate in the outer surface of the membrane. I indicated this chemical asymmetry in molecular diagrams in later papers by filling in part of the protein as a zigzag line because I was thinking of the dimensions of the repeating unit. A lipid bilayer is ~5 nm thick, and a fully spread denatured protein monolayer is ~1.2–1.5 nm thick. The repeating unit needed to contain two bilayers and four monolayers of protein, according to Finean's analysis. The overall thickness of the lipid would be ~10 nm. This left ~7.0–8.5 nm to be

[76]

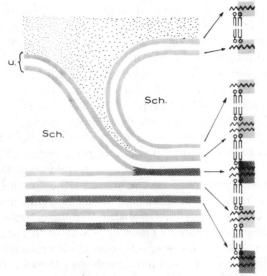

Figure 20. Composite diagram superimposing Finean's molecular diagram on myelin diagram, including outer mesaxon, as seen in Fig. 15. Chemical asymmetry of unit membranes in compact myelin is indicated by darker stippling for cytoplasmic surfaces. Single bilayer of free Schwann cell membrane with its polar surfaces directed outward and covered by monolayers of nonlipid is indicated. [From Robertson (119).]

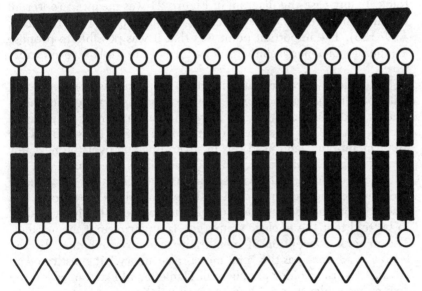

Figure 21. Diagram of unit membrane with outer nonlipid monolayer inked in to show chemical asymmetry.

occupied by protein. In my initial analysis, I supposed that this protein would be fully spread out like protein in a monolayer at an air-water interface and hence ~1.5 nm thick. This much protein would fit comfortably into the repeating unit with a little room left over for some carbohydrate residues, which I supposed would be covalently bound to the outer protein monolayer. I wasn't thinking about glycolipids then, but of course, they too could play a role.

Of course, one could suppose that the protein molecules were composed of α-helices, β-sheets, or random coils, and depending on the exact thickness of the polypeptide chains, more than one chain could possibly be fitted into two or even all four of the monolayers. If the protein was integral, each molecule could be thicker. Today we know that the two principal myelin proteins are the Folch-Lees proteolipid protein (53) and the basic protein (83). Both of these seem to be integral, most of the mass of the Folch-Lees protein seems to be in the external surface of the glia cell membrane, and most of the basic protein mass probably is in the cytoplasmic surface. Together these two proteins make up ~85% of the protein of CNS myelin. We still do not know exactly how these proteins are arranged in the myelin membrane and how much they increase the thickness of the bilayer, although Blaurock (7, 8) has some recent evidence from X-ray diffraction studies of myelin. However, at the time, the concept of integral proteins had not been introduced, and I was thinking only of how protein monolayers could be fitted into the structure without penetrating the lipid; it seemed likely that the molecules were spread out into layers about one polypeptide chain thick. In any case the diagram in Figure 21 was meant to represent this all rather crudely, as explained in various papers (119–123, 125, 129–133). The important point was that it was possible to propose a general molecular architectural pattern of the triple-layered structure.

The triple-layered structure had been found in plasma membranes, mitochondria, Golgi membranes, and endoplasmic reticulum membranes, and it now appeared to be the fundamental repeating unit in the myelin sheath. I began to refer to it in 1957 (115–116) as the "unit membrane" and used this in several papers over the next few years (117, 119, 125, 127, 129). This perhaps ungrammatical, but graphic, term came into widespread use, and I find it very useful today in making a clear distinction between a single membrane and a compound one such as the "nuclear membrane." It helps a student to be told that the nuclear membrane is a pair of unit membranes. The general conception of membrane structure embodied by the term was advanced as the "unit membrane theory." It constituted an advance over the older paradigm of Danielli and Davson, called the "pauci-molecular theory," because it added three new features. First, it restricted the lipid core to a single bilayer; second, it introduced the new idea of chemical asymmetry; and third, it was applicable in a rigorous sense as a generality to all biological membranes. However, the generalization had to await more evidence, as detailed later.

[78]

This seemed a fairly satisfying argument, but I was well aware that it was a very flimsy one because it all depended importantly on the correctness of the X-ray diffraction analysis of Finean. We all knew very well that Finean's analysis was only one of several possibilities because the X-ray phase problem had not been solved and his particular choice of phases was therefore arbitrary. This meant that the repeating units could be any of the ones specified earlier by Schmitt et al. [Fig. 18; (148)]. I thought that we might get some useful information on this problem by looking at lipid model systems. I had met Dr. Margaret McFarlane in London and knew that she had some very pure preparations of lipids of the type that occurred in membranes. She kindly gave me some highly purified samples of phosphatidylcholine, phosphatidylethanolamine, and some others. I also made some crude ethanol-ether (50:50 vol/vol) extracts of rat brain. I dried samples from these various preparations and treated the residues like pieces of tissue, fixing them in OsO_4 and $KMnO_4$, embedding them in epoxy, and cutting thin sections for study. Right away I saw a layered structure (as in Fig. 22) consisting of dense and light strata repeating periodically at ~4 nm. Geren had done a similar thing earlier with a phospholipid (62) and noted that the repeat period correlated very closely with the repeat that was measured in such systems by X-ray diffraction. It was quite clear that we were looking at lipid bilayers in a smectic fluid crystalline state with the molecular pattern in cross section shown in Figure 23a. The problem was that we didn't know whether the dense strata represented the lipid polar heads (as in Fig. 23b) or the carbon chains (as in Fig. 23c). Clearly, on symmetry grounds, it had to be one or the other. The problem was to identify a single bilayer in isolation and see how the densities were distributed. It was clear that if the nonpolar carbon chains were responsible for the dense strata, a single bilayer should appear in cross section as a single dense stratum; if the polar heads were responsible, then each bilayer should appear in transection as a pair of dense strata making a triple-layered structure similar to but thinner than a unit membrane. I shall never forget the joy I felt late one night in London when I first saw what I thought was a single bilayer in isolation. I wanted to run down the hall and find somebody to show it to, but I was apparently alone in the building. It was triple-layered like the unit membrane but only ~6 nm thick. I immediately knew that this effectively solved the X-ray phase problem. This first observation was made on the crude alcohol-ether brain extract I had made. This preparation turned out to contain scattered micelles that appeared in cross section as stacks of bilayers of unequal width. Each whole micelle appeared like an isolated spicule tapering at both ends, perhaps a few hundred nanometers long and 50–100 nanometers thick in the middle. At high power the middle part was made of a lattice of repeating dense and light strata, like the ones in Figure 22. However, at the tapering ends of the micelles, the last one turned out to be a single pair of dense strata with a uniformly wide light zone in the middle, making a triple-layered structure similar to but thinner than a unit membrane; none

Figure 22. Appearance of egg lecithin fixed with OsO_4, embedded in epoxy resin, and sectioned with glass knife. Repeat period is ~4 nm. × 1,200,000.

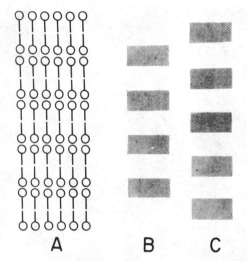

A   B   C

Figure 23. Diagram showing smectic fluid crystal of lipid in *A*, assumed to underlie Fig. 22, and two possible interpretations of dense and light strata in *B* and *C*.

of the micelles tapered to a single dense stratum (122). It was then immediately clear to me that this triple-layered structure must be a single bilayer. This meant that the polar heads of the lipid molecules were responsible for the dense strata [as in Fig. 23b; (122)]. Although it seemed convincing to me, I realized that it might not be convincing to someone else, so I repeated the procedure with the very pure preparations from Dr. McFarlane. Here I found areas in which the bilayers of the smectic fluid crystals were hydrated, and I could clearly see several individual bilayers, all of which appeared triple layered. One of the original micrographs showing this is reproduced in Figure 24. Although I had not consciously and systematically set out to do the experiment, as soon as I saw these triple-layered structures, I knew that in effect I had done the experiment reported by Schmitt et al. (148) in 1941 in which they had shown by X-ray diffraction methods that a smectic fluid cystalline system of lipids could be hydrated reversibly, splitting off individual bilayers (as in Fig. 25, reproduced from their paper). This gave me great confidence in my analysis. Figure 26 shows the only three plausible molecular interpretations of the triple-layered pattern (models A–C). Because I now knew that both the protein and the lipid polar heads showed densities in electron micrographs of sections, it was possible to eliminate models B and C; model B would show three dense strata and two light zones for each bilayer, and model C would show simply a single dense band. This assumed, of course, that there was enough resolution in the micrographs to detect the five-layered structure in model B. However, even if this were not so, the image would be a single dense band, not the triple-layered structure seen.

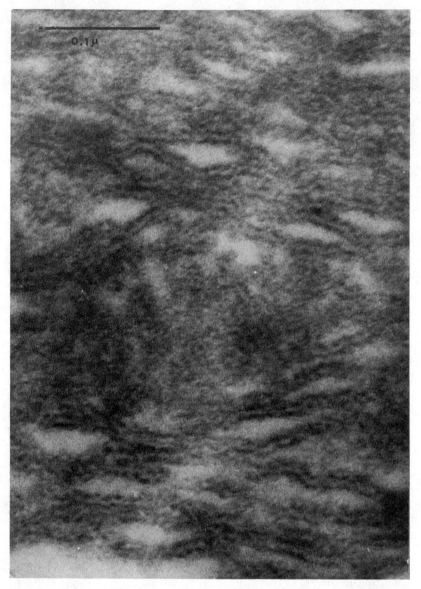

Figure 24. Part of preparation of egg lecithin like that in Fig. 22 but partially hydrated. Note that individual triple-layered structures <10 nm thick are seen. ×245,000.

Thus I felt that model A was the only plausible one. At this point I believed I was no longer dealing with a theory of membrane molecular organization but with an established fact. I began to refer to the "unit membrane concept" rather than the "unit membrane theory."

In 1958 I went to the European Regional Electron Microscopy Congress in West Berlin and reported these findings. I met Walther Stoeckenius there and found that he had done a similar study of lipid

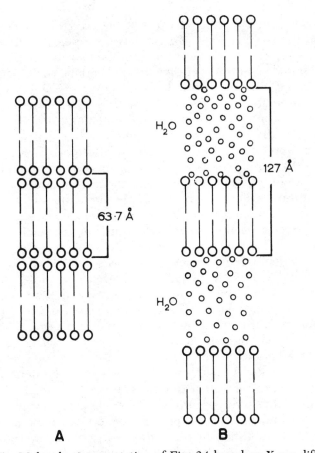

Figure 25. Molecular interpretation of Fig. 24 based on X-ray diffraction experiments. [From Schmitt et al. (148).]

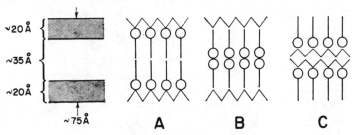

Figure 26. Diagrams of 3 (*A–C*) possible interpretations of the triple-layered unit membrane structure.

model systems but had come to a totally different conclusion. As I interpreted his pictures, he had observed single bilayers in isolation and had even succeeded in coating them with a thin layer of protein. He showed pictures of a triple-layered structure like mine. However, he had concluded that each of these triple-layered structures was two bilayers covered with monolayers of protein. He evidently adopted this interpretation because be believed that the dense strata seen in the smectic fluid crystals must represent the lipid carbon chains, since this was where the double bonds were known to be concentrated and it was known from the work of Von Criegee (166, 167) that OsO_4 reacted with double bonds. Stoeckenius was basing his interpretation mainly on the chemistry of the reaction of OsO_4 with bilayers and was misled; my interpretations were based mainly on morphology, not only of bilayers but of nerve fibers. We still do not know precisely why the density of the polar heads becomes higher than that of the carbon chain region of bilayers, but there are some plausible reasons and there is no doubt at all about the observed fact that this is true. We can rationalize the findings by saying that, while some osmium is deposited in the centers of bilayers as a result of the Criegee reaction, more is deposited in the polar regions for two reasons. First, OsO_4 reacts directly with some of the polar groups. Perhaps as importantly, OsO_3 is a product of the Criegee reaction and it is relatively hydrophilic. Hence it tends to be driven out of the hydrophobic core of the bilayer and deposited preferentially in the polar regions. These are rationalizations to explain the facts, but even if they are not correct, the facts remain. The polar heads are regularly more dense than the carbon chains in preparations of the kind we were using.

Stoeckenius and I had a public discussion on the floor at one of the plenary sessions at that Congress on this fundamental disagreement about the interpretations of our micrograph models. I presented my findings in a symposium at the meeting, and my views were finally published in the *Proceedings* volume in 1960 (122). Stoeckenius published his findings and interpretations in 1959 (157). Some time later Stoeckenius did some experiments based on Vittorio Luzatti's work with Husson on controlled lipid-phase changes in model systems monitored by X-ray diffraction (89). These workers in France found that it was possible to get lipid molecules to orient in hexagonal cylindrical columns in which the lipid polar heads were together with little or no water or, alternatively, with the polar heads surrounding columns of water. Stoeckenius fixed such lipid phases with OsO_4, sectioned them, and found that the polar heads were stained but the carbon chains were not. He published these findings (158) in tandem with a paper by Luzatti and Husson (89) in 1962, and this constituted elegant support of my position in our previous debate.

At about this time, what seemed to me a curious phenomenon occurred. Somehow the word got around that I thought all biological membranes were the same; "the same" was soon translated to "identical." To this day there are people who believe that this is what I was saying. This, of course, completely distorted my meaning and missed the point. I was saying that all membranes were constructed on the same architectural principle. Perhaps I did not point out explictly enough that this architectural principle could include very wide variations in chemical species to account for the obvious diversity of different kinds of membranes, but I thought that this was obvious to anyone. Perhaps my use of the word *unit* caused this misunderstanding. In fact, in a paper I wrote in 1951 entitled "The Unit Membrane" (124), I said the following (which I quote verbatim from page 88, line 6):

"The attachment of substances such as RNA granules to the unit membrane does not alter the general conception nor does wide variation in the molecular species composing the lipid core or the non-lipid monolayers. There is ample room for specificity within this structural pattern."

During this period a group in the microbiology department across the street at University College Medical School (a separate institution for reasons too complicated to review here) was working on influenza virus and I was given a sample of the pure virus to examine by EM. I fixed some of the viruses with $KMnO_4$ and examined sections. To my surprise I found the unit membrane structure at their surfaces. I found this disturbing because I did not know that the virus contained lipid. I asked about this, was told that lipid was present, and was given some figures. Just after receiving the figures, I happened to be seated with Hugh Davson at lunch in the faculty dining room. I told him about what I had found and about by concern that there might not be lipid in the virus particles. However, we looked at the chemical data and together proceeded to calculate whether or not there was enough lipid. We concluded that there was almost exactly enough lipid to make a bilayer at the surface of each particle.

This incident may serve to indicate that I was acquainted with Davson and we got along well. I did not at first know Jim Danielli, but I met him on the occasion of presenting an invited paper at a symposium organized by the Biochemical Society in London in 1958. It was published in 1959 (119), and I believe it is my most quoted paper (135). At this meeting there was a rather large audience, and I gave about a forty-minute presentation in which I developed the theme of the unit membrane concept for the first time in complete detail. Danielli was in the audience, and at the end of my talk he rose to make some lengthy comments. The essence of what he said, as I remember it, was that he believed that most of what I was looking at and calling cell membranes represented "soap bubbles." I recall vividly that his opening statement was to the effect that I had just illustrated to the society that EM was then about where histochemistry had been years earlier. I attempted to answer his criticisms as politely as possible, but I did not regard them as very friendly comments. Later on I got to know Danielli and found him a very nice person whom I grew to like, but on that occasion my reaction was very negative. This experience may perhaps explain the irritation I have sometimes manifested when people refer to the unit

[85]

membrane model as the "Danielli" model. The least they could do is call it the "Danielli-Davson" model. But even that is not accurate. The point is that the pauci-molecular model could not and did not specify that the bilayer was the universal structure, not to mention saying anything about chemical asymmetry. To be sure, Danielli (32–35) and Danielli and Davson (36) had mentioned the bilayer earlier as one of the possible structures, but there was no way that the earlier evidence could, in any rigorous way, justify putting forward the bilayer as a general structure. This is quite well illustrated in their 1935 paper in which the core of the membrane is presented in a diagram as a "lipoid" one of indefinite thickness. There was in fact some evidence for restriction of the lipid to one bilayer in the red blood cell (RBC), but the evidence was equivocal and there was no evidence at all for any generalization beyond the red cell.

The work of Gorter and Grendel in 1925 (69) had suggested that the RBC membrane contained a single lipid bilayer, and the capacitance measurements of Fricke at about the same time (55, 56) had suggested that the thickness of the RBC membrane was much less than 100 Å. The measurements of Waugh and Schmitt in 1940 (170) with an instrument they called the analytical leptoscope could be interpreted as supporting the presence of only one bilayer, but that was not the only possible interpretation, and in any case, it dealt with only one particular membrane. It should be mentioned here that Schmitt, Bear, and Ponder (146, 147) found that there was radially positive intrinsic birefringence in the RBC membrane. This is consistent with the presence of a bilayer but contributes nothing about the number of bilayers or any generalization.

Thus, while some of these various studies could have been interpreted as evidence for the single bilayer in the RBC, they certainly did not support this as a general structure. Even such a conclusion for the RBC was doubtful. For instance, the techniques used by Gorter and Grendel were suspect because it was obvious by the 1950s that the method they had used to extract lipids was not adequate to get them all and the method they had used to calculate the area of the cells was flawed. The capacitance measurements, although correct, were not supportive of any generalization to other membranes because Cole and Curtis (29) had published measurements of membrane capacitance that varied from <1 $\mu F/cm^2$, which Fricke had obtained, to as much as 6 $\mu F/cm^2$ for various other membranes. This seemed to mean either that any measurements of thickness based on capacitance measurements alone were suspect or that membranes varied greatly in thickness. This ambiguity in capacitance measurements remained a problem until Lord Rothschild discovered what was wrong. Capacitance values in membranes are always calculated in terms of membrane area, and if the area figure used was wrong, the capacitance value would be wrong. Victor Rothschild showed that in marine eggs there were many infoldings of the surfaces of the eggs that made area calculations based on the superficial surface area of the egg too low (143). My work (113) (referred to previously), the work of Porter and Palade (101), and the later work of Franzini-Armstrong and Porter (54) on the transverse tubule systems of muscle gave a similar explanation for the erroneous values of muscle capacitance. The point is that in the late 1950s there was no way that the generality that I made about the bilayer could be made with any certainty except as I had done it. The unit membrane concept was the first one to establish the lipid bilayer as the general core structure of all biological membranes. This has stood the test of time, for there is now essentially no debate about any other general core structure.

The General Cell

At about this time, I proposed (119, 126) that cell structure could be considered in general terms, utilizing the unit membrane concept. I had established that all membranes had the same essential architectural pattern, but I believed that all membranes in cells were highly specific in their molecular composition and that they were chemically asymmetric, with sugar groups on their outside surfaces. So I proposed that eukaryotic cells in general could be thought of usefully as a three-phase system, as indicated in Figure 27. Here the cell is depicted with *1)* a nucleocytoplasmic endomatrix phase, which is everywhere bounded by *2)* a continuous unit membrane phase, which separates it from *3)* an exomatrix phase, which is somewhere at some time continuous with the outside world, as indicated in the diagram. Apart from mitochondria, the exomatrix is considered to represent the content of all cytoplasmic organelles, including the

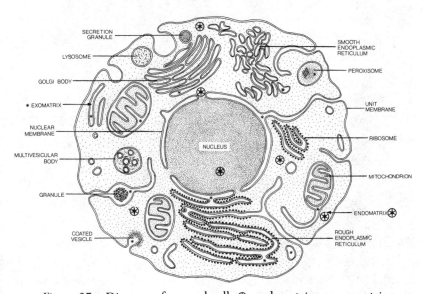

Figure 27. Diagram of general cell: ⊕, endomatrix; ∗, exomatrix.

[87]

endoplasmic reticulum, Golgi apparatus, lysosomes, endosomes, peroxisomes, secretion granules, and so forth. Mitochondria are now regarded as primitive bacteria-like symbiotic organisms dwelling in the exomatrix phase, consisting of an endomatrix phase separate from that of the cell and bounded by its own separate single unit membrane.

When I first advanced this concept of cell structure, I was uncertain about the whole matter of membrane fusion. I thought that it was possible to explain all the phenomena of pinocytosis, phagocytosis, secretion, and indeed all the phenomena that can be classified as forms of endo- or exocytosis by supposing that all the membranes of the cell are continuous, even including the mitochondrial outer membrane, and that a wisp of the exomatrix material, too tenuous to be seen by light microscopy and too unstable to be preserved in any cell fractionation procedure, was always present, connecting the various intracellular membrane compartments. Isolated unit membrane–limited vesicles seen by electron microscopy, I thought, could be explained as artifacts of preparation, simply representing break-

Figure 28. First electron micrograph of node of Ranvier. It is now clear that this is at branch point of axon, but at the time the structure was not at all clear. OsO_4 fixed. ×14,500.

down products of small unit membrane–limited tubules. The main reason I liked this was that it permitted one to understand readily how the molecular individuality of the membranes of the various compartments and the plasma membrane could be maintained. However, it soon became apparent that this viewpoint was wrong, and I modified it. This, however, requires one to imagine some specific mechanisms for accomplishing membrane fusion while maintaining the asymmetric distribution of membrane lipids and proteins that are known to exist. One also must invoke special structures in the cytoskeleton to guide and motivate intracellular vesicles to go to the right place and fuse with the correct membranes. All this is, in fact, not too difficult to imagine. There is ample evidence of the existence of special transmembrane receptor proteins that permit certain membranes to fuse selectively when brought into contact, and there is evidence accumulating to explain how the cytoskeleton could guide and motivate vesicles to the right locations. Thus clathrin cages of the cytoskeleton enclose specific vesicles in the endomatrix, and actin and myosin complexes with other proteins can readily be invoked to move vesicles in an orderly manner. Microtubules and their associated proteins play a role in this and participate with the other cytoskeletal elements in determining cell shape. As we gain more insight into how the cytoskeleton operates, all this becomes more understandable. The most difficult problem for me remains that of maintaining the chemical asymmetry, order, and specificity of the lipid and protein components of membranes during the multiple fusions of endo- and exocytosis. This is surely one of the major membrane enigmas that we need to solve today.

The Node of Ranvier

After moving to London in 1955, I continued to work on the motor end plate in lizard muscle and one day I recorded the micrograph reproduced in Figure 28. I found this picture very puzzling at the time, although I soon realized that it is a node of Ranvier at the origin of a branch of an axon. At that point the only micrographs of nodes of Ranvier that had been published were some that Gasser had included in his report at the Cold Spring Harbor Symposium mentioned earlier (59). The techniques he used at that time were inadequate to show many details, and thus his pictures had not been much more informative than light micrographs. I thought my picture must be of a node, but I could not interpret it, so I set out deliberately to obtain some micrographs of nodes. First I tried cutting longitudinal sections of myelinated nerves (such as the frog sciatic) and could not find nodes. Then I teased out some individual fibers or bundles of fibers and embedded and sectioned them. In such preparations I could see nodes by light microscopy before I prepared sections for electron microscopy, so I knew where they were. I soon found a few nodes, and once I had seen the relationships, I immediately returned to the whole nerve fibers and had no trouble at all finding nodes in both longitudinal and cross sections. I have always thought this experience was interesting in showing how difficult it is for the human mind to make the transition from a rather crude idea of a structure that is being sought to looking for something that is very specific in one's mind. I knew very well what a node looked like by light microscopy, but having never seen one with certainty by electron microscopy, I simply could not pick up the essential cues to recognize what I was observing. Once I knew exactly what to look for, there was not the least difficulty in finding nodes in sections, but before that it was quite impossible for me. Figures 29 and 30 show two of the first cross sections of nodes that I could interpret. I realized that the axon was surrounded by minute fingerlike processes of the two adjacent Schwann cells that formed a collar around the axon beneath the basement lamina and that there was a space ~10–15 nm wide between the membranes of these Schwann cell elements and the axon membrane. I immediately sent word to John Young that I had something to show him. He came to the bench beside the Siemens microscope and eagerly examined the pictures with a hand lens as I told him what I thought they meant. He studied them closely for some time and then stepped back and took my hand and shook it, congratulating me. This gave me a feeling of intense satisfaction

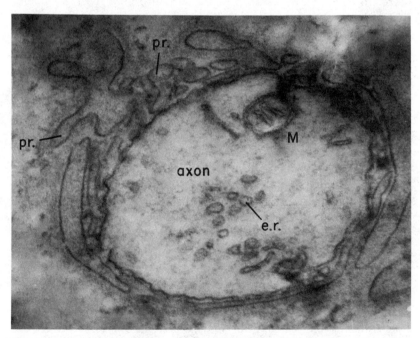

Figure 29. Early cross section of node of Ranvier showing fingerlike processes of Schwann cell (pr.) making collar around nodal axon. Axon contains mitochondrion (M) and some vesicular and tubular components of endoplasmic reticulum (e.r.). ×60,000.

[90]

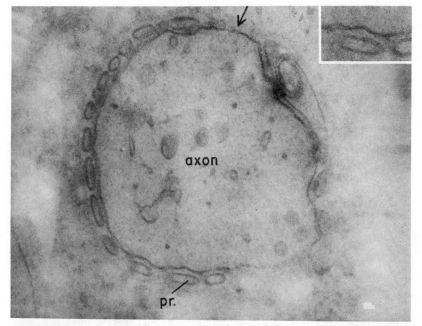

Figure 30. Transection of node as in Fig. 29 but with Schwannian nodal collar reduced to single layer of processes (pr.). One region of axon membrane is completely bare (*arrow*). Unit membrane patterns of axon and Schwann cell membranes can be delineated. It is apparent (*inset*) that there is space of variable width but at least 10 nm wide between axon and Schwann cell membranes. KMnO$_4$ fixed; epoxy embedded. ×60,000.

because he was one of the people who had studied nodes very thoroughly by light microscopy in both the peripheral and central nervous systems (74).

I then determined the structure of the node of Ranvier in longitudinal section and found that there is a collar of repeated mesaxons in the juxtaterminal myelin region (Fig. 31). I also noted that here the Schwann cell membrane and the axon membrane were in very intimate contact with closure of the 10- to 15-nm gap that was present elsewhere between these two membranes. It seemed particularly important that this gap was present between the axon membrane and the membranes of the Schwann cell nodal processes (see Figs. 29, 30). The gap also was present in the internode beyond the juxtaterminal myelin region. These relationships are shown in the longitudinal section of the juxtaterminal myelin at a node fixed in permanganate in Figure 31 and in diagrammatic form in Figure 32. I interpreted these relationships in terms of the spiral myelin concept by constructing the model shown in Figure 33. I made this model by spreading a thin layer of plasticene on a sheet of aluminum foil with a rolling pin. When I had flattened the Plasticene to ~1–1.5 mm, I covered it with another sheet of aluminum foil and rolled it

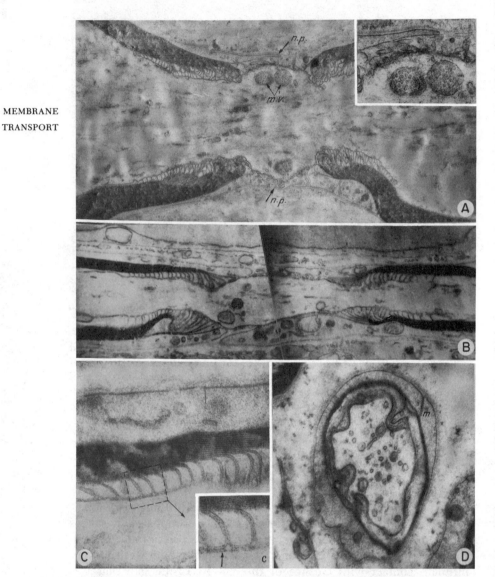

Figure 31. Juxtaterminal myelinated region of node of Ranvier. *Inset* in *C* shows axon and Schwann cell unit membranes in intimate contact. KMnO$_4$ fixed; epoxy embedded. *A,* ×10,000 (*inset* ×21,000); *B,* ×11,000; *C,* ×65,000 (*insets,* ×110,000); and *D,* ×21,700.

into tight contact with the rolling pin. Then I cut the end of this thin sheet with a knife (*A*). I let this represent the mesaxon and wrapped it around a long cylinder of Plasticene, representing the axon (*A, B*). I then rolled the mesaxon sheet tightly about the axon and tightened the terminal part of the node with my hands; this made longitudinal ridges (*B, C*). I then cut a longitudinal section of the model node (*D*) and observed the mesaxons cut in longitudinal section in the juxta-terminal myelinated region (*E, F*). I cut across the model in the internodal region and, of course, saw the outer and inner mesaxons related to the myelin sheath just as I had seen them in many sections of internodal myelin (not shown). I thought that this model fully explained the relationships of the membranes observed at the node except, of course, the fingerlike processes of the Schwann cell making the collar about the axon at the node. Furthermore, I realized that the multiple mesaxons in the longitudinal section at the node explained the structure that had been described earlier by Nageotte called the "spinous bracelet." In addition, the longitudinal ridges seen in my model in the juxtaterminal myelin region had been seen and described by light microscopists. This was surprising because my model mesaxon was a rigid structure and it was clearly this feature that led to the ridges. Perhaps this was just a coincidence, since the myelin membranes of the mesaxon would not be regarded as rigid structures by current thinking about membrane structure; perhaps this was more than coincidence. In any case, for this and other reasons, I think there is room for some revision in our current conceptions of membrane fluidity. Be that as it may, the most important relationshhips were adequately explained by the model.

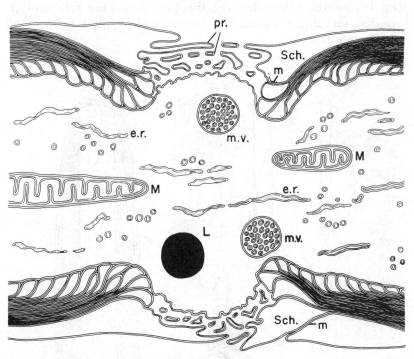

Figure 32. Diagram of longitudinal section of node of Ranvier, based on observations of permanganate-fixed material, emphasizes unit membrane relationships. Basement lamina is very faint in such preparations and was not included in this diagram. L, dense lipid body; M, mitochondria; m, mesaxon; m.v., multivesicular body; pr., Schwann cell nodal processes. [From Robertson (116a).]

The Schmidt-Lanterman Cleft

In the process of working out the structure of nodes, I also determined the structure of Schmidt-Lanterman (S-L) clefts. The only electron micrographs in the literature, of which I was aware at the time, showed practically no detail of S-L clefts, and there was some doubt occasionally expressed by neurocytologists about their being real structures; some regarded them as artifacts. However, in KMnO$_4$-fixed myelin, I soon observed the relationships seen in Figure 34. This micrograph is very poor by modern standards, but it is one of the originals in my 1958 paper (118) showing the membranes in the cleft, and it contains the essential details. It was clear that the cleft simply represents an intrusion of Schwann cytoplasmic matrix into the myelin sheath separating the myelin lamellae by pushing the myelin membranes apart at the major dense lines. This thin layer of Schwann cytoplasm followed a long helical pathway connecting the outer nucleated Schwann cytoplasm with the thin layer of cytoplasm next to the axon. The relationships were as shown in the diagram in Figure 35. In this case my paper was published long before any other showing the key membrane relationships required for understanding the structure, leaving (I believe) no priority questions.

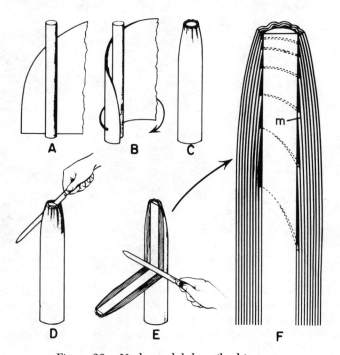

Figure 33. Node model described in text.

The Crayfish Synapse, Continued

While I was working out the structure of peripheral nerves and the unit membrane concept, I continued to work on crayfish synapses. In the second journal paper I ever published (111), I included a micrograph of a synaptic process in the median-to-motor giant synapse that showed an important membrane relationship. This was one of the first thin sections I had obtained with OsO_4–fixed material embedded in methacrylate. It showed the pre- and postsynaptic membranes in close apposition, measuring ~15 nm in overall thickness. However, it did not show the unit membrane structure, and therefore, I could not interpret it beyond saying that the micrograph showed that the pre- and postsynaptic membranes were intact, as Ramón y Cajal had predicted and as my first paper had established (108). Looking at the picture today, I wish I had at least given a more thorough description, though I could not have drawn any more conclusions. The picture shows an important difference between the nonsynaptic axon membranes apposed to the Schwann cell membranes and the pre- and postsynaptic axon membrane. The space

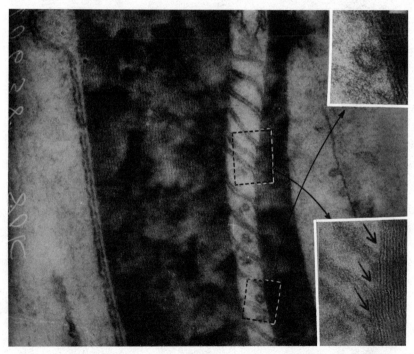

Figure 34. First published electron micrograph of Schmidt-Lanterman cleft showing key unit membrane relationships. Note unit membrane–limited vesicular profiles in Schwann cytoplasmic layers between paired membranes within the cleft. $KMnO_4$ fixed; epoxy embedded. ×57,000; *upper inset* ×171,000; *lower inset* ×131,000. [From Robertson (118).]

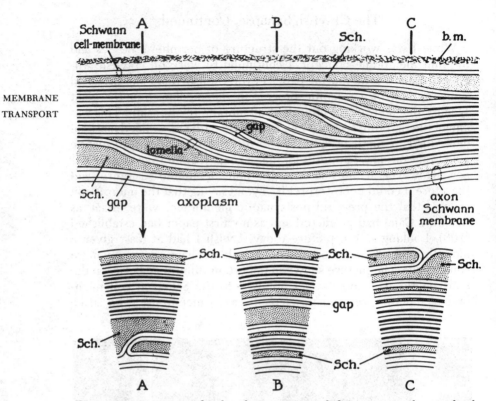

Figure 35. Diagram of Schmidt-Lanterman cleft as seen in longitudinal section (*top*) and three cross sections (*bottom*).

between the Schwann cell membrane and the axon membrane was at least twice as thick as the clear space between the pre- and postsynaptic axon membranes. In the former the overall combined thickness of the two membranes was ~30 nm, in contrast to ~15 nm at the synapse. The problem was that the membranes each appeared only as single dense strata <10 nm thick. In retrospect I can now say that this was so because only the cytoplasmic halves of the membranes were visible in the micrographs, but this is hindsight. At the time I did not understand this. I was reluctant to describe things I did not understand; perhaps there is some reason to this, but I think now that it is better for a morphologist to give a thorough description of what is apparent even if it is not understandable at the time. Good clear description has a value of its own. I think I may have been unconsciously reacting here to the perjorative sense in which "descriptive morphology" was viewed by some of the budding "molecular biology" community of the day. In any case, as soon as I had looked at these synapses in $KMnO_4$–fixed material (Fig. 36), I knew how to interpret the micrographs. There was simply no space visible between the membranes because they were in such intimate contact.

[96]

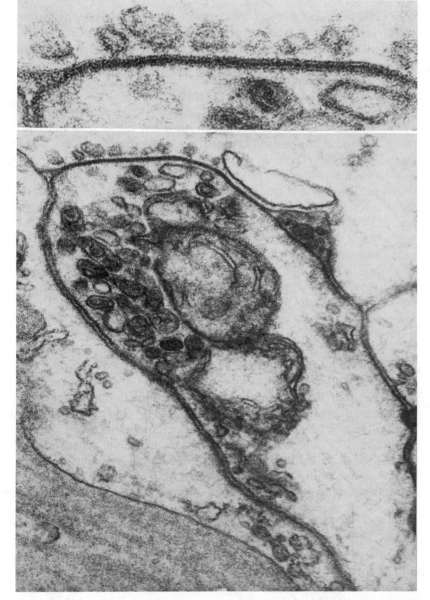

Figure 36. Electron micrograph of postsynaptic process in contact with
median giant fiber in crayfish giant fiber synapse. Schwann cell is seen (*right*
and *left*), and synaptic process axoplasm contains mitochondrion and simple
and complex vesicles. Note axon and Schwann cell membranes are separated
by gap >10 nm, but no gap is seen where pre- and postsynaptic membranes
are in contact (*inset*). Here unit membranes are in intimate contact. ×64,000;
inset (*top*), ×168,000.

[97]

THE HARVARD PERIOD

The Synaptic Disk (Gap Junction)

In 1960 I decided to leave University College and to take a position at Harvard as an assistant professor of neuropathology in the Department of Neurology and Psychiatry in the research laboratory at McLean Hospital (the psychiatric division of Massachusetts General Hospital) in Belmont, Massachusetts.

At about the same time, Ed Furshpan and David Potter joined Steve Kuffler's group in the pharmacology department at Harvard Medical School. Ed and I occasionally saw one another despite our being located on opposite sides of the Charles River and my being even beyond the Harvard campus. One day he told me that he and Furakawa, a postdoctoral fellow from Japan working with him, had found evidence, while studying goldfish medullae by microelectrode techniques, that there was an electrical synapse on the Mauthner cell (58). He didn't know where it was exactly, but I believe he suggested it might be on the lateral dendrite. The lateral dendrite receives numerous synaptic endings of VIII nerve fibers that were described in 1934 by Bartelmez (2) as the distinctive "club" endings, and it seemed likely, as Ed and I discussed the problem, that these endings were the electrical transmitters, although this was by no means certain. These endings were illustrated very well in a widely used drawing by Bodian (10, 11), reproduced in Figure 37. At the time two Harvard Medical students, David Stage and Tom Bodenheimer, were working with me in my new EM laboratory at McLean Hospital. I told them about my conversation with Ed and expressed my belief that the club endings on the lateral dendrite might be electrical endings. If so, I said that we should be able to find the pre- and postsynaptic unit membranes in close contact in these endings in the kind of relationship found in crayfish giant synapses. We proceeded to fix goldfish brain by perfusion with permanganate and in due course found exactly what we were looking for. This did not produce that rare feeling of exhilaration that I had experienced before with completely new discoveries because I was not at all surprised, but we were all very happy. Figure 38 from our original paper (139) is a light micrograph of a cross section of the lateral dendrite of the Mauthner cell showing several club endings. Figure 39 is a low-power electron micrograph of one of the endings fixed in formaldehyde followed by OsO_4 and stained with uranyl acetate. We found that the club endings were characterized by disk-shaped regions, ~0.3–0.5 μm in diameter, scattered over the extensive contact areas of the synaptic endings on the lateral dendrite in which the pre- and postsynaptic membranes were in very close contact (*arrows* in Fig. 39 and *upper right inset* enlargement). In cross sections these disks were separated by regions about the same sizes as the disks, in which the pre- and postsynaptic membranes were

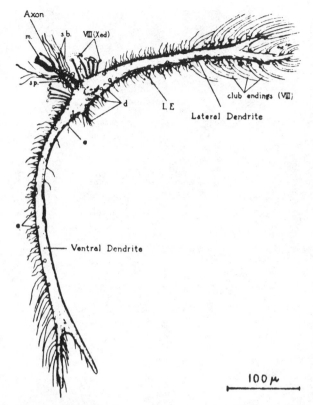

Figure 37. Diagram of Mauthner cell showing lateral dendrite with many
club endings. [From Bodian (11).]

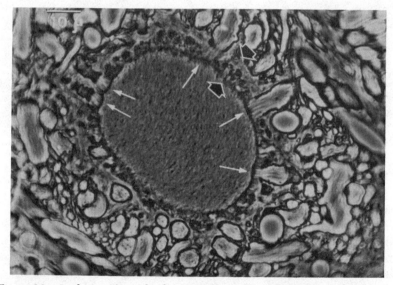

Figure 38. Light micrograph of cross section of lateral dendrite of Mauthner
cell in goldfish medulla. ×900. [From Robertson et al. (139).]

[99]

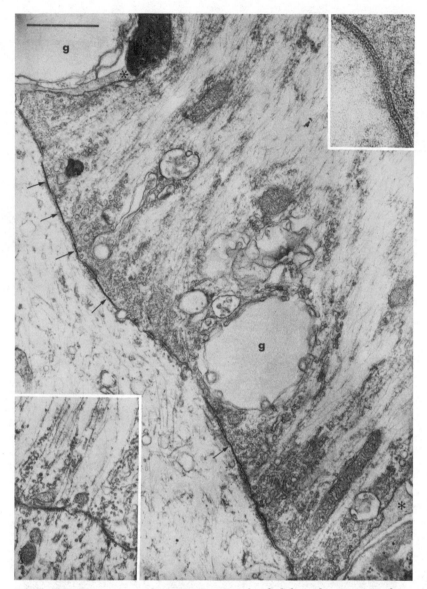

Figure 39. Low-power electron micrograph of club ending on Mauthner cell lateral dendrite. Preparation fixed in formaldehyde-OsO₄. ×12,000; *insets,* ×80,000.

separated by the usual 10- to 15-nm gap seen in other synapses. Furthermore, there were accumulations of synaptic vesicles in the terminal axon and there was some increase in density of the pre- and postsynaptic membranes in these intervening regions (*lower left inset* in Fig. 39). In the disks, which we began calling "synaptic disks," we saw some very interesting substructures, first in the permanganate-fixed material, but also in formaldehyde-OsO₄–fixed material.

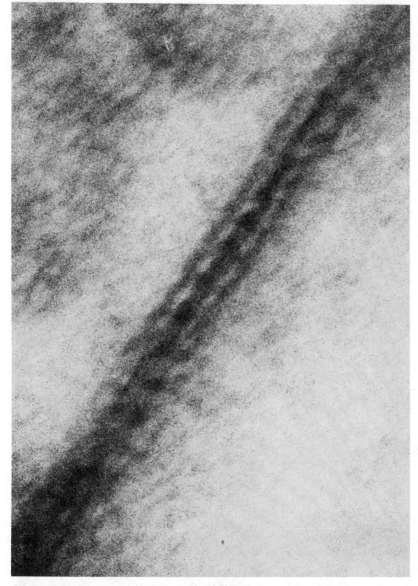

Figure 40. High-power cross section of synaptic disk in preparation fixed
with KMnO$_4$. ×1,200,000. [From Robertson (128).]

Figure 40 shows my original (1963) high-power electron micrograph
of a synaptic disk in cross section (128). In cross section there were
dense spots repeating at a period of ~9 nm between the two membranes, and in some places there were vague densities running across
the two membranes roughly opposite the dense spots to dimples in
the cytoplasmic surfaces that seemed to repeat at ~9 nm, as in Figure
40. In frontal views of the membranes, the disk showed an almost

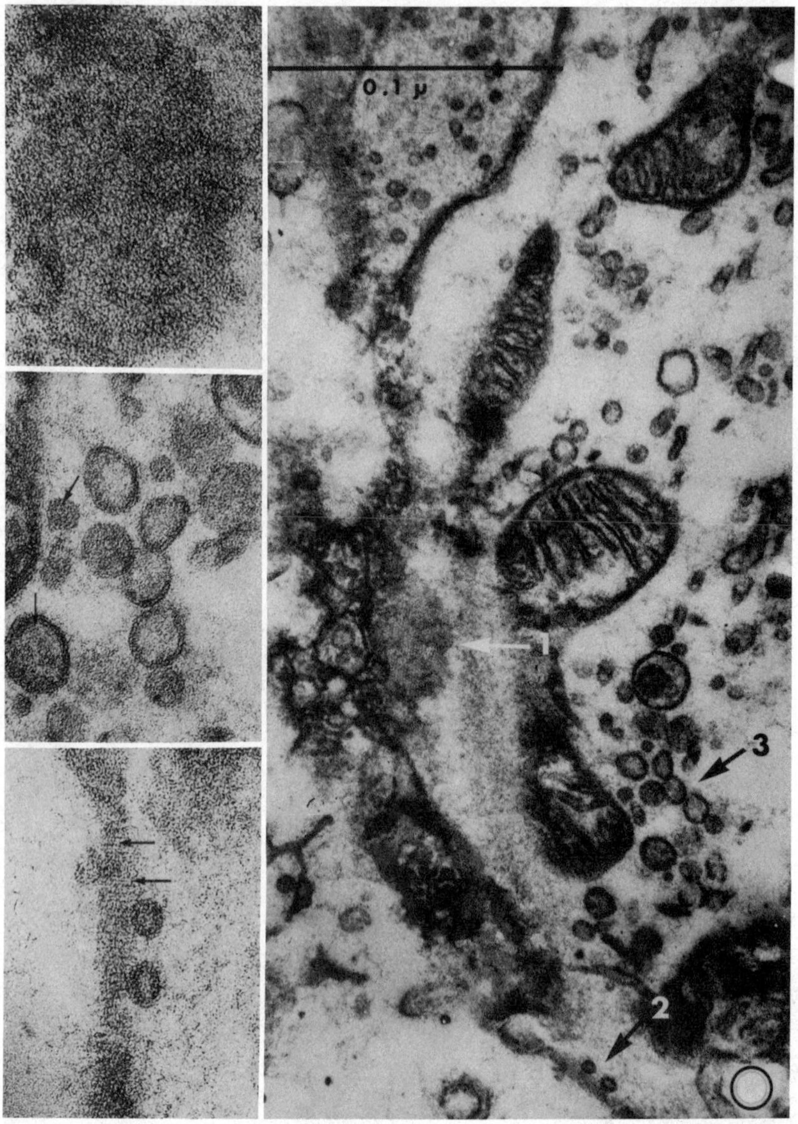

Figure 41. Oblique section of club ending showing synaptic disk in frontal view (*center*), enlarged (*upper left*) to show substructure. Some synaptic vesicles are enlarged (*middle left*), and an obliquely sectioned synaptic disk (2) is enlarged (*lower left*). Sections are KMnO$_4$ fixed and epoxy embedded. ×27,300. *Insets: top,* ×78,000; *middle,* ×58,500; *bottom,* ×91,000. [From Robertson (128).]

crystalline pattern of dense bordered facets spaced ~9 nm apart, as in Figure 41 (*upper left inset*) (128). The facets were closely packed with dense, straight boundaries ~2 nm wide. Each facet had a central dense spot ~2 nm in diameter. We wondered whether or not the transverse densities and the scollops in the cytoplasmic surfaces represented transmembrane channels. However, it seemed attractive to account for the transverse densities as image-overlap artifacts produced by the lattice as a result of a slight degree of membrane tilt in the sections, so we interpreted the dense spots as being localized mainly in the external surfaces of the membranes.

In 1962, just after we had completed our work on the synaptic disk, I was invited to give a lecture at the University of Michigan in Ann Arbor. I presented the synaptic disk work and a general review of membrane structure, which was published a considerable time later in a review article (129). Maynard Dewey and Lloyd Barr were in the audience and invited me to visit their lab after the lecture. We had a lively discussion about synaptic disk structure, and they showed me some micrographs of close contacts between unit membranes of smooth muscle cells that they thought were mediating electrical coupling between smooth muscle cells measured with microelectrodes. I believed that Dewey and Barr were seeing the same structure that we had seen in the club endings, even though they had not seen the substructure that we had seen in the synaptic disk. Shortly after we had talked, they published a paper (44) in *Science* that appeared in 1962, giving the name "nexus" to the kind of close contact that we had all seen. I presented the work at the meeting of the newly formed American Society for Cell Biology in 1962, but the proceedings were not published and this abstract was thus lost like my earlier one on the transverse tubules of muscle. However, Dewey and Barr had not seen the synaptic disk substructure, so this didn't really matter. In any case, the full papers from my lab (128, 139) and from theirs (45) appeared in 1963 and 1964; all the work was done completely independently.

THE LANTHANUM TRACER TECHNIQUE

Shortly after the synaptic disk work, I received a telephone call from Jerry Lettvin at MIT. He told me about some work he had been doing with lanthanum ions on the electrical properties of nerve membranes. He said that lanthanum seemed to lock up membranes so that no ions could get through them and suggested that I try fixing some membranes in lanthanum permanganate. He said he had a postdoctoral fellow from Chile in his lab who could make some of the material, which, he incidentally pointed out, was explosive. He also had another Chilean postdoctoral fellow, Samy Frenk, who wanted to participate. I too had a Chilean postdoctoral fellow, Carlos Doggenweiler, and agreed to try this if he wanted to work with Frenk, which he did. We tried the lanthanum permanganate as a fixative on several tissues and it worked! In addition to beautifully clear-cut unit membranes, we got a remarkable staining of the intercellular material in the gaps between the membranes (46). Jean Paul Revel and Morris Karnofsky (across the river at the Harvard Medical School quadrangle), presumably following up on this work,

[103]

devised a way to introduce lanthanum ions into a glutaraldehyde fixative. They applied their method to many tissues and saw structures like the synaptic disks in many locations such as liver, intestinal epithelium, kidney, and others. They came to my lab and showed me their material before they published it in 1967 (105). They had taken advantage of the staining of intercellular membrane gap material and had surveyed a number of tissues. In many places they had seen close contact between membranes in disk-shaped regions, which in frontal view showed the same kind of paracrystalline hexagonal arrays of facets as the ones we had seen in the synaptic disk. We all agreed that these structures must be the same as the synaptic disk, but it was clear that the term *synaptic disk* was not general enough, and they didn't want to use the term *nexus*. They had seen a small gap ~2 nm wide between the membranes, which had not appeared in our material, and had begun to call the structure a "gap junction." I didn't like the term, but they used it in their paper and it became widespread and is still used. I objected because I thought the term put the emphasis incorrectly; it was not the gap that was important but the place where there was no gap. However, the structure is now known as the gap junction despite an abortive attempt by George Palade and his colleagues (149), with my support, to get people to call the structure a communicating junction or "macula communicans." In 1965 Benedetti and Emmelot (3) described a membrane structure isolated from liver that looked, in negative-stain preparations, very much like our pictures of the synaptic disk. Although they referred to our papers and those of Dewey and Barr, they seemed to regard the structure they found as representative of the kind of globular substructure that Lucy and Glauert (87), Lucy (85, 86), and Sjoestrand (155) had discussed. In 1968 these authors published another paper (4) relating all these structures and unfortunately applied the term *tight junction* to them, but this time they related them more directly to the synaptic disk (128, 139) and the junctions described by Revel and Karnofsky (105). In retrospect we all now realize that they were looking at essentially the same structure we had all seen, which is now called the gap junction. This history is the main reason I prefer to avoid the term *tight junction*, although it is now generally understood to refer to an *occluding junction* (47), as explained subsequently.

THE DUKE PERIOD

In 1966 Dan Tosteson, now Dean of Harvard Medical School, who was then chairman of the Department of Physiology and Pharmacology at Duke and chairman of a search committee to find a chairman for the anatomy department, invited me to come to Durham, N.C., to consider the position. I also found Bill Anlyan, the Dean at Duke, very sympathetic and helpful, and I accepted the job. It was a very challenging and interesting experience for me to try to build up a new anatomy department, and I soon learned most of the essential tricks of the trade from Dan and from Phil Handler, the chairman of biochemistry. With their help I was able to get a new building on the drawing boards and in due course managed to build up a department of which I am proud. My research undoubtedly suffered, but this was a price I was willing to pay and the whole thing has been fun. Today I have ~23 full time faculty members, all but one of whom I chose and appointed as assistant professors. I perhaps chauvinistically, but sincerely, rate my department as second to none. In fact, the department will become three new departments when I retire as chairman in June, 1988: the Department of Cell Biology, the Department of Neurobiology, and the Department of Biological Anthropology and Anatomy.

During the late 1960s, some efforts were made by various people to break down the unit membrane paradigm. Most of these had to do with placing protein molecules within the bilayer in various ways, like the models proposed by Lenard and Singer (84), Wallach and Zahler (169), Chapman and Wallach (27), and Sjoestrand (154). The most extreme model was that of Benson (5, 6), who conceived of the membrane simply as a monolayer of globular protein molecules with lipid molecules tucked here and there at random. There was no way that a general model such as this could be reconciled with the body of facts about membrane structure, and the other arguments seemed almost as weak, so I paid little attention to them. Ed Korn (80) attacked the unit membrane model on various grounds in an article in *Science* in 1966, but it seemed unconvincing. Nevertheless, I replied to many of his arguments in a review article some time later (133). In general it seemed to me that all these models could easily be proven or rejected by X-ray diffraction or other experimental techniques, and I believed the available evidence excluded most of them. In any case, the main points—the ubiquity of the bilayer and the concept of chemical asymmetry—were secure and still remain so.

In 1969 a review paper was published by Walther Stoeckenius and Don Engleman (159), describing the various reasons why the bilayer structure is the correct general one for biological membranes. The review was quite comprehensive, and I agreed in general with its conclusions. However, the term *unit membrane* was downplayed to such an extent that it seemed as if the authors considered it hardly relevant. Throughout the paper they repeatedly referred to the bilayer model as the "Danielli" model. In the end it seemed to me that the idea was conveyed that the work that I had done to establish the bilayer as a general membrane paradigm had already been done by Danielli before I started. Davson was hardly mentioned. Although my work was not ignored, it was certainly not presented in proper perspective. Perhaps the reader will understand from what I have said about my early interaction with Danielli and my understanding of how the field actually developed that I found this review, despite its many excellent qualities, quite distorted. Its conclusions supported my views completely, but the experimental foundations of these views were not accurately conveyed. This kind of article can sometimes be published without a critical editorial review, and this one, I would venture to guess, probably was not reviewed at all.

In 1963 Moor and Muehlethaler (92) published a paper giving results of the application of a new technique with an instrument that they had developed (93), based on a method first introduced by Steere in 1957 (156). Steere had found that he could visualize virus particles that had been frozen in water by putting them in a vacuum chamber on a cold stage, etching away some of the ice, and shadow casting them with metal, using the metallic shadow-casting technique that had been introduced much earlier by Williams and Wyckoff (174). Moor and Muehlethaler developed an elaborate apparatus that essentially was a standard shadow-casting machine with a vacuum chamber containing a microtome and a cold stage. They were able to mount a piece of frozen tissue on the cold stage and keep it frozen while evacuating the chamber. The stage was perfused with liquid nitrogen to keep the specimen frozen, and it also contained a heater and a feedback mechanism to alternate heating with cooling so the specimen temperature could be kept constant at any chosen value between room temperature and about −180°C. They cooled the microtome knife with liquid nitrogen in a similar way and proceeded to cut away some of the surface of the tissue, discarding the sections shaved off. They then deposited metal on the exposed fresh surface either immediately after making a pass with the microtome knife or after a minute or so, holding the specimen temperature at about −100°C to permit some sublimation of ice to occur, essentially as Steere had done before. It was soon realized that the knife did not cut the tissue but simply fractured it away. They observed membranes in cross fracture and found that they showed the unit membrane pattern. This pleased me, and I didn't pursue the matter, partly because I did not have the apparatus, partly because I was fully occupied with other things, and partly because at first glance nothing new seemed to have been revealed. I also thought the technique was so prone to artifact production that it would take more time than I wanted to invest to evaluate it. However, Daniel Branton, after a period of work in the laboratory of Moor and Muehlethaler, developed the technique in his laboratory at Berkeley and soon, alone or with colleagues, published a series of papers (13–17) in which the very important concept was advanced that membranes tend to fracture along the bilayer into two halves that do not etch away, as does the surrounding water table. It was concluded that these were the hydrophobic surfaces of the two lipid monolayers making the central bilayer of the membrane. On these etched-membrane inner surfaces, rather large particles, ~10 nm or more in diameter, were seen and interpreted as protein molecules.

This, of course, caught my attention, and I immediately built an apparatus that would allow me to do some crude freeze fracturing in my simple Kinney vacuum evaporator. I soon was able to produce particles like those that Branton and his colleagues were seeing, but I could accentuate the particles by doing the procedure in a relatively poor vacuum. If I took the trouble to work at higher vacuum, many of the particles disappeared. Mistakenly ignoring the fact that all the particles did not disappear, I concluded that they represented artifacts produced by deposition of material from the vacuum chamber on the cold fractured membrane surfaces. So at that time I did not accept the interpretation of these particles as protein molecules. After a number of years, we now know that, while some such particles are indeed, as I believed, artifacts not only of deposition of gas molecules but also of plastic deformation (140), some of them do indeed represent protein molecules disposed across the bilayer.

John Singer, not to my surprise, apparently believed that Branton's original interpretation was literally correct and used this concept to support the advancement of the fluid mosaic model of membrane structure with Nicholson (151) in 1972 (see also ref. 150). At first I resisted this conception because I did not believe there was enough hard evidence for it. There was, as I have indicated, considerable doubt about the nature of intramembrane particles (38, 133, 134, 136, 137), and the biochemical evidence involving labeling proteins in membranes did not seem convincing to me. Bretscher (19–21) reported on labeling experiments that at face value seemed in keeping with the fluid mosaic ideas, but I did not find his papers convincing because he made a fundamental assumption that I was sure was wrong. Nevertheless, his conclusions were later confirmed (160).

At this point I found myself in the position that Thomas Kuhn described in his book *The Structure of Scientific Revolutions* (82). I did not realize exactly what was going on at the time, but I was in fact in the impossible position of one who dares to defend an old paradigm when the field is ready for a new one. I was naive enough to think that the fluid mosaic model simply represented a refinement of the unit membrane model and that, after its details had been discussed and evaluated, any of the new features that it emphasized would simply be incorporated into the unit membrane concept as an evolution of normal science. One of the main features of the model was fluidity, but after all, there was nothing new about this idea, which had been accepted as a feature of membranes since the work of Chambers and others in the 1930s (25, 26), as discussed in the 1945 edition of *The Physical Chemistry of Cells and Tissues* by Hoeber (76), which I had used as a text as a graduate student. To be sure, the work of Frye and Edidin (57) and of Cone (31) put a new emphasis on this feature of membranes and added some new quantitative and graphic data. Even though there had never been any doubt in my mind that membranes were in a sense fluid, it seemed to me almost as though this idea came to the scientific public as some great new revelation. Also the idea of mosaic patches of protein and lipid in a kind of "fluid mosaic" model had been proposed in the 1930s by Collander and Baerlund (30) and was discussed in detail in Hoeber (p. 234 and 277 in ref. 76). It was discounted on surface-tension arguments by Davson and Danielli (37) because it was said that such a membrane would self-destruct, to translate into modern jargon: the surface tension of the lipid part would be so different from the protein part that the membrane would be unstable and fall apart. To be sure, this kind of fluid mosaic model was not exactly what was being proposed. The Singer-Nicholson model proposed that single protein molecules were embedded in isolation in a naked lipid bilayer (151). Under these conditions of course, the surface-tension argument would not apply and the overall surface tension would be that of a lipid bilayer. In any case I completely underestimated the degree to which the scientific community was ready for a change of paradigm, so I continued to examine each piece of evidence critically and discarded it if I found a flaw in the arguments. The early double-labeling experiments published by Bretscher (19–21), as mentioned, did not convince me because the interpretations seemed to depend critically on the assumption that no molecular rearrangement occurred when erythrocyte ghost membranes were produced. I regarded this as a false assumption for some of the reasons mentioned subsequently. I could see nothing about the Singer arguments that I could regard as more than theoretical possibilities. However, in 1974, a double-labeling experiment was published by Whiteley and Berg (170a) that I could not break down, and at that point I accepted publicly the transmembrane-protein concept. However, it was now too late. Most people in the field had been persuaded by Singer's arguments, and because I had opposed them, I think the scientific public simply discarded the whole unit membrane concept and put it out of mind. I was trapped in the change of paradigm and thoroughly discredited. I can look at this dispassionately now and realize how wrong I was. The fact that the Singer-Nicholson model did not change the two main features of the unit membrane model, the ubiquity of the bilayer and the idea of chemical asymmetry, and the fact that several of the main features of the Singer-Nicholson model are incorrect will eventually become clear. But at the moment the pendulum is still swinging very much in the same direction that was started by Singer and Nicholson. In fact they deserve considerable credit for what they did because the unit membrane model was not adequate to deal with many membrane functions. The concept of transmembrane proteins is extremely powerful and quite rightly pervades every aspect of modern membrane biology.

In 1977 I participated in a biennial membrane symposium that has been organized and run since 1965 by Professora Liana Bolis, a pharmacologist whose eclectic interests and international life makes her unclassifiable as to nationality. She lives most of the time in Geneva, and the symposium has been held in Crans-sûr-Sierre, Switzerland, the last ten times. At the meeting in 1977, I presented some of my work on cystalline membranes; Professor Ernst Helmreich of the University of Würzburg was in the audience. After the session he expressed great interest in my micrographs and suggested that I visit him in Würzburg and give a seminar. I did this and he offered to get me a Humboldt Prize. My old friend Pfizi Lynen was President of the Humboldt Society and indeed had been at the symposium where we met, so I agreed to accept the prize and spend a year in Würzberg. The prize was worth about $30,000 tax-free in both the U.S. and W. Germany, so I took the only sabbatical leave I have ever had. Dody and I had a marvelous year living in the 12th-century village of Waldbüttelbrün near Würzberg. I didn't accomplish much beyond getting acquainted with Ernst Helmreich and learning from him and his delightful wife Milein what great people live in Germany. On one occasion I went to a symposium in Sicily that was organized by Bolis, who is Professor of Pharmacology in Messina, and a group of us stayed at the home of a friend of hers where we had one of those parties one night that I suppose the Sicilians are expert at giving. The group was extremely congenial, and after a feast of Italian food and wine, someone started a conga line; Figure S3 shows the group. (I don't remember all the names, but Liana Bolis is at the head of the line in the left center with her back to the camera. Clockwise, Ralph Straub is sixth, Joe Hoffman is seventh, I am tenth, Professor Monnier is thirteenth, Murdock Ritchie is fourteenth, and Mrs. Monnier is sixteenth.) I had had a coronary thrombosis the summer before, but Joe Hoffman, Murdock Ritchie, Ralph Straub, and I climbed halfway up Mt. Etna one day while we were there.

This brings us to the present, in which I find myself out of step with much of the scientific community because I do not see evidence that should be in hand for the most popular interpretation of another important concept. Perhaps I am wrong, but I can only read the evidence as I see it and so am compelled to make interpretations that are not in line with those of many. This problem concerns the fundamentals of how transmembrane channels are constructed molecularly. Some people seem to regard this as a very easy problem. They take the maximum size of a molecule or ion that can go through a transmembrane channel and conclude that there is a hydrophilic protein-lined, sewer pipe-like structure of that diameter filled with water that exists as a permanent stable structure in the membrane with some kind of gate at the membrane surface that opens and closes to permit molecules to pass through. I take the position that

Figure S3. Group of scientists at play in Sicily in 1979.

if this were so, somebody would be able to get some heavy metal stain into one of the channels and visualize it with the EM. I say this not only for theoretical reasons but because of an experiment that I have frequently done. I often use tobacco mosaic viruses (TMV) as test objects for various reasons, and I routinely do a negative-stain preparation (18, 71) with either uranyl acetate or phosphotungstic acid to assure myself that I have the right concentration of virus. The TMV is a protein particle 18 nm in diameter that has a sewer pipe hole down its center lined with protein in which is embedded a coiled strand of RNA (94). The RNA strand is not in direct contact with the lumen of the hole; it is covered by protein. The lumen of this hole is 4 nm in diameter, and it is clearly hydrophilic because it invariably becomes filled with negative stain and is readily seen without any particular care with microscopy. Even though this channel is complicated by containing an RNA strand embedded in the protein and the hole is larger than the usual postulated transmembrane channel, it is protein lined, and I believe it is analogous enough to make the point that an aqueous pore of molecular dimensions traversing a membrane should be demonstrable. This should, in my opinion, be true even if it is only 12–20 Å in diameter, the size postulated by some (22, 23, 66, 90) for the gap junction channel, because uranyl and phosphotungstate ions are small enough to enter such channels. To be sure, there have been claims that such transmembrane pores have been demonstrated by filling them with heavy metal stain; I dealt with one such claim (66, 67) by showing that what was seen was a simple optical artifact (136, 137). I have never

[110]

seen any such demonstrations that I found acceptable, and I have tried very hard to produce such a picture myself. So I am driven to take the position that the nature of transmembrane channels must be different from the popular concept of a simple water-filled pore 1.5–2 nm in diameter.

As a graduate student in my lab, Guido Zampighi, now a professor at UCLA, chose as his thesis topic the structure of the gap junction. I have already referred to the paper we did with Corless (176) in which we analyzed what could be seen in thin sections or negative-stain preparations of gap junctions in relation to the problem of channels. Guido and I wrote an earlier paper on this point that seemed to me to be quite compelling (177). This reported that isolated liver gap junctions tend to break up into fragments when dialysed against EDTA. Some of the fragments contain only a small number of the repeating subunits of the junction, and some of these are seen edge-on in negative-stain preparations. It is easy to see in such preparations that the negative stain does not penetrate the channels.

It became clear, as Zampighi's work progressed, that he needed to be able to do the kind of analysis on isolated purified gap junctions that Nigel Unwin and Richard Henderson (73, 163) had done on the purple membrane of *Halobacterium halobium*. We were not, at that time, equipped to do this at Duke, so I suggested that he go to the MRC laboratory in Cambridge and do the analysis with Unwin. Unwin was agreeable to this, and they proceeded with a minimal-dose computer analysis of the gap junction structure in negative stain to a resolution of ~1.8 nm (164, 178). They produced some very interesting evidence regarding the arrangement of the gap junction protein in the bilayer as a transmembrane protein, which gave them a very appealing idea about how a channel might be opened and closed. They found that there are six protein subunits composing each channel unit, called a connexon, and that the subunits were either arranged almost perpendicular to the plane of the membrane or each was relatively tilted, depending on the concentration of detergent in the preparation. Their evidence suggested that this tilt, occurring simultaneously in all six subunits, might by a cooperative twisting motion open or close the channel, as indicated by their model reproduced in Figure 42 from the cover of the February, 1980, issue of *Nature* (164). The picture shows six connexons, each of which consists of six subunits. Each subunit is made of a roughly cylindrical piece of wood that is comma-shaped in cross section. Six of these are fitted together to make a connexon, and six connexons are shown in the figure. In three of the connexons, the subunits are perpendicular to the plane of the membrane and there is no opening between them. In the other three alternating connexons, the subunits are each slightly tilted and the resulting twist in the whole structure produces a small hole in the center that runs

Figure 42.   Diagram from cover of *Nature.*

through the connexon from one side of the membrane to the other. I think this model suggests a very likely mechanism, and I agree in general with their conclusions.

There is just one point of emphasis that I would put somewhat differently. They found no evidence that negative stain entered the channel. Figure 43, taken from their work (164), is a composite of two forms of one membrane of the junction, showing the electron-density distribution across the bilayer (*arrowheads*) including one subunit. Note that the negative stain (*stippling*) is concentrated in both forms in pools in indentations of the outer surface of the membrane, but it does not extend across the bilayer. In their papers they attribute this fact to their not having sufficient resolution to detect the stain, if it were there. To me the important point is that stain is not found at a resolution of 1.8 nm. If one bears in mind that, at a resolution of 1.8 nm, structures much smaller than this can be detected, it becomes clear that there may simply be no stain there to detect. Unwin et al. (162) have proceeded with a similar analysis of isolated junctions embedded in ice. This has already revealed many fascinating facts about the disposition of the connexon subunits across the bilayers in forms that are presumably open and closed, and the results give support to the general idea illustrated in the Unwin-Zampighi model. This mechanism seems more and more attractive. However, the exact nature of the channel in the hydrophobic part of the bilayer has not yet been completely clarified and will not be until the resolution is carried to the higher levels that Unwin and his collaborators are likely to achieve in the near future.

Because the problem of how channels are constructed is so important, I cannot resist a few speculations in anticipation of the solutions

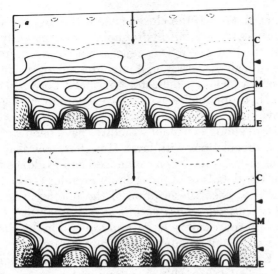

Figure 43. Diagram from Unwin and Zampighi (164).

that will come in the near future. The gap junction channel clearly
consists of some kind of protein structure defining a pathway that
can, when open, permit free flow of ions and small molecules. In this
state it behaves as though it were a cylindrical structure 10–20 Å in
diameter lined with hydrophilic amino acids. But if it were like this
when prepared for observation by EM methods, it should be easily
demonstrable with heavy metals; this has never been accomplished
convincingly. One plausible explanation is that the structure is in a
closed state when prepared, behaving as though the channel were
only a potential opening lined by hydrophobic amino acids. This
would explain the failure of the EM to reveal it. This implies a
structure that can change its properties greatly during preparation,
as the result of subtle molecular alterations. Such a structure is
plausible. In Figure 43 the central pool of negative stain between
the external surfaces of the membranes penetrates the bilayer appreciably, depending on its exact definition. Thus the hydrophobic
region is less than the thickness of the bilayer. The water-free region
in the middle of the bilayers has been estimated from nuclear
magnetic resonance data to be only ~2.6 nm thick. Thus the hydrophobic part of the channel might be only ~2.6 nm long. One kind of
plausible channel structure is a β-pleated sheet rolled into a cylindrical β-barrel. This could conceivably be made of the six connexon
proteins of the channel. The spacing between amino acid residues in
a β-pleated sheet is 3.5 Å. Thus a sliding change in the relative
positions of amino acid residues in each of the six polypeptide chains
of the six connexons could change local net hydrophobicity to hydrophilicity. This mechanism is quite comparable with the one postulated by Unwin and Zampighi and similar to one postulated by

[113]

Greenblatt et al. (70). If the closed state is the lowest energy state, the channel would become hydrophobic spontaneously when removed from the living cell. This general kind of structure would be compatible with all the data, and it might be called an "amphipathic" channel. Such a channel, delicately balanced between net hydrophilicity and hydrophobicity, would have a definite biological advantage by providing a useful safety factor. As indicated, it will be some time yet before enough channel structures have been worked out to replace such speculations with facts. However, now that one of the gap junction proteins has been sequenced and evidence is accumulating about the details of other channel proteins (161), this should occur soon. The recent description of the photosynthetic reaction center in *Rhodopseudomonas veridis* at 3-Å resolution by Dersenhofer et al. (39) represents a large step forward in this direction.

In the 1950s the unit membrane structure was best seen in material fixed with $KMnO_4$, but today it is easily seen in glutaraldehyde-OsO_4–fixed material that is block-stained with uranyl acetate before embedding. A few years ago I used one of the polyglutaraldehyde-embedding methods introduced independently by Pease and Peterson (98, 99) and by Heckman and Barnett (72) to show the triple-layered structure very clearly in erythrocyte membranes, as in Figure 12. This preparation was fixed in glutaraldehyde and embedded in polyglutaraldehyde by the GACH method (72). Sections were cut and stained with uranyl and lead salts. The membrane is somewhat thicker (~100–120 Å) because the lipid is largely retained, and there is a tendency for the two halves of the bilayer to separate during the sectioning. However, this technique gives a very clear representation of the triple-layered structure. The fact that the same triple-layered structure is demonstrable this way, as seen previously with $KMnO_4$ or OsO_4, makes arguments against the unit membrane concept, as put forward by Chapman and Wallach (27), seem irrelevant. Recently I put together my thoughts about membrane structure in a model diagrammed in Figure 44. This model attempts to show all the essential points about membrane structure that seem, on current evidence, to have general significance. It shows the bilayer core of the membrane with the asymmetry of the lipids indicated by showing the polar heads of the inner monolayer as filled circles. This is meant to indicate that the amino lipids phosphatidylethanolamine and phosphatidylserine are located in the inner monolayer (9, 68, 91, 104, 141, 142, 165). The polar heads are left open in the outer monolayer to indicate that the choline-containing and sphingolipids phosphatidylcholine and sphingomyelin are confined to the outer monolayer. Cholesterol is shown confined to the outer monolayer (24). The fact that the carbon chains in the outer monolayer tend to be longer and less unsaturated than those in the inner monolayer is not indicated. A few glycolipids are shown in the outer monolayer. Some transmembrane proteins are shown with the polypeptide chains in the

[114]

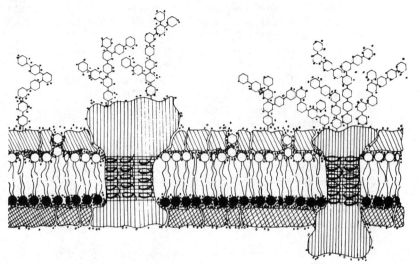

Figure 44. Cross section diagram of current model of unit membrane, sometimes called *hydrophobic barrier model.*

core of the bilayer, diagrammed as relatively hydrophobic α-helices. Sugar residues are shown on some of the external proteins. The polar lipid surfaces on both sides are shown completely covered with either protein or sugar residues. I conceive of this model as possessing fluidity of two kinds. I believe many of the lipid molecules are free to rotate and translate, although generally not to flip-flop from one side to the other (81). Also I believe some of them are bound to protein molecules on the surfaces mainly by head group interactions. I see the proteins as having varying degrees of mobility. Some are relatively free to move about, but some are anchored—most likely by attachments on the cytoplasmic side to cytoskeletal elements and on the outside to extracellular proteins such as fibronectin (28, 51, 100). The model possesses fluidity, but it is a controlled and regular fluidity. To emphasize the feature of this model that I consider to be the most important one, I call this the *hydrophobic barrier model.* I do not expect this to be adopted as a new paradigm. The field is not yet ready for this, and after all, it is not that different from the old unit membrane model.

BIBLIOGRAPHY

1. ARVANITAKI, A. Effects evoked in an axon by the activity of a contiguous one. *J. Neurophysiol.* 5: 89–108, 1942.
2. BARTELMEZ, G. W. Mauthner's cell and the nucleus motorius tegmenti. *J. Comp. Neurol.* 25: 87: 128, 1915.
3. BENEDETTI, E. L., and P. EMMELOT. Electron microscopic observations on negatively stained plasma membranes isolated from rat liver. *J. Cell Biol.* 26: 299–304, 1965.

4. BENEDETTI, E. L., and P. EMMELOT. Hexagonal array of subunits in tight junctions separated from isolated rat liver plasma membranes. *J. Cell Biol.* 38: 15–24, 1968.

5. BENSON, A. A. Plant membrane lipids. *Annu. Rev. Plant Physiol.* 15: 1–16, 1964.

6. BENSON, A. A. On the orientation of lipids in chloroplast and cell membranes. *J. Am. Oil Chem. Soc.* 43: 265–270, 1966.

7. BLAUROCK, A. E. The spaces between membrane bilayers within PNS myelin as characterized by X-ray diffraction. *Brain Res.* 210: 383–387, 1980.

8. BLAUROCK, A. E. X-ray and neutron diffraction by membranes: How great is the potential for defining the molecular interactions? In: *Progress in Protein-Lipid Interactions*, edited by A. Watts and J. J. DePont, 1986, vol. 2, p. 1–43.

9. BLOJ, B., and D. B. ZILVERSMIT. Asymmetry and transposition rates of phosphatidylcholine in rat erythrocyte ghosts. *Biochemistry* 15: 1277–1283, 1976.

10. BODIAN, D. The structure of the vertebrate synapse. A study of the axon ending on Mauthner's cell and neighboring centers in the goldfish. *J. Comp. Neurol.* 68: 117–159, 1937.

11. BODIAN, D. Introductory survey of neurones. *Cold Spring Harbor Symp. Quant. Biol.* 17: 1, 1952.

12. BODIAN, D., AND H. A. HOWE. Experimental studies on intraneural spread of poliomyelitis. *Bull. Johns Hopkins Hosp.* 68: 248–267, 1941.

13. BRANTON, D. Fracture faces of frozen membranes. *Proc. Natl. Acad. Sci. USA* 55: 1048–1056, 1966.

14. BRANTON, D. Fracture faces of frozen myelin. *Exp. Cell Res.* 45: 703–707, 1967.

15. BRANTON, D. Freeze-etching studies of membrane structure. *Philos. Trans. R. Soc. Lond. B Biol. Sci.* 261: 133–138, 1971.

16. BRANTON, D., AND D. W. DEAMER. Membrane structure. In: *Protoplasmatologia*, edited by M. Alfert, H. Baner, W. Sandritter, et al. New York: Springer-Verlag, 1972, p. 1–70.

17. BRANTON, D., and R. B. PARK. Subunit chloroplast lamellar. *J. Ultrastruct. Res.* 19: 288–303, 1967.

18. BRENNER, S., and R. W. HORNE. A negative staining method for high resolution electron microscopy of viruses. *Biochim. Biophys. Acta* 34: 103–110, 1959.

19. BRETSCHER, M. S. Major human erythrocyte glycoprotein spans the cell membrane. *Nature Lond.* 231: 229–232, 1971.

20. BRETSCHER, M. S. Phosphatidyl-ethanolamine: differential labelling in intact cells and cell ghosts of human erythrocytes by a membrane-impermeable reagent. *J. Mol. Biol.* 71: 523–528, 1972.

21. BRETSCHER, M. S. Membrane structure: some general properties. *Science Wash. DC* 181: 622–629, 1973.

22. BRINK, P. R., and M. M. DEWEY. Evidence for fixed charge in the nexus. *Nature Lond.* 285: 101–102, 1980.

23. CASPAR, D. L. D., D. A. GOODENOUGH, L. MAKOWSKI, and W. C. PHILLIPS. Gap junction structures. I. Correlated electron microscopy and X-ray diffraction. *J. Cell Biol.* 74: 605–628, 1977.

24. CASPAR, D. L. D., and D. A. KIRSCHNER. Myelin membrane structure at 10 Å resolution. *Nature Lond.* 231: 46–52, 1971.

25. CHAMBERS, R., and M. J. KOPAC. The coalescence of living cells with oil drops. *J. Cell. Comp. Physiol.* 9: 331–345, 1937.
26. CHAMBERS, R., and H. POLLACK. Micrurgical studies in cell physiology. IV. Colorimetric determination of the nucleus and cytoplasmic pH in the starfish egg. *J. Gen. Physiol.* 10: 739–755, 1927.
27. CHAPMAN, D., and D. F. H. WALLACH. Recent physical studies of phospholipids and natural membranes. In: *Biological Membranes*, edited by Dennis Chapman. Academic: New York, 1968, p. 125–202.
28. CHEN WEN-TIEN, E. HASEGAWA, T. HASEGAWA, C. WEINSTOCK, and K. M. YAMADA. Development of cell surface linkage complexes in cultured fibroblasts. *J. Cell Biol.* 100: 1103–1114, 1985.
29. COLE, K. S., and H. J. CURTIS. Bioelectricity: electric physiology. In: *Medical Physics*, edited by Otto Glasser. Chicago, IL: Year Book, 1950, vol. II, p. 82–90.
30. COLLANDER, R., and H. BAERLUND. Permeabilitaets studuien an chara ceratophylla. *Acta Bot. Fenn.* 11: 1–114, 1933.
31. CONE, R. A. Rotational diffusion of rhodopsin in the visual receptor membrane. *Nature Lond.* 236: 39–42, 1972.
32. DANIELLI, J. F. The thickness of the wall of the red blood corpuscle. *J. Gen. Physiol.* 19: 19–22, 1935.
33. DANIELLI, J. F. Some properties of lipoid films in relation to the structure of the plasma membrane. *J. Cell. Comp. Physiol.* 7: 393–408, 1936.
34. DANIELLI, J. F. Protein films at the oil-water interface. *Cold Spring Harbor Symp. Quant. Biol.* 6: 190–195, 1938.
35. DANIELLI, J. F. *Colston papers.* vol. VII, 1954.
36. DANIELLI, J. F., and H. A. DAVSON. A contribution of the theory of permeability of thin films. *J. Cell. Comp. Physiol.* 5: 495–508, 1935.
37. DAVSON, H., and J. F. DANIELLI. *The Permeability of Natural Membranes.* New York: Hafner, 1943.
38. DEAMER, D. W., and N. YAMANAKA. Freeze-fracture particles in protease treated membranes. *Biophys. J.* 15: 110–111, 1975.
39. DEISENHOFER, J., O. EPP, K. MIKI, R. HUBER, and H. MICHEL. Structure of the protein subunits in the photosynthetic reaction centre of *Rhodopseudomonas viridis* at 3 Å resolution. *Nature Lond.* 318: 618–624, 1985.
40. DEL CASTILLO, J., and B. KATZ. Quantal components of the end-plate potential. *J. Physiol. Lond.* 124: 560–573, 1954.
41. DEL CASTILLO, J., and B. KATZ. Biophysical aspects of neuromuscular transmission. *Prog. Biophys.* 6: 121–170, 1956.
42. DE ROBERTIS, E., and H. S. BENNETT. Some features of the submicroscopic morphology of synapses in frog and earthworm. *J. Biophys. Biochem. Cytol.* 1: 47–58, 1955.
43. DE ROBERTIS, E., and F. O. SCHMITT. An electron microscope analysis of certain nerve axon constituents. *J. Cell Comp. Physiol.* 31: 1–24, 1948.
44. DEWEY, M. M., and L. BARR. Intercellular connection between smooth muscle cells: the nexus. *Science Wash. DC*, 137: 670–672, 1962.
45. DEWEY, M. M., and L. BARR. A study of the structure and distribution of the nexus. *J. Cell Biol.* 23: 553–585, 1964.
46. DOGGENWEILER, C. F., and S. FRENK. Staining properties of lanthanum on cell membranes. *Proc. Natl. Acad. Sci. USA* 53: 425–430, 1965.

47. FARQUHAR, M. G., and G. E. PALADE. Junctional complexes in various epithelia. *J. Cell Biol.* 17: 375–412, 1963.

48. FATT, P. Biophysics of junctional transmission. *Physiol. Rev.* 34: 674–710, 1954.

49. FATT, P., and B. KATZ. Chemo-receptor activity at the motor end-plate. *Acta Physiol. Scand.* 29: 117–125, 1953.

50. FERNANDEZ-MORAN, H. The submicroscopic organization of vertebrate nerve fibers. *Exp. Cell Res.* 3: 282–359, 1952.

51. FIERSCHBACHER, M. D., E. C. HAYMAN, and E. RUOSLAHTI. The cell attachment determinant in fibronectin. *J. Cell. Biochem.* 28: 115–126, 1985.

52. FINEAN, J. B. Recent ideas on the structure of myelin. In: *Biochemical Problems of Lipids*. London: Butterworths, 1956, p. 127–131.

53. FOLCH, J., and M. LEES. Proteolipids, a new type of tissue lipoproteins. *J. Biol. Chem.* 191: 807–817, 1951.

54. FRANZINI-ARMSTRONG, C., and K. R. PORTER. Sarcolemmal invaginations constituting the T system in fish muscle fibers. *J. Cell Biol.* 22: 675–696, 1964.

55. FRICKE, H. The electric capacity of suspensions with special references to blood. *J. Gen. Physiol.* 9: 137–152, 1925.

56. FRICKE, H. The electric impedance of suspensions of biological cells. *Cold Spring Harbor Symp. Quant. Biol.* 1: 117–124, 1933.

57. FRYE, L. D., and M. EDIDIN. The rapid intermixing of cell surface antigens after formation of mouse-human heterokaryons. *J. Cell Sci.* 7: 319–335, 1970.

58. FURSHPAN, E. J., and T. FURAKAWA. Intracellular and extracellular response of the several regions of the Mauthner cell of the goldfish. *J. Neurophysiol.* 25: 732–771, 1962.

59. GASSER, H. S. Addendum to paper by B. Frankenhaeuser. *Cold Spring Harbor Symp. Quant. Biol.* 17: 32–36, 1952.

60. GASSER, H. S. Properties of dorsal root unmedullated fibers on two sides of the ganglion. *J. Gen. Physiol.* 38: 709–728, 1955.

61. GEREN, B. B. The formation from the Schwann cell surface of myelin in the peripheral nerves of chick embryos. *Exp. Cell Res.* 7: 558–562, 1954.

62. GEREN, B. B., and F. O. SCHMITT. The structure of the nerve sheath in relation to lipid and lipid-protein layers. EMSA *J. Appl. Phys.* 24: 1421, 1953.

63. GEREN, B. B., and F. O. SCHMITT. The structure of the Schwann cell and its relation to the axon in certain invertebrate nerve fibers. *Proc. Natl. Acad. Sci. USA* 40: 863–871, 1954.

64. GLAUERT, A. M., G. E. ROGERS, and R. H. GLAUERT. A new embedding medium for electron microscopy. *Nature Lond.* 178: 803, 1956.

65. GLAUERT, A. M., G. E. ROGERS, and R. H. GLAUERT. Araldite as an embedding medium for electron microscopy. *J. Biophys. Biochem. Cytol.* 4: 191–194, 1958.

66. GOODENOUGH, D. A. Channels traversing two junctional membranes and intervening "gap." *J. Cell Biol.* 71: 334–335, 1976.

67. GOODENOUGH, D. A. In vitro formation of gap junction vesicles. *J. Cell Biol.* 68: 220–231, 1976.

68. GORDESKY, S. E., and G. V. MARINETTI. The asymmetric arrangement of phospholipids in the human erythrocyte membrane. *Biochem. Biophys.*

Res. Commun. 50: 1027–1031, 1973.

69. GORTER, E., and R. GRENDEL. Bimolecular layers of lipid on the chromocytes of the blood. *J. Exp. Med.* 41: 439–443, 1925.

70. GREENBLATT, R. E., Y. BLATT, and M. MONTAL. Hypothesis: the structure of the voltage-sensitive sodium channel. *FEBS Lett.* 193: 125–134, 1985.

71. HALL, C.E. Electron densitometry of stained virus particles. *J. Biophys. Biochem. Cytol.* 1: 1–12, 1955.

72. HECKMAN, C. A., and R. J. BARRNETT. GACH: a water-miscible, lipid-retaining embedding polymer for electron microscopy. *J. Ultrastruct. Res.* 42: 156–179, 1973.

73. HENDERSON, R. The structure of the purple membrane for *Halobacterium halobium*. Analysis of the X-ray diffraction pattern. *J. Mol. Biol.* 93: 123–138, 1975.

74. HESS, A., and J. Z. YOUNG. The nodes of Ranvier. *Proc. R. Soc. Lond. B Biol. Sci.* 140: 301–320, 1952.

75. HILL, A. V. The abrupt transition from rest to activity in muscle. *Proc. R. Soc. Lond. B Biol. Sci.* 136: 399–435, 1949.

76. HOEBER, R. *The Physical Chemistry of Cells and Tissues.* Philadelphia: Blakiston, 1945.

77. HUXLEY, A. F. The Croonian lecture 1967. The activation of striated muscle and its mechanical response. *Proc. R. Soc. Lond. B Biol. Sci.* 178: 1–27, 1971.

78. HUXLEY, A. F., and R. E. TAYLOR. Function of Krause's membrane. *Nature Lond.* 176: 1068, 1955.

79. JOHNSON, G. E. Giant nerve fibers in crustaceans with special reference to *Cambarus* and *Palaemonetes. J. Comp. Neurol.* 36: 323–373, 1924.

80. KORN, E. D. Structure of biological membranes. *Science Wash. DC* 153: 1491–1498, 1966.

81. KORNBERG, R. D., and H. M. McCONNELL. Inside-outside transitions of phospholipids in vesicle membranes. *Biochemistry* 10: 1111–1120, 1971.

82. KUHN, T. *The Structures of Scientific Revolutions* (2nd edition). Chicago, IL: University of Chicago Press, 1970.

83. LAATSCH, R. H., M. W. KIES, S. GORDON, and E. C. ALVORD. The encephalomyelitic activity of myelin isolated by ultracentrifugation. *J. Exp. Med.* 115: 777–778, 1962.

84. LENARD, J., and S. J. SINGER. Protein conformation in cell membrane preparations as studied by optical rotatory dispersion and circular dischroism. *Proc. Natl. Acad. Sci. USA* 56: 1828–1835, 1966.

85. LUCY, J. A. Globular lipid micelles and cell membranes. *J. Theor. Biol.* 7: 360–373, 1964.

86. LUCY, J. A. Theoretical and experimental models for biological membranes. In: *Biological Membranes—Physical Fact and Function.* London: Academic, 1968, p. 233–288.

87. LUCY, J. A., and A. M. GLAUERT. Structure and assembly of macromolecular lipid complexes composed of globular micelles. *J. Mol. Biol.* 8: 727–748, 1964.

88. LUFT, J. H. Permanganate—a new fixative for electron microscopy. *J. Biophys. Biochem. Cytol.* 2: 799–801, 1956.

89. LUZATTI, V., and F. HUSSON. The structure of the liquid-crystalline phases of lipid-water systems. *J. Cell Biol.* 12: 207–220, 1962.

90. Makowski, L., D. L. D. Caspar, W. C. Phillips, and D. A. Goodenough. Gap junction structures. II. Analysis of the X-ray diffraction data. *J. Cell Biol.* 74: 629–645, 1977.

91. Miljanich, G. P., P. P. Nemes, D. L. White, and E. A. Dratz. The asymmetric transmembrane distribution of phosphatidylethanolamine, phosphatidylserine and fatty acids of the bovine retinal rod outer segment disk membrane. *J. Membr. Biol.* 60: 249–255, 1981.

92. Moor, H., and K. Muhlethaler. Fine structure in freeze-etched yeast cells. *J. Cell Biol.* 17: 609–628, 1963.

93. Moor, H., K. Muhlethaler, H. Waldner, and A. Frey-Wyssling. A new freezing ultramicrotome. *J. Biophys. Biochem. Cytol.* 10: 1–14, 1961.

94. Namba, K., and G. Stubbs. Structure of tobacco mosaic virus at 3.6 angstroms resolution: implications for assymmetry. *Science Wash. DC* 231: 1401–1406, 1986.

95. Palade, G. E. (assisted by S. L. Palay). Electron microscope observations of interneuronal and neuromuscular synapses. *Anat. Rec.* 118: 335–336, 1954.

96. Palay, S. L. (assisted by G. E. Palade). Electron microscopy study of the cytoplasm of membranes. *Anat. Rec.* 118: 336, 1954.

97. Palay, S. L. Synapses in the central nervous system. *J. Biophys. Biochem. Cytol. Suppl.* 2: 193–202, 1956.

98. Pease, D. C., and R. G. Peterson. Polymerizable glutaraldehyde urea mixtures as polar, water-containing embedding media. *J. Ultrastruct. Res.* 45: 124–148, 1972.

99. Peterson, R. G., and D. C. Pease. Myelin embedded in polymerized glutaraldehyde urea. *J. Ultrastruct. Res.* 41: 115–132, 1972.

100. Piersschbacher, M. D., E. G. Hayman, and E. Ruoslahti. The cell attachment determinant in fibronectin. *J. Cell. Biochem.* 28: 115–126, 1985.

101. Porter, K. R., and G. E. Palade. Studies on the endoplasmic reticulum. III. Its form and distribution in striated muscle cells. *J. Biophys. Biochem. Cytol.* 3: 269–300, 1957.

102. Reger, J. F. Electron microscopy of the motor end-plate in intercostal muscle of the rat. *Anat. Rec.* 118: 344, 1954.

103. Reger, J. F. Electron microscopy of the motor end-plate in rat intercostal muscle. *Anat. Rec.* 122: 1–15, 1955.

104. Renooij, W. L., M. G. van Golde, R. F. A. Zwaall, and L. L. M. van Deenen. Topological asymmetry of phospholipid metabolism in rat erythrocyte membranes: evidence for flip-flop of lecithin. *Eur. J. Biochem.* 61: 53–58, 1976.

105. Revel, J. P., and M. J. Karnowski. Hexagonal array of subunits in intercellular junctions of the mouse heart and liver. *J. Cell Biol.* 33: C7–C12, 1967.

106. Robertson, J. D. Addendum to neurones of arthropods by C. A. G. Wiersma (*Abstract*). *Cold Spring Harbor Symp. Quant. Biol.* 17: 161, 1952.

107. Robertson, J. D. The Ultrastructure of Two Invertebrate Synapses. Boston, MA: MIT Press, 1952, Ph.D. thesis.

108. Robertson, J. D. Ultrastructure of two invertebrate synapses. *Proc. Soc. Exp. Biol. Med.* 82: 219–223, 1952.

109. Robertson, J. D. Electron microscope observations on a reptilian

myoneural junction. *Anat. Rec.* 118: 346, 1954.

110. ROBERTSON, J. D. The ultrastructure of a reptilian myoneural junction. *J. Appl. Physics* 25: 1466–1467, 1954.

111. ROBERTSON, J. D. Recent electron microscope observations on the ultrastructure of the crayfish median-to-motor giant synapse. *Exp. Cell Res.* 8: 226–229, 1955.

112. ROBERTSON, J. D. The ultrastructure of adult vertebrate peripheral myelinated fibers in relation to myelinogenesis. *J. Biophys. Biochem. Cytol.* 1: 271–278, 1955.

113. ROBERTSON, J. D. Some features of the ulrastructure of reptilian skeletal muscle. *J. Biophys. Biochem. Cytol.* 2: 369–379, 1956.

114. ROBERTSON, J. D. The ultrastructure of a reptilian myoneural junction. *J. Biophys. Biochem. Cytol.* 2: 381–394, 1956.

115. ROBERTSON, J. D. The cell membrane concept. *J. Physiol. Lond.* 140: 58–59, 1957.

116. ROBERTSON, J. D. New observations in the ultrastructure of frog peripheral nerve fibers. *J. Biophys. Biochem. Cytol.* 3: 1043–1948, 1957.

116a.ROBERTSON, J. D. The ultrastructure of the myelin sheath near nodes of Ranvier. *J. Physiol. Lond.* 135: 56P–57P, 1957.

117. ROBERTSON, J. D. Structural alterations in nerve fibers produced by hypotonic and hypertonic solutions. *J. Biophys. Biochem. Cytol.* 4: 349–364, 1958.

118. ROBERTSON, J. D. The ultrastructure of Schmidt-Lanterman clefts and related shearing defects of the myelin sheath. *J. Biophys. Biochem. Cytol.* 4: 39–46, 1958.

119. ROBERTSON, J. D. The ultrastructure of cell membranes and their derivatives. *Biochem. Soc. Symp.* 16: 3–43, 1959.

120. ROBERTSON, J. D. The molecular biology of cell membranes. *Molecular Biology*, edited by D. Nachmansohn. New York: Academic, 1960, p. 87–151.

121. ROBERTSON, J. D. The molecular structure and contact relationships of cell membranes. *Prog. Biophys.* 10: 343–417, 1960.

122. ROBERTSON, J. D. A molecular theory of cell membrane structure. *Proc. Int. Conf. Electron Microsc., 4th.* Berlin: Springer, 1960, vol. XX, p. 159–171.

123. ROBERTSON, J. D. Cell membranes and the origin of mitochondria. *Proc. Reg. Neurochem. Symp. 4th*, edited by S. S. Kety and J. Elkes. Oxford, UK: Pergamon, 1961, p. 497–530.

124. ROBERTSON, J. D. Ultrastructure of excitable membranes and the crayfish median giant synapse. *Ann. NY Acad. Sci.* 94: 339–389, 1961.

125. ROBERTSON, J. D. The unit membrane. In: *Electron Microscopy in Anatomy*, edited by J. D. Boyd, F. R. Johnson, and J. D. Lever. London: Arnold, 1961, p. 74–99. (Proc. Symp. Anat. Soc Gr. Brit. Ultrastruct. Cells.)

126. ROBERTSON, J. D. The membrane of the living cell. *Sci. Am.* 206: 64–72, 1962.

127. ROBERTSON, J. D. The unit membrane of cells and mechanisms of myelin formation. *Res. Publ. Assoc. Res. Nerv. Ment. Dis.* 40: 94–158, 1962.

128. ROBERTSON, J. D. The occurrence of a subunit pattern in the unit membranes of club endings in Mauthner cell synapses in goldfish brains. *J. Cell Biol.* 19: 201–221, 1963.

129. ROBERTSON, J. D. Unit membranes: a review with recent new studies

of experimental alterations and a new subunit structure in synaptic membranes. In: *Cellular Membranes in Development*, edited by Michael Locke. New York: Academic, 1963, p. 1–81. (Proc. Symp. Soc. Stud. Dev. Growth, XXII.)

130. ROBERTSON, J. D. Design principles of the unit membrane. In: *Principles of Biomolecular Organization*, edited by G. E. W. Wolstenholme and M. O'Connor. London: Churchill, 1966, p. 357–408. (Ciba Found. Symp.)

131. ROBERTSON, J. D. Granulo-fibrillar and globular substructure in unit membranes. *Ann. NY Acad. Sci.* 137: 421–440, 1966.

132. ROBERTSON, J. D. The organization of cellular membranes. In: *Molecular Organization and Biological Function*, edited by J. M. Allen. New York: Harper & Row, 1966, p. 65–106.

133. ROBERTSON, J. D. The structure of biological membranes, current status. *Arch. Intern. Med.* 129: 202–228, 1972.

134. ROBERTSON, J. D. On the nature of intramembrane particles. *Proc. Electron Microsc. Soc. Am., 35th*, edited by J. W. Bailey. Baton Rouge, LA: Claitor's, 1977, p. 680–683.

135. ROBERTSON, J. D. Citation classic: the ultrastructure of cell membranes and their derivatives (*Abstract*). *Current Contents* Jan. 1982: 20.

136. ROBERTSON, J. D. A review of membrane structure with perspectives on certain transmembrane channels. In: *Demyelinating Disease and Basic and Clinical Electrophysiology*, edited by S. G. Waxman and J. M. Ritchie. 1981, p. 419–477.

137. ROBERTSON, J. D. The nature and limitations of electron microscopic methods in biology. In: *Physiology of Membrane Disorders*, edited by T. Andreoli, J. Hoffman, and D. Fanestil. New York: Plenum, 1986, p. 59–80 (2nd ed.).

138. ROBERTSON, J. D. The early days of electron microscopy of nerve tissue and membranes. *Int. Rev. Cytol.* 100: 129–196, 1987.

139. ROBERTSON, J. D., T. S. BODENHEIMER, and D. E. STAGE. The ultrastructure of Mauthner cell synapses and nodes in goldfish brains. *J. Cell Biol.* 19: 201–221, 1963.

140. ROBERTSON, J. D., and J. A. VERGARA. An analysis of the structure of intramembrane particles of the mammalian urinary bladder. *J. Cell Biol.* 86: 514–528, 1980.

141. ROTHMAN, J. E., and E. P. KENNEDY. Assymmetrical distribution of phospholipids in the membrane of bacillus megaterium. *J. Mol. Biol.* 110: 603–618, 1977.

142. ROTHMAN, J. E., and J. LENARD. Membrane asymmetry. The nature of membrane asymmetry provides clues to the puzzle of how membranes are assembled. *Science Wash. DC* 195: 743–753, 1977.

143. ROTHSCHILD, V. The membrane capacitance of the sea urchin egg. *J. Biophys. Biochem. Cytol.* 3: 103–110, 1957.

144. SCHMIDT, W. J. Doppelbrechung und Feinbau der Markescheide der Nervenfasern. *Z. Zellforsch. Mikrosk. Anat.* 23: 657–676, 1936.

145. SCHMIDT, F. O., R. S. BEAR, and G. L. CLARK. X-ray diffraction studies in nerve. *Radiology* 25: 131–135, 1935.

146. SCHMITT, F. O., R. S. BEAR, and E. PONDER. Optical properties of the red cell membrane. *J. Cell. Comp. Physiol.* 9: 89–92, 1937.

147. SCHMITT, F. O., R. S. BEAR, and E. PONDER. The red cell envelope considered as a Wiener mixed body. *J. Cell. Comp. Physiol.* 11: 309–

313, 1937.

148. SCHMITT, F. O., R. S. BEAR, and K. J. PALMER. X-ray diffraction studies on the structure of the nerve myelin sheath. *J. Cell. Comp. Physiol.* 18: 1–41, 1941.

149. SIMIONESCU, M., N. SIMIONESCU, and G. E. PALADE. Segmental differentiations of cell junctions in the vascular epithelium. *J. Cell Biol.* 67: 863–885, 1975.

150. SINGER, S. J. The molecular organization of membranes. *Annu. Rev. Biochem.* 43: 805–833, 1974.

151. SINGER, S. J., and G. L. NICHOLSON. The fluid mosaic model of the structure of cell membranes. *Science Wash. DC* 175: 720–731, 1972.

152. SJOESTRAND, F. S. The lamellated structure of the nerve myelin sheath as revealed by high resolution electron microscopy. *Experientia* 9: 68–69, 1953.

153. SJOESTRAND, F. S. The ultrastructure of the retinal rod synapses of the guinea pig eye. *J. Appl. Physics* 24: 1422, 1953.

154. SJOESTRAND, F. S. A new repeat structural element of mitochondrial and certain cytoplasmic membranes. *Nature Lond.* 199: 1062–1064, 1963.

155. SJOESTRAND, F. S., and J. RHODIN. The ultrastructure of the proximal convoluted tubules of the mouse kidney as revealed by high resolution electron microscopy. *Exp. Cell Res.* 4: 426–456, 1953.

156. STEERE, R. A. Electron microscopy of structural detail in frozen biological specimens. *J. Biophys. Biochem. Cytol.* 3: 45–60, 1957.

157. STOECKENIUS, W. An electron microscope study of myelin figures. *J. Biophys. Biochem. Cytol.* 5: 491–500, 1959.

158. STOECKENIUS, W. Some electron microscopical observations on liquid-crystalline phases in lipid-water systems. *J. Cell Biol.* 12: 221–229, 1962.

159. STOECKENIUS, W., and D. M. ENGELMAN. Current models for the structure of biological membranes. *J. Cell Biol.* 42: 613–646, 1969.

160. THOMPSON, T. E., and C. HUANG. Composition and dynamics of lipids in biomembranes. In: *Physiology of Membrane Disorders* (2nd ed.), edited by T. E. Andreoli, J. F. Hoffman, D. D. Fanestil, and S. G. Schultz. New York: Plenum, p. 25–44, 1986.

161. UNWIN, N. Is there a common design for cell membrane channels? *Nature Lond.* 323: 12–13, 1986.

162. UNWIN, P. N. T., and P. D. ENNIS. Two configurations of a channel-forming membrane protein. *Nature Lond.* 307: 609–613, 1984.

163. UNWIN, P. N. T., and R. HENDERSON. Molecular structure determination by electron microscopy of unstained crystalline specimens. *J. Mol. Biol.* 94: 425–440, 1975.

164. UNWIN, P. N. T., and G. ZAMPIGHI. Structure of the junctions between communicating cells. *Nature Lond.* 283: 545–549, 1980.

165. VERKLEIJ, A. J., R. F. A. ZWAAL, B. ROELOFSEN, P. COMFURIUS, D. KOSTELIJN, and L. L. M. VAN DEENEN. The asymmetric distribution of phospholipids in the human red cell membrane: a combined study using phospholipases and freeze-etch electron microscopy. *Biochim. Biophys. Acta* 323: 178–193, 1973.

166. VON CRIEGEE, R. Osmium Saure-ester als Mischerproducte beir Oxydation. *Ann. Chim.* 522: 75–96, 1936.

167. VON CRIEGEE, R., B. MARCHAND, and H. WANNOWIUS. Zur Kentnis der

organischer Osmium-Vergindungen. *Ann. Chim.* 550: 99–133.

168. WADE, R. H., and A. BRISSON. Three-D structure of luminal plasma membrane protein from urinary bladder. *Proc. Electron Microsc. Soc. Am., 41st.* San Francisco, CA: San Francisco Press, 1983, p. 436–437.

169. WALLACH, D. F. H., and P. H. ZAHLER. Protein conformations in cellular membranes. *Proc. Natl. Acad. Sci. USA* 56: 1552–1559, 1966.

170. WAUGH, D. F., and F. O. SCHMITT. Investigations of the thickness and ultrastructure of cellular membranes by the analytical leptoscope. *Cold Spring Harbor Symp. Quant. Biol.* 8: 233–241, 1940.

170a. WHITELY, N. W., AMD H. C. BERG. Amidination of the outer and inner surfaces of the human erythrocyte membrane. *J. Mol. Biol.* 87: 541–561, 1974.

171. WIERSMA, C. A. G. Giant nerve fiber system of the crayfish. A contribution to comparative physiology of synapse. *J. Neurophysiol.* 10: 23–38, 1947.

172. WIERSMA, C. A. G. Synaptic facilitation in the crayfish. *J. Neurophysiol.* 12: 267–275, 1949.

173. WIERSMA, C. A. G., and W. SCHALLEK. Influence of drugs on response of a crustacean synapse to pre-ganglionic stimulation. *J. Neurophysiol.* 11: 491–496, 1948.

174. WILLIAMS, R. C., and R. W. G. WYCKOFF. Application of metallic shadow-casting to microscopy. *J. Appl. Physiol.* 17: 23–33, 1946.

175. YOUNG, J. Z. Fused neurons and synaptic contact in the giant nerve fibers of cephalopods. *Philos. Trans. R. Soc. Lond. B Biol. Sci.* 229: 465, 1939.

176. ZAMPIGHI, G., J. M. CORLESS, and J. D. ROBERTSON. On gap junction structure. *J. Cell Biol.* 86: 190–198, 1980.

177. ZAMPIGHI, G., and J. D. ROBERTSON. Fine structure of synaptic discs separated from the goldfish medulla oblongata. *J. Cell Biol.* 56: 92–105, 1973.

178. ZAMPIGHI, G., and P. N. T. UNWIN. Two forms of isolated gap junctions. *J. Mol. Biol.* 135: 451–464, 1979.

IV

Transport Pathways:
Water Movement Across
Cell Membranes

ARTHUR K. SOLOMON

M Y involvement with water transport across plasma membranes has its roots in the achievements of two great modern physiologists, Merkel Jacobs and August Krogh, the first concerned with the human red cell, and the second with amphibian epithelia. Merkel Jacobs completed the body of his research before World War II, and he never worked with radioactive or stable isotopes, which were not discovered until the 1930s. Naturally, when Jacobs (23) made the first quantitative measurement of red cell water permeability in 1932, he measured net flux. By 1935, Krogh was able to report on the use of D_2O, which had been discovered by Harold C. Urey in 1931, to measure the water permeability of frog skins. Krogh gained significantly from the close-knit scientific community in Copenhagen and enlisted the cooperation of von Hevesy, who was at the Institute of Theoretical Physics, and Hofer, who was at the Institute of Chemical Physics (20). Von Hevesy had won the Nobel prize for discovering tracers by showing that radium D, which he had been unable to separate from lead, could be used as a tracer for lead.

My involvement in water transport grew directly from my experience with radioactive and stable isotopes. I was trained as a physical chemist, but as soon as I had received my Ph.D. in 1937 under George B. Kistiakowsky's direction, I went as a postdoctoral fellow to the Cavendish Laboratory at Cambridge University to work on radioactivity under the direction of Lord Rutherford. Artificial radioactivity had been discovered in 1934 by Curie and Joliot, and I wanted to learn what contributions artificial radioactivity could make to physical chemistry. Plans were for the cyclotron at the Cavendish to be completed by the time I arrived, so that radioactive isotopes would be available; I expected that the postdoctoral phase of my education would begin by characterizing these isotopes. I arrived just in time for Lord Rutherford's funeral in Westminster Abbey. Thus virtually my first experience with British science was attendance with everyone at the Cavendish at the solemn service for Rutherford's interment in the Abbey.

When I got to Cambridge, the cyclotron had barely been started; there was a large magnet, partially complete, and a few pieces scattered around the lab. I was put to work with three physics colleagues, Robert Latham, Albert Kempton, and Donald Hurst, to build the cyclotron. It was an extraordinarily useful experience from which I learned a tremendous amount. It taught me how to design the instruments required to carry out my subsequent research. It also taught me that if I wanted radioactive isotopes, I had to make them myself; one of these was ^{82}Br. We used this ^{82}Br in collaboration with Nick Werthessen and the sterol chemist E. Friedmann to synthesize radioactive cholesterol dibromide in 1939, which must have been one of the earliest examples of incorporation of radioactivity into compounds of biological interest (15). At that time, the only tracer for carbon was ^{11}C, with its 20.7 min half-life. This short half-life, as I was later to learn in excruciating detail, made carbon tracer measurements difficult. I spent some time bombarding a carbon target with deuterons from the cyclotron in the fruitless hope of making a more suitable carbon tracer. When Ruben and Kamen (48) announced their discovery of ^{14}C in 1940, I learned that the radiation was so soft that it could not have penetrated the walls of the Geiger counter I had built to detect it.

With the outbreak of the war clearly imminent, I returned in the fall of 1939 to Harvard to take up an appointment as a Research Fellow in Physics and Chemistry to work on a project that would utilize my recently acquired skills at Cambridge. J. B. Conant, the famous organic chemist who had recently become Harvard's president, had the idea that glucose was synthesized in the body by the end-to-end fusion of two lactic acid molecules. He had enlisted my professor, Kistiakowsky, and the professor of biological chemistry at the Medical School, A. Baird Hastings, to form a team to put this idea to experimental test, using ^{11}C as a tracer. By the fall of 1939, when I joined the team, Harvard had built its own cyclotron, and Richard Kramer, in Kistiakowsky's laboratory, had worked out a way to synthesize lactic acid from $^{11}CO_2$ in two hours. My assignment was to transmute boron 10 into carbon 11 in the cyclotron, construct the Geiger counter, and build the amplifiers and scalers required for the radioactivity assay. The other members of the team were all biochemists: Birgit Vennesland, Friedrich Klemperer, and Jack Buchanan. Notwithstanding Kramer's rapid synthesis, the constraints of radioactive decay were imposing. Allowing one hour for the rat to metabolize the lactic acid, another hour to purify the glycogen from the liver, time for transit from cyclotron to laboratory, and time to make the counts, we were lucky to finish 5 h after I had stopped bombarding the boron target in the cyclotron. At the end of 5 h, radioactive decay by itself had reduced the activity of the sample to 0.004% of the initial radioactivity. As it turned out, about one experiment in three succeeded and, as it also turned out, President Conant's idea was wrong. Much more important was the result of our control experiment. At that time, the perceived biochemical wisdom was that carbon molecules passed in only one direction in metabolism, from glucose to lactate and thus to CO_2 but never backward up the glycolytic chain, notwithstanding Wood and Werkman's demonstration (78) of travel in the reverse direction in bacteria. Our control experiments, which we were very slow to publish (64) because we first had to convince ourselves of their validity, showed, for the first time in a mammalian system, that HCO_3^- injected into the rat could make its way backward up the glycolytic chain to produce significant ^{11}C concentrations in the glycogen we purified from the liver.

WATER MOVEMENT ACROSS MEMBRANES

Baird Hastings was a syndic of the Harvard University Press, where he spoke for the scientific community. One day he asked if I knew any books that would make science more intelligible to an educated lay audience. When I was at the Cavendish, I had written three such articles on the cyclotron and related matters, which had been published in *Discovery*, a magazine edited by C. P. Snow, then a fellow of Christ College. It occurred to me that with additional chapters these articles might be expanded into the book that became *Why Smash Atoms?*, published by the Harvard University Press in 1940.

By mid-1941 the American scientific community had begun to play an active role in the war effort, and peacetime scientific endeavors were largely abandoned. I wanted to join my British scientific colleagues, and in late August 1941 I went to England to become a British scientific civil servant. The rest of the war was spent overseeing the design and construction of antiaircraft radar, initially for the British Army, but subsequently for use in the warships of the British Navy. While I was in England, Allen Lane decided to publish *Why Smash Atoms?* as a Penguin book. It was in press when the bomb fell, and thus became the first book of its kind to be available to the British public. At the close of the war, Professor Hastings invited me to start the Biophysical Laboratory at the Harvard Medical School, whose initial charge was to help the medical scientific community learn how to use isotopic tracers in medical research.

By today's standards, times were primitive when the Biophysical Laboratory set up shop on July 1, 1946. We no longer had to make our own isotopes in a cyclotron we had built because they could be obtained from the nuclear reactor at the Oak Ridge National Laboratory. However, the industrial production of scientific instruments was not yet under way and for the first two or three years after the war, every counter used for the detection of radioactivity in the Harvard medical community was made in our laboratory. We also had to make our own scalers and amplifiers. The medical community was deeply interested in tracers for Na^+ and K^+, whose short half-lives of 15.0 h and 12.4 h meant that we had to arrange for weekly air shipments of a great deal of lead and significant amounts of radioactivity up and down the east coast. It soon turned out that what Oak Ridge called Na^+ and K^+ contained contaminants, so we routinely purified the isotopes on cation-exchange columns. We followed the decay of the purified $^{24}Na^+$ for 13 half-lives to obtain the accurate value of its half-life (57), which was necessary for our calculations. As a result of these several endeavors we were able, by 1954, to make a simultaneous determination, by isotope dilution, of the total exchangeable Na^+ and K^+ content of humans (1).

It had been decided at the outset that the Biophysical Laboratory should also have the capacity to measure stable isotopes, a decision that was to lead me directly to studies on the properties of water in biological systems. The instrument of choice was a mass spectrometer because it would not only permit measurements of deuterium but also of stable ^{18}O, which is the only tracer for oxygen, since the longest radioactive oxygen half-life is 2 min. It was also desirable to have the capacity to determine deuterium, even though radioactive hydrogen, ^3H, could be obtained from Oak Ridge. The ^3H radiation was so soft (19 keV) that there was no practicable method of measuring it routinely because scintillation counters had not yet been developed. Because mass spectrometers were not available commercially, the only way to get one was to build it, which we proceeded to do. At that time, the total body water in humans had not been determined accurately, and the professor of surgery at the Peter Bent Brigham Hospital, Dr. Francis Moore, was anxious to measure it by isotope dilution. We measured the D_2O content in samples of serum with the mass spectrometer and combined our measurements with the results obtained in his laboratory by the densimetric technique; these measurements led to the determination of the total body water in humans, published in 1950 (50).

WATER FLUXES IN RED CELLS

One final step was still required to channel my scientific interest into studies on membrane transport, and that was a painful one. The research projects in the Biophysical Laboratory were divided about equally between a biochemical arm, run by Chris Anfinsen, who was later to get a Nobel prize, and the biophysical section, of which I was in charge. Daniel Steinberg, a young M.D. who also wanted a Ph.D. degree, had begun his Ph.D. research under my direction, starting out with studies on the use of radioactive isotopes in biological problems. One day Dan told me that he wanted to move to biochemistry and complete his dissertation under Chris Anfinsen's direction. Since Chris and I were very close friends, there was no personal problem in arranging the transfer, but Dan's request was a great shock to my ego. After a great deal of consideration, I concluded that Dan's desire to move was driven by a desire to devote his research to problems of fundamental biological interest, rather than to making improvements, however important, in the measurement of results. On reflection, our laboratory seemed like a scientific chauffeur who made it possible to complete a journey whose direction has been set by others. Clearly, if I expected to attract graduate students of the quality of Dan Steinberg, I had to set the scientific course myself.

The intellectual problem was to choose a biologically significant area of research to which I could reasonably hope to make useful contributions. My attributes were a rigorous education in physical chemistry, considerable hands-on experience in measuring radioactive and stable isotopes, and the ability to design scientific instruments as they were needed for my research. The search narrowed down to two possibilities, the study of the transmission of the nervous impulse or the measurement of transport across the cell membrane. When I had been in England during the war, I had become very friendly with Alan Hodgkin and Andrew Huxley, particularly Andrew. It was not difficult to reach the conclusion that it would be the height of folly to enter a field in which Hodgkin and Huxley were already firmly ensconced.

Progress in individual scientific fields follows an S-shaped curve. In the earliest days progress is slow, as the fundamental principles that shape a field are slowly enunciated. This phase is followed by a period of almost explosive growth when these established principles are exploited. After a period of some years during which all of the easy (and a great many of the hard) experiments are done, growth slows to a walk, and new advances are only achieved with inordinate effort. It seemed to me that by the 1940s membrane transport had reached this later phase. Physical chemical concepts had been applied with brilliance, and essentially all the information that could be extracted from net flux experiments had been obtained. I reasoned that the use of radioactivity to measure unidirectional fluxes would power the next intellectual advance and start the membrane transport field again on a period of explosive growth. Thus I launched myself into membrane transport. I decided to study red cells, which could easily be obtained in a suspension, free from other cells, without requiring any kind of dissection for which I was not fitted.

When I started to study cation exchange across the red cell membrane, the general principle that cation fluxes drive water transport in epithelial tissues was not yet apparent and so, strictly speaking, these studies have not earned a place in a chapter on water transport. Nonetheless, the studies are of far-reaching importance in membrane transport, and their enunciation has taught me a wry lesson that may be of general interest. We found, first, that cardiac glycosides such as ouabain are very effective inhibitors of unidirectional cation fluxes across the red cell (63); second, that these effects on red cell cation flux are correlated with severe damage to heart muscle (63); and third, that under equilibrium conditions, rather than net uptake conditions, there is a 1:1 exchange of K^+ influx with Na^+ efflux (16). The article describing the first result was published simultaneously with that of Glynn (17), yet it is never credited in the literature; the second result appeared seven years before the paper of Repke that is normally credited (see ref. 51, p. 89) but that neglect has recently been rectified. The reason for these omissions is that these papers were omitted from the first review and, having escaped that first review, were forever lost in the literature. The lesson is that unless scientists seize the high ground when reviews are written, they court obscurity. The third result has not been generally accepted because it runs counter to Post and Jolly's observation (41) that three Na^+ leave the cell for every two K^+ that enter, when there is net uptake by K^+-depleted and Na^+-repleted cells, a result that has been broadly confirmed. I am convinced that the conflict reflects the difference between equilibrium exchange, which we measured, and net uptake, which Post measured, a difference whose roots probably lie in the basic stoichiometry of the process. Clearly, even in science it is necessary to speak out firmly, and surely more often than once.

When Charles Paganelli came as a graduate student in 1952, the laboratory's experience with red cell cation fluxes led us to try a more difficult problem, the measurement of red cell permeability to water. In 1935, von Hevesy, Hofer, and Krogh (20) had found that, when water was driven across the frog skin by an osmotic pressure gradient, the resultant flux was much greater than that measured by the diffusion of D_2O. Koefoed-Johnsen and Ussing (30) in a landmark paper in 1953 had carefully dissected conventional diffusion across a frog skin, measured by tracers such as D_2O, from hydraulic conductivity, driven by an osmotic pressure gradient and measured as net flux. Although Jacobs (23) had measured the hydraulic conductivity of the red cell, there had been no measurement of red cell diffusional flux. Diffusional fluxes had been measured across other single-cell membranes using D_2O, particularly by Prescott and Zeuthen (42) in Copenhagen, who had measured fluxes across frog and *Xenopus* egg membranes and found permeability coefficients of the order of 0.1×10^{-3} cm·s^{-1}. Because Jacobs (24) had pointed out that the red cell water permeability was greater than that of any other known cell, we felt quite sure that red cell diffusional flux measurements would require rapid reaction techniques.

Hartridge and Roughton (19) had developed the rapid reaction method in 1923 to measure the rate of reaction of hemoglobin (Hb)

[130]

with O_2. Two solutions, one containing COHb and the other HbO_2, were rapidly mixed in a chamber whose effluent flowed into an observation tube where the extent of the reaction was measured photometrically. The elapsed time between mixing and observation was obtained from the ratio of the distance between mixing chamber and observation port, divided by the velocity of the flowing solution. Alterations of this distance permitted Hartridge and Roughton to observe the reaction profile and thus to resolve reaction times as short as 10 ms. Subsequently, the rapid reaction technique was modified and improved, particularly by Milliken and Chance (see ref 4). The rapid reaction method was generally applied to reactions whose advancement could be monitored spectrophotometrically so that detection was essentially instantaneous on the reaction time scale. Spectrophotometric determination of the D_2O concentration in red cell suspensions was not practicable and so we were forced to use the much more cumbersome technique of filtration to separate the suspending medium from the cells. In 1931, Dirken and Mook (10) had combined the rapid reaction method with pressure filtration to separate plasma from suspensions of red cells in order to make direct measurements of the distribution of HCO_3^-, Cl^-, CO_2, and O_2 between red cells and plasma, with a resolution of ~100 ms. As it turned out, the diffusion of water across the red cell membrane was so fast that we needed a resolution of ~1 ms, which we obtained in the apparatus shown in Figure 1.

Even with the highest gas pressures we dared employ in driving the solutions through our apparatus, we were only able to obtain filtered samples of 10 to 50 mg, whose analysis for ^3H_2O content posed significant technical problems. In those more primitive times, tritium could only be obtained in the gas phase in sealed ampoules; we converted it to water by equilibration in vacuo for three days in the presence of platinum oxide. Because there were no scintillation counters and the 19-keV β-particles from ^3H would not penetrate any known counter window, Charles Robinson, a physicist in our laboratory, devised a method (47) to synthesize the water into ^3H-CH_4 in the gas phase by a Grignard reaction and then to use the ^3H-CH_4 as a filling gas in a proportional counter using the apparatus shown in Figure 2A. Using this method Paganelli was able to analyze four samples spaced equally between 2 and 8 ms after mixing, as shown in Figure 2B, and to make the first determination (38) of the diffusional permeability (P_d) of the human red cell, $P_d = 5.3 \pm 1.4 \times 10^{-3}$ cm$\cdot$s^{-1}.

In the next paper in this series, when Villegas, Barton, and I (72) measured P_d for dog and beef red cells, we uncovered an error in the Paganelli and Solomon (38) paper. In addition to the four experimental points Paganelli had obtained with the rapid reaction apparatus, we had included a zero-time point, based on the red cell hematocrit, determined by centrifugation. The determination of the

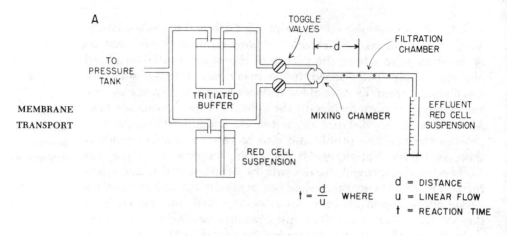

A

TO PRESSURE TANK

TOGGLE VALVES

TRITIATED BUFFER

RED CELL SUSPENSION

MIXING CHAMBER

FILTRATION CHAMBER

EFFLUENT RED CELL SUSPENSION

$$t = \frac{d}{u} \quad \text{WHERE}$$

d = DISTANCE
u = LINEAR FLOW
t = REACTION TIME

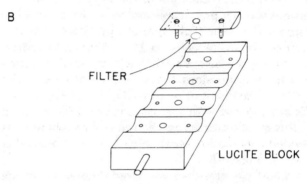

B

FILTER

LUCITE BLOCK

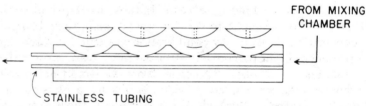

FROM MIXING CHAMBER

STAINLESS TUBING

Figure 1. A: continuous flow apparatus for measuring diffusional permeability of water in human red cell. B: filtration chamber in continuous flow apparatus. Millipore filter closes aperture in observation tube and fluid is collected from hole in cap above filter.

advancement of the reaction from its time course in the rapid reaction apparatus gives absolute values, while the hematocrit measured by centrifugation is a relative value that depends on the speed of the centrifuge and other extraneous factors; this meant that the inclusion of the zero-time point was incorrect and introduced a substantial error. Subsequently, Ted Barton, one of the authors of the Villegas et al. (72) paper, repeated the determination of P_d in humans on an improved apparatus. His value, $P_d = 4.2 \pm 1.2 \times 10^{-3}$

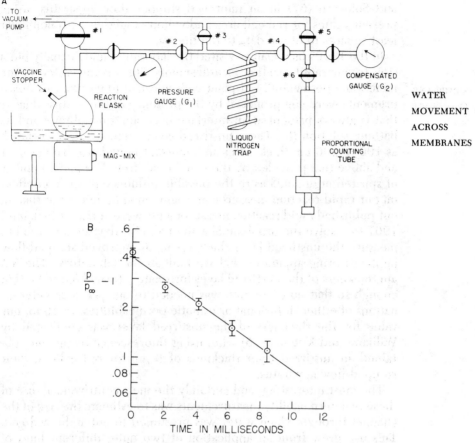

WATER
MOVEMENT
ACROSS
MEMBRANES

Figure 2. A: apparatus for synthesizing ^{3}H-CH$_4$ from ^{3}H$_2$O in the gas phase and then filling the proportional counter. B: linear relationship showing uptake of ^{3}H$_2$O by red cell. p/p$_\infty$, Ratio of counts per minute in the sample, divided by the infinite time value.

cm·s^{-1} (after conversion to present values for red cell volume and area and 20°C) agrees very well with the present average value of $4.2 \pm 0.5 \times 10^{-3}$ cm·s^{-1}, which includes the averaged results of 10 determinations by 9 groups of investigators (11), many using proton nuclear magnetic resonance, the present method of choice.

At just about the same time, Victor Sidel and I measured the hydraulic conductivity of the red cell by 90° scattered light in a continuous flow apparatus (56). These measurements were much easier than the P_d measurements because the time scale covered the range from 50 to 250 ms. The value we obtained for the hydraulic conductivity, $L_p = 0.95 \times 10^{-11}$ cm^3·dyn^{-1}·s^{-1}, was larger by ~50% than that previously obtained by Jacobs, using the time taken for red cells to hemolyze as a measure of the permeability coefficient. Our most recent value for L_p, and we believe the most accurate one, is

[133]

$1.8 \pm 0.1 \times 10^{-11}$ cm$^3 \cdot$dyn$^{-1} \cdot$s^{-1} at 25°C obtained by Terwilliger and Solomon (67) in an improved stopped-flow apparatus, using present values for red cell area and volume together with nonlinear least squares to fit the data to the theory.

In the 1960s Jack Dainty visited the laboratory and virtually put a stop to our rapid reaction measurements. He was interested in the membrane permeability of plant cells and had found that his measurements were compromised by the presence of the unstirred layer that is always present at an interface between a membrane and its bathing solution (8). These unstirred layers, which can be as much as 100 to 200 μm thick, offer an important impedance barrier over and above that provided by the membrane itself. We had a number of spirited arguments as to the possible influence of such an effect on our rapid reaction measurements and when he left, I felt that all our published rapid reaction measurements were at risk. It took until 1967 to resolve the problem. Sha'afi et al. (54) devised a method to measure the unstirred layer thickness in an improved stopped-flow light-scattering apparatus that we built in the laboratory. The 5.5 μm thickness of the unstirred layer fortunately turned out to be thin enough so that no correction was needed to our previous determinations of either diffusional or osmotic permeabilities. In 1985, our value for the thickness of the unstirred layer was confirmed by Williams and Kutchai (76) who, using fluorescence quenching, obtained an unstirred layer thickness of 8 μm for red cells in their stopped-flow apparatus.

The most interesting, and certainly the most controversial, use of these two permeability measurements was to estimate the size of the channel through which water was presumed to enter the red cell. This use grew from an application of two quite different lines of reasoning, each discussed in the chapter by Pappenheimer in this volume. Pappenheimer [see his chapter in this volume and Pappenheimer, Renkin, and Borrero (40)] devised a technique to measure the size of the channels in the capillary membrane by combining the results of osmotic flow measurements with those of diffusion. According to the classical laws of fluid dynamics, the coefficient that relates flow through a small tube to the pressure difference that drives the flow is dependent on the fourth power of the tube's radius. According to Fick's law, diffusional flow through a small tube is related to its conjugate force by the Stokes-Einstein coefficient, which depends on the area of the tube. In principle, as Pappenheimer, Renkin, and Borrero (40) have described, if these laws are applicable to flows in channels as small as those in capillary membranes, the radius of the channel can be obtained from the ratio of the hydraulic conductivity to the Stokes-Einstein coefficient. Koefoed-Johnsen and Ussing (30) came to a similar conclusion from a more general derivation, which they made to account for effects of vasopressin on water permeability in the frog skin.

It was already a bold step to suggest that Poiseuille's law, which had been devised to describe flows in very small glass capillaries with diameters as small as 13 μm, could be applied to flow through the very much smaller capillary channels. We had the temerity to propose that these same concepts could be applied to an even smaller channel and to suggest that water crossed the red cell membrane through an equivalent pore. The pore was considered the equivalent of a right circular cylinder of uniform diameter whose constraints to diffusion and osmotic flow were described by the equations Renkin (44) had shown to describe these flows through cellophane membranes with radii as small as 15 Å. Based on the permeability coefficients available to Paganelli and me, the equivalent pore radius was computed to be 3.5 Å; with the best present values for these coefficients, the equivalent pore radius is now computed (59) as 6.5 ± 0.6 Å, which is still only a few times larger than the 1.5 Å radius of the water molecule. Even though we had been careful to point out that the equivalent pore radius was to be considered as a construct and an effective radius that "provided a consistent description of the cell's permeability to water," the proposal that such a small pore permeated the red cell membrane provoked a storm.

The ratio of the two permeability coefficients of the red cell is conventionally given (see ref. 58) as the dimensionless ratio, P_f/P_d, in which the L_p has been transformed to P_f, the filtration coefficient, which has the same units as the diffusion coefficient. There is a fundamental difference between the relative motion of water molecules in osmotic flow and in diffusion. As Onsager (36) pointed out, "*viscous flow is a relative motion of adjacent portions of a liquid. Diffusion is a relative motion of its constituents.*" If one imagines a membrane in which the pore radius can be made smaller at will, the first effect will be a diminution of P_f relative to P_d so the P_f/P_d ratio will diminish with the pore radius and finally go to unity when the pores have disappeared and the only movement across the membrane is that resulting from water dissolution in the membrane fabric. When thin lipid membranes became available, it was possible to test this expectation and we all watched with great interest, none perhaps with more concern than I, to see whether this fundamental tenet of the theory would pass experimental tests. The experiments are technically very difficult because they are plagued with problems from the unstirred layer, which clings particularly tenaciously to the interface between the black lipid membrane and the solutions that bathe it. Finally, Cass and Finkelstein (3) in a carefully controlled series of experiments showed that $P_f/P_d = 1$ for black lipid membranes and I breathed a sigh of relief.

If water in these narrow aqueous channels behaves like bulk water, the activation energies for diffusion and osmotic flow through the channel should be very nearly the same as those for bulk water. In order to see if this is the case, we measured the activation energies for water flow through the red cells of humans and dogs. The self-diffusion coefficient for water is given in classical terms by the

Stokes-Einstein relation, $D_w = RT/(6\pi\eta_w N_{Av} a_w)$ in which D_w is the diffusion coefficient, RT is the gas constant $\times$ absolute temperature, η_w is the viscosity of water, N_{Av} is Avogadro's number, and a_w is the radius of the water molecule. Wang (75) who determined the activation energy for self-diffusion in water, also observed that the product $D_w\eta_w/T$, which contains all the temperature-dependent terms in the Stokes-Einstein coefficient, was itself independent of temperature between 5°C and 25°C, varying between 1.58 and 1.61 $\times 10^{-7}$ dyn$\cdot$deg^{-1}. Wang concluded, therefore, that the self-diffusion of water consists of movement of individual water molecules, notwithstanding the structural properties of bulk water. A similar coefficient for the diffusional permeability of water across the red cell membrane is given by $P_d\eta_w/T$, which can be computed from the data of Vieira et al. (71) for dog red cells. The ratio $P_d\eta_w/T$ is also temperature independent between 7°C and 37°C, varying between 7.79 and 7.70 $\times 10^{-10}$ dyn$\cdot$cm$^{-1}\cdot$deg^{-1}. The equivalent pore radius in dog red cells, which, unlike human red cells, are freely permeable to glucose is about twice as large as that of humans (72). The activation energies for diffusion are 4.9 $\pm$ 0.3 and 6.0 $\pm$ 0.2$\cdot$kcal$\cdot$mol^{-1} for dogs and humans, close to Wang's figure of 4.8 kcal$\cdot$mol^{-1} for self-diffusion. Similarly, the activation energies for osmotic flow are 3.7 $\pm$ 0.7 and 3.3 $\pm$ 0.4 kcal$\cdot$mol^{-1} for dogs and humans, very close to the activation energy of 4.2 kcal$\cdot$mol^{-1} for the viscosity of bulk water. Even after considerable reflection it is difficult to understand why the properties of water in these narrow channels are so similar to those in bulk water, but the similarity of the activation energies leaves little doubt. Perhaps the hydrogen bonds of the hydrophilic faces of the amino acids that line the channel have conformations not dissimilar to those of bulk water.

The demonstration that equivalent pores had a real existence was also a product of black lipid membrane research. Holz and Finkelstein (21) used the antibiotics amphotericin B and nystatin to form aqueous channels in black lipid membranes that contained cholesterol, and found that the permeability of a number of hydrophilic nonelectrolytes decreased smoothly with increases in solute dimensions. Sha'afi et al. (53) subsequently obtained a similar result in their determination of the sieving properties of the human red cell membrane. When these two sets of data were overlaid on the same graph (62), as shown in Figure 3A, it became apparent that red cell permeation was virtually identical to that of the amphotericin B and nystatin pores. Furthermore the equivalent pore radius, calculated from the ratio of the osmotic to the diffusional permeability to water, is 4.3 Å for the amphotericin B–treated membrane and 4.6 Å for the nystatin treated membrane, similar to the 6.5-Å equivalent pore radius for the red cell. These experiments showed that the permeability properties of amphotericin B pores in thin lipid bilayers approximated those of the human red cell.

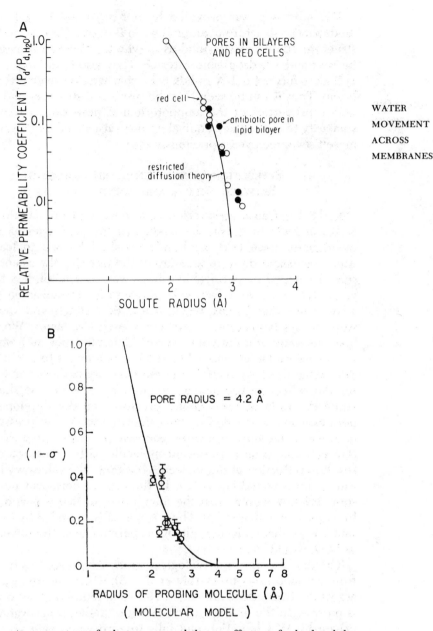

Figure 3. A: comparison of relative permeability coefficients for hydrophilic nonelectrolyte permeation of red cell membranes (*open circles*) and nystatin and amphotericin B pores in black lipid bilayers (*filled circles*). Curve for restricted diffusion theory has been computed from the Renkin equations (44). [From Solomon et al. (61).] *B:* $(1 - \sigma)$ for hydrophilic nonelectrolytes in human red cells, fitted to the Renkin equations (44). [From Goldstein and Solomon (18).]

The next step was provided by deKruijff and Demel (9), who modeled the structure of amphotericin B pores. They found that the structure could be described as a cylinder that had a hydrophilic interior and a hydrophobic exterior. They also found that two such cylinders formed a 4 Å radius pore that would traverse the membrane. Thus it can be seen that the presumed structure and permeability properties of the amphotericin B pore bear a remarkable similarity to the red cell equivalent pore visualized by Paganelli and myself seventeen years previously (38).

REFLECTION COEFFICIENTS AND COUPLING BETWEEN SOLUTE AND SOLVENT FLUXES

Karl Sollner, an Austrian scientist who had emigrated to the United States in 1937 to pursue his studies on the properties of artificial membranes, once told me how astonished he was to learn that American scientists were ignorant of the fact that the osmotic pressure developed by permeable solutes was, in general, less than the van't Hoff pressure, a fact well known to his European colleagues. Thus it was that Jacobs, whose scientific instincts and knowledge were always impeccable, never considered this factor. We did not become aware of it until Staverman's difficult paper (65), published in an obscure Dutch journal in 1951, was brought to our attention. Staverman used irreversible thermodynamics to derive an equation for the reflection coefficient, denoted by σ, which he showed to equal the ratio of the osmotic pressure actually developed by a permeant solute, divided by the calculated van't Hoff pressure. We understood the term reflection coefficient to arise from the notion that collision of an impermeant molecule with a membrane would result in reflection of the molecule back into the solution whence it came, and σ would equal 1.0. In the case of permeant molecules, some fraction would cross the membrane, so that reflection would be imperfect with $\sigma < 1.0$. The σ is normally bounded by limits of 0 and 1.0, so that molecules almost as permeable as the solvent, such as D_2O, would have $\sigma \sim 0$.

The first use of σ in a biological study, I believe, was in a paper from our laboratory by Durbin et al. (13), who measured water flow across the frog gastric mucosa in 1956 and found that net water flux is passive. In the absence of an osmotic gradient, net water flux is driven by HCl secretion and falls to zero when acid secretion is stopped. The study was also concerned with the size of the aqueous pores in the mucosa and showed that $(1 - \sigma) = A_s/A_w$. The expression, A_s/A_w, previously derived by Renkin (44), is the ratio of the hindrance of solute molecule filtration through a pore, A_s, to the hindrance to water filtration, A_w. The equality of the ratio to $(1 - \sigma)$ is important because it relates the thermodynamic parameter, σ, to geometrical dimensions of solute, solvent, and membrane channel. In a subsequent paper, Durbin (12) verified this equality in cellulose membranes with pore radii of 20 to 80 Å. Durbin also showed, in an appendix to the gastric mucosa paper, that the treatments of Koefoed-Johnsen and Ussing (30) and of Pappenheimer (39) gave equivalent results for estimations of the equivalent pore radius of the gastric mucosa.

The real impetus for the use of the reflection coefficient in biological studies came from the use of irreversible thermodynamics in 1958 by Kedem and Katchalsky (26) to formulate two equations that were immediately applicable to laboratory experiments. Aharon Katchalsky gave me a preprint of that article, which I studied en route to a membrane transport meeting in Czechoslovakia that we were both to attend. I was so impressed that I invited him to give a course on the topic in the Harvard biophysics program; it was this course that gave rise to the book by Katchalsky and Curran (25), *Nonequilibrium Thermodynamics in Biophysics*, published in 1964 and still in print more than twenty years later. Kedem and Katchalsky (27) had returned to the same topic in 1961 in a particularly lucid article that made their equations accessible to a broad group of scientists working on membrane transport problems. It was Katchalsky's tragic fate to be murdered in the course of a terrorist attack on the airport in Lod, Israel.

The Kedem-Katchalsky equations that follow are those applied to transport across a biological membrane in which only solvent and a single solute can move,

$$J_v = L_p \Delta\pi_i - \sigma L_p \Delta\pi_s \qquad (1)$$

$$J_s = (1 - \sigma)\bar{c}_s J_v + \omega\Delta\pi_s \qquad (2)$$

in which J_v is volume flux across the membrane, $\Delta\pi_i$ is the difference in osmotic pressure caused by the impermeable solute; $\Delta\pi_s$ is the difference in osmotic pressure caused by the permeable solute; J_s is the permeable solute flux; $\bar{c}_s$ is the mean solute concentration in the membrane, usually taken as the mean between inside and outside concentrations; and ω is the solute permeability coefficient.

The Kedem-Katchalsky equations are particularly useful because of the illumination they cast on coupling of solute and solvent transport across the membrane. In Equation 1, if $\Delta\pi_s = 0$, the hydraulic conductivity, L_p, is the ratio of the volume flux to its conjugate force, $\Delta\pi_i$ (either the osmotic pressure difference of an impermeable solute or, alternatively, applied hydrostatic pressure). If $\Delta\pi_i = 0$, the volume flux will be driven by the osmotic pressure difference of the solute. If the solute is impermeable, $\sigma = 1$ and the flux will be the same as for an impermeable solute; for more permeable solutes, σ becomes <1.0 and there is a smaller flux just as Staverman showed. In Equation 2, when $J_v = 0$, the permeability coefficient (ω) is the ratio of the solute flux to its conjugate force, the osmotic pressure gradient of the solute. However, when $J_v > 0$, there is volume flow, which propels solute across the membrane by a process called solvent drag that can be envisioned as a flux of entrained solute molecules carried along in the solvent stream. Solvent drag is proportional to $(1 - \sigma)$; when $\sigma = 1$, the membrane is impermeable to solute and there can be no solvent drag. When $\sigma \sim 0$, as for D_2O, the membrane can barely distinguish D_2O from H_2O and the tracer concentration inside the pore approximates that in the solution from which it came.

The reflection coefficient may be viewed as a composite of two different frictions, that between solute and membrane as illustrated in Equation 1 and that between solute and solvent as illustrated in Equation 2. It is easy to understand that if $\sigma = 1$, solvent and solute cannot traverse the membrane through the same pathway. Conversely, Equation 2 shows that, if σ is significantly <1, both solute and solvent traverse the same route, a consequence of great importance when we consider how hydrophilic nonelectrolytes cross cell membranes.

In 1960, Goldstein and I (18) devised the zero-time method to determine the reflection coefficient for the interactions of a number of hydrophilic nonelectrolytes with the red cell membrane. Once a permeant solute has been introduced into a red cell suspension in a stopped-flow apparatus, solute begins to enter the cell. In general, there is no practicable technique for measuring the intracellular solute concentration instantaneously, but at zero time ($t = 0$) the concentration is known. Our method depends on the use of Equation 1 and the determination of $J_{v,t=0}$ by extrapolation from later observations. A series of experiments was carried out at constant $\Delta\pi_i$, and $\Delta\pi_s$ was varied systematically to determine $J_{v,t=0}$ as a function of $\Delta\pi_s$. The $\Delta\pi_s$ for which $J_{v,t=0} = 0$ was then determined by interpolation and σ was given by the ratio $\Delta\pi_i/\Delta\pi_s$. Having established values of σ for a number of small hydrophilic nonelectrolytes, we then used the equation $(1 - \sigma) = A_s/A_w$ and fitted our values of σ to Renkin's equations for the hindrance of filtration through cellulose membranes, as shown in Figure 3B. These data gave an equivalent pore radius of 4.2 Å and provided an independent confirmation of the equivalent pore radius we had previously derived from the P_f/P_d ratio.

As it has subsequently developed, one of our σ values, $\sigma_{urea} = 0.62 \pm 0.02$, has provided the fuel for a long-running controversy. The reason for the controversy is that my view that urea and water cross the red cell membrane by the same aqueous channel has drawn increasing fire. It is generally conceded that values of σ_{urea} significantly <1.0 provide very strong thermodynamic evidence that water and urea share a common channel. Therefore, those who believe urea crosses the membrane independently of water would like to show that $\sigma_{urea} \sim 1$. Thus, in 1983, Levitt and Mlekoday (33) found $\sigma_{urea} = 1$, but they reported that their data was fit almost as well by a value of 0.75. All the other experimental determinations of σ_{urea} give values <1. In 1975, Owen and Eyring (37) found $\sigma_{urea} = 0.79 \pm$

0.02. We have made two more determinations, one by Chasan and Solomon (5) in 1985 giving 0.70 ± 0.02 and another one, in 1987 by Toon and Solomon (69), of 0.65 ± 0.03. We continue to believe that our three separate measurements are valid. If so, they indicate that urea and water share, at least in part, a common pathway across the red cell membrane; if not, the pathways are separate. And there the matter rests.

Water Permeability in Intestine and Kidney

The necessity to learn enough physiology to be able to teach medical students was responsible for the enlargement of my research interests to include the intestine. My initial appointment in the Harvard Medical School was in the department of biological chemistry, but in the late 1940s or early 1950s it was decided that I was better fitted to physiology, and Gene Landis, the chairman, felt that as a member of the department I should participate in the Medical School teaching. It was not instantly obvious that there was any area of physiology to which my talents would be appropriate and Gene decided that I should teach intestinal physiology, which I then began to study intensively.

At that time, the prevailing view of the mechanism of intestinal absorption was the fluid circuit theory of Visscher and his group [see Ingraham et al. (22)]. Visscher and his colleagues (73) had used radioactive Na^+ and Cl^- tracers to determine that absorption of these ions from the intestinal lumen could not be due to simple passive diffusion and they advanced the fluid circuit theory as the explanation. The theory proposed that the mucosal membrane had a mosaic structure and that two separate flows took place through different parts of the membrane, one flow containing solute and water, and the other flow containing water alone; active transport of water was an essential feature of the theory. I found the theory difficult to believe, and even more difficult to explain comprehensibly to the medical students. The more I had to try to understand the theory so that I could teach it, the surer I became that it was wrong. It seemed to me that Ussing's concepts, which defined active and passive transport (70), would be applicable to intestinal absorption and I was determined to see if this was the case.

The research Peter Curran carried out in my laboratory during his senior year at Harvard College made such a strong impression that I urged him to become a graduate student. Obdurate to these insistent pleas, he decided to enter medical school and I was surprised and delighted when, one year later, he decided that research was his métier and came back as a graduate student. A study of intestinal absorption appeared to be an ideal topic for his dissertation. One of the problems that faces a physical chemist who has turned to physiology is the question of how to tell whether tissue is alive or dead. This problem never arose in studies on red cells, which are very

[141]

tough and only need a supply of glucose to live for hours. The position at which I have now arrived, after all these years, is the purely pragmatic one that holds a tissue to be alive if it preserves the function I want to measure. I had not yet come to this conclusion when we decided that the rat intestine was the tissue of choice for our intestinal transport studies, so we perfused the intestine in situ and only cut an ileal segment at its two ends to insert the perfusing pipettes. Thus the intestinal blood supply was maintained and we could safely assume that the intestinal segment was in a normal physiological environment.

All of the measurements necessary to define flux across the ileal segment could be obtained from analysis of the perfusion fluid. We used hemoglobin concentration to measure net water movement. One-way fluxes of Na^+ and Cl^- were determined with $^{24}Na^+$ and $^{36}Cl^-$ and net fluxes of these ions were determined by chemical analyses. We used mannitol as an indifferent solute to adjust the osmolality of the perfusion fluid and measured its loss from the system by chemical analysis. By varying the composition of the perfusion fluid, we could impose defined ion and osmotic pressure gradients across the system and observe the resultant flows across the ileal membrane.

We first found that net water flux was correlated with net Na^+ flux, but we could not be absolutely sure of the explanation until we also took account of the leakage of mannitol out of the perfusion fluid. When this was done, we were able to show [Curran and Solomon (7)] that the net solute flux across the membrane drove net water flux, as illustrated in Figure 4. When the net solute flux went to zero, so did the water flux, thus proving that water movement across the ileal epithelium is entirely passive. We also found that both Na^+ and Cl^- transport are active and depend on the lumenal [NaCl], as shown in Figure 4B. With these relatively simple concepts, we were able to account for virtually all the classical measurements of salt and water movement across the ileal membrane that had been reported in the literature.

These studies led to a continuing interest in epithelial transport in the intestine in which Curran played a leading role, prior to his untimely death from a heart attack at age 40. Stanley Schultz, who worked closely with Curran in our laboratory, has continued to make important contributions in this field, as also have others who worked in the laboratory at that time, such as Ernie Wright and Michael Field.

Gradually my interests in physiology broadened to include the kidney and I became fascinated with the elegant studies of Richards and his group (see ref. 45), who were able to probe kidney function with meticulous micropuncture techniques, and I immersed myself in the literature. I remember that one day Baird Hastings brought Richards to visit the laboratory and I was brash enough to tell him

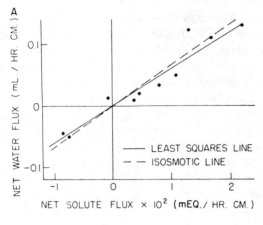

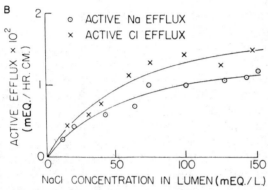

Figure 4. A: net water flux across intestinal ileum in rat as a function of net solute flux. [From Curran and Solomon (7).] B: active component of Na$^+$ and Cl$^-$ fluxes across intestinal ileum in rat. [From Curran and Solomon (7).]

that I hoped to study the proximal tubule. He murmured something polite and walked on. He didn't tell me what I was not bright enough to work out for myself—that it was a formidable enterprise and that I would have to mobilize all my resources to succeed. Gerhard Giebisch, then at Cornell and now at Yale, who entered the field at about the same time, did it right. He apprenticed himself to Phyllis Bott, who was the last member of Richards' group to do micropuncture and learned micropuncture techniques in her laboratory. By this time, a number of young M.D.s who wanted postdoctoral research experience had begun to join my laboratory. I reasoned, wrongly, that their medical school education would have taught them enough about surgical techniques so that they would only have to read the papers of the Richards group in order to emulate their technique. This was a mistake that almost ruined the laboratory.

Richards and Walker (46) devised the technique to perfuse the proximal tubule of the *Necturus maculosus*, an amphibian that is

[143]

distinguished by having unusually large tubules. In *Necturus* kidneys, the first segment of the proximal tubule is visible on the top surface of the kidney. The tubule then descends into the body of the kidney and pursues a convoluted 14.0 mm path until the distal end of the proximal tubule emerges at the surface again where it can usually

be found lying parallel to the initial segment, as illustrated in Figure 5A. The tubule has a 0.14 mm diam, wide enough for puncture by a pipette of 20 μm diam, which Richards and Walker used to inject droplets of mercury at both ends of the tubule. This procedure, which is a great deal easier to describe than to do, blocks off the entire length of the proximal tubule, whose contents can then be accessed by another pipette, much as if it were a length of rat ileum. Or so I reasoned, as I confidently expected to do in the kidney what we had done in the intestine.

The project was started by Joe Shipp and Irwin Hanenson, who soon enough mastered the technique and were able to isolate solutions of known initial composition between droplets of mercury. We used ^{14}C-inulin, to which the proximal tubule is impermeable, to monitor the changes in water we confidently expected to see as NaCl was absorbed out of the tubule. However, we were totally unable to observe any fluid movement, and I was at my wits' end. The project was joined by Guillermo Whittembury, Erich Windhager, and Hans Schatzmann. Schatzmann had come for a postdoctoral year from Switzerland where he had previously discovered that ouabain specifically inhibited red cell cation transport. The size of the group meant that a very large segment of the laboratory was now committed to studies on the kidney. We remained unable to make the system work until Schatzmann substituted oil for the mercury in the tubules, as well as introducing other important modifications in our technique (55). I think it was the mercury that had poisoned the system, but whatever the cause, we were delighted and immediately started to do the experiments we had planned from the outset. Those were very trying days when everything that we did failed and tension ran high in the lab. I remember that Schatzmann used to swear in Schweizerdeutsch, relieving his feelings and making all the rest of us laugh. Subsequently, many of those involved, such as Windhager, Whittembury, and Khuri, went on to do further kidney research and Schatzmann devoted his attention to Ca^{2+}.

One of our first observations was that of Schatzmann et al. (49), who showed 27% net water absorption from a tubule filled with isosmolar NaCl, in agreement with the earlier determinations of Walker and Hudson (74). Addition of 10^{-4} M ouabain to the lumen of the proximal tubule inhibited water absorption to 10%, which suggested that water absorption was driven by the active transport of Na^+. Windhager et al. (77) next addressed the mechanism of water transport. They first varied the luminal NaCl concentration, using mannitol to maintain constant osmolality, and showed that net water

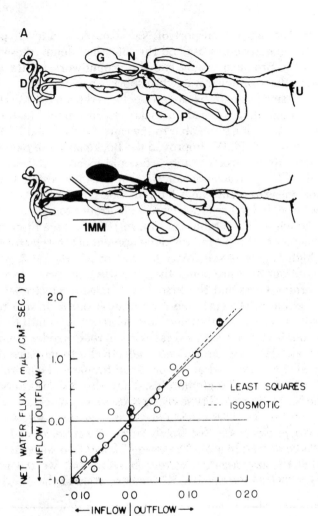

Figure 5. A: technique of stopped-flow microperfusion in the *Necturus* kidney. G, glomerulus; N, neck; P, proximal tubule; D, distal tubule; U, ureter. Dark fluid in tubule is oil. [From Shipp et al. (55).] *B*: net water flux as a function of net solute flux out of lumen of *Necturus* proximal tubule. Full line is the least squares line and the dashed line is the prediction for isosmotic flux. [From Windhager et al. (77).]

flux was dependent on NaCl concentration. They then measured the total net solute flux out of the lumen under the same conditions. Net water flux was linearly dependent on net solute flux and fell to zero at zero solute flux, as shown in Figure 5*B*. Measurements of the potential difference (77) showed that Na$^+$ was transported up an electrochemical potential gradient. These results show that water movement out of the proximal tubule lumen is a passive process

[145]

driven by the active transport of Na^+. Coupled with our previous results in the gastric mucosa and the intestinal ileum, these experiments in the proximal tubule laid the groundwork for the general conclusion that epithelial water transport is passive.

The very small size of the samples collected from the tubules, 1 μl or less, meant that we had to make improvements in the flame photometer in order to be able to measure the Na^+ and K^+ content of the collected fluid. We improved the Beckman flame photometer, particularly its electronics, until we could measure (60) as little as 200 pmol of K^+ to an accuracy of 8%. Our most interesting adventure in cation measurement came later in the series of kidney papers when we decided to use ion-selective electrodes to measure the Na^+ and K^+ activities in the tubule in situ. In 1957 George Eisenman and his colleagues (14) reported the development of ion-selective glasses with a high degree of selectivity for Na^+ or K^+. By 1962, when we carried out our measurements, the glass was only available in lumps from Corning Glass and the first step in making electrodes was to blow a bubble in the glass and then draw it out to form a tube. In successive steps we drew thinner and thinner tubes until they were small enough so that we could fashion microelectrodes from them. Khuri et al. (28) used the Na^+ microelectrodes to measure the Na^+ activity in situ at the end of the proximal tubule and confirm that it agreed with that in the glomerulus. They also did the experiments with the K^+ electrodes. These electrodes were particularly delicate and we were lucky if we could get as many as three measurements from a single electrode. We finally brought our series of K^+ measurements to an end because the ratio of failures to successes was so great that the investigators were exhausted, but we did not stop before we had shown that the K^+ concentration at the distal end of

---

Figure 6. A: time course of ^{203}Hg-pCMBS uptake by red cell membrane proteins under conditions in which only the water flux inhibition site is labeled. Cells were treated with N-ethylmaleimide and 4,4'-diisothiocyanostilbene-2,2'-disulfonic acid to block sulfhydryl sites that do not affect water or urea transport. Curves have been fitted by nonlinear least squares to the equation for bimolecular association with a negligible back reaction. For the band 3 site, $\tau = 15 \pm 2$ min, in fair agreement with the kinetics computed according to data from Toon and Solomon (68) for the water flux inhibition site for which $\tau = 28 \pm 5$ min under the same conditions. We ascribe the other peaks for which $\tau = 98 \pm 7$ min ($\approx$46 kDa peak) and $\tau = 78 \pm 65$ min (band 4.5) to a slow leakage of pCMBS into the cell through the anion-exchange channel. [Data from D. Ojcius and A. K. Solomon (unpublished observations).] B: similar to A but conditions have been chosen so that only the urea flux inhibition site is labeled and $\tau = 38 \pm 11$ min, in agreement with the figure of $\tau = 21 \pm 19$ min computed from Toon and Solomon's kinetic data for pCMBS inhibition of urea flux under the same conditions (68).

[146]

the proximal tubule was 1.8 ± 0.1 times more concentrated than in the glomerulus.

LOCUS OF RED CELL AQUEOUS CHANNEL

It was Sha'afi and his colleagues (see ref. 2) who first suggested that the red cell aqueous channel was associated with a specific membrane protein, band 3, the anion-exchange protein. In 1970, Macey and Farmer (35) had found that the mercurial sulfhydryl

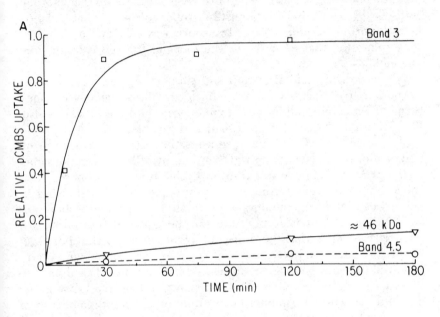

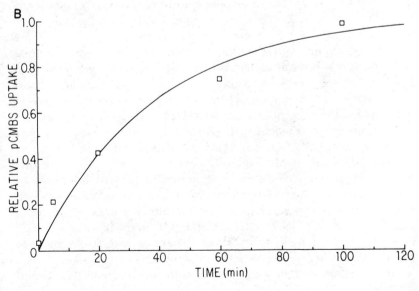

reagent *p*-chloromercuribenzene sulfonate (pCMBS) inhibited osmotic water flow across the red cell membrane by about 80% and was also an effective inhibitor of urea flux. Sha'afi and Feinstein (52) characterized a number of other sulfhydryl reagents that inhibited water transport, including *p*-chloromercuribenzoate (pCMB), which they found to be localized on band 3, as determined by polyacrylamide gel electrophoresis. In order to obtain satisfactory gels, they had found it necessary to block other sulfhydryl groups with *N*-ethylmaleimide (NEM) and mersalyl, neither of which had any effect on normal water transport. When Brown et al. (2) found that another sulfhydryl reagent, 5,5'-dithiobis(2-nitrobenzoic acid), that had also been reported to inhibit water transport, was localized on band 3, they suggested that band 3 was the locus of the red cell aqueous channel.

We seized on this suggestion and supported it enthusiastically, proposing (61) that band 3 was the locus of an aqueous channel through which not only water and anions entered the cells, but also nonelectrolytes such as urea. Knauf and Rothstein (29) had previously shown that pCMBS induced a massive cation leak in the red cell membrane and we suggested that their pCMBS site was the same as the pCMBS site for water transport inhibition and proposed that the cation leak went through the same aqueous channel. Band 3 comprises 25% of the membrane proteins and is present in $0.5–0.6 \times 10^6$ dimers. If it is responsible for the aqueous transport it is necessary to show that there are sufficient copies to transport the water. We have computed (61) that 2.7×10^5 aqueous channels are required to accommodate the measured flux of water, based on the hindrance to nonelectrolyte diffusion shown in Figure 3*A*; band 3 is one of the few integral membrane proteins of which sufficient copies are present. As might have been expected, this proposal engendered a good deal of discussion and the ensuing controversy has enlivened the past several years of research in the laboratory.

The dust has not settled and it is too soon to know what the final answer will be, but one aspect is of general interest and deserves a place in this account. Kopito and Lodish (31) have sequenced band 3 and report that it contains a total of six cysteine residues. Five of these six sulfhydryl groups react with NEM, which has no effect on water transport and blocks further reaction of these five sulfhydryl groups with pCMBS (43, 61), leaving the sixth sulfhydryl group as the candidate for pCMBS inhibition of water and urea transport. Macey (34) has reported that the kinetics for pCMBS inhibition of urea transport are much faster than those for water inhibition, which indicates that the sites for these two inhibitory processes are separate. Toon and Solomon (68) have characterized the binding affinities of each site, which differ by three orders of magnitude and the kinetics that differ by two orders of magnitude.

These differences in kinetics are sufficiently great that David

[148]

Ojcius and I have been able to label the sites separately in red cells in which five of the sulfhydryl groups on band 3 have been blocked with NEM, using ^{203}Hg-pCMBS and polyacrylamide gel autoradiography. We find, as shown in Figure 6, that each site is saturable with the appropriate kinetics and that each site is on band 3 (D. Ojcius and A. K. Solomon, unpublished observations). What surprised us was the stoichiometry; there are three times as many water transport inhibition sites as urea transport inhibition sites. There is one band 3 molecule for each site so that four band 3 molecules are required to accommodate the three water transport inhibition sites plus the one urea transport inhibition site. The observation that both the water and urea transport inhibition sites are on band 3 suggests that band 3 plays a role in the transport of both water and urea. The stoichiometry suggests that band 3 may be present in the membrane as a tetramer.

One other recent observation contributes to the picture. We reasoned that we could obtain an unequivocal answer to the question of whether band 3 was responsible for the aqueous channel by reconstituting purified band 3 into phospholipid vesicles and determining whether there was pCMBS-inhibitable water flux. In preparatory experiments, David Ojcius and I reconstituted a mixture of all the red cell membrane proteins into vesicles, but were unable to find any evidence of pCMBS inhibition of water transport, though glucose transport in this preparation was normal.

Next we turned to white ghosts [prepared essentially according to the method of Steck and Kant (66)], which are known to have water permeability very similar to red cells (6, 32). To our surprise, we found that pCMBS did not inhibit water transport in these ghosts and that normal urea permeability was entirely absent. When we made ghosts very gently, without repeated washing so that they exhibited a pink color, we retained normal pCMBS-inhibitable water transport and normal urea transport. It is too soon to know the exact cause of these manifestations, but the experiments suggest that interactions between the cytoskeleton and the membrane transport protein(s) are requisite for normal water and urea transport, possibly to keep a tetrameric assembly of band 3 molecules in its physiological conformation. These are exciting prospects because they may relate the transport of water across the cell membrane to profound interactions with structures that extend throughout the cytoplasm.

BIBLIOGRAPHY

1. ARONS, W. L., R. J. VANDERLINDE, and A. K. SOLOMON. The simultaneous measurement of exchangeable body sodium and potassium utilizing ion exchange chromatography. *J. Clin. Invest.* 33: 1001–1007, 1954.
2. BROWN, P. A., M. B. FEINSTEIN, and R. I. SHA'AFI. Membrane proteins

related to water transport in human erythrocytes. *Nature Lond.* 254: 523–525, 1975.

3. CASS, A., and A. FINKELSTEIN. Water permeability of thin lipid membranes. *J. Gen. Physiol.* 50: 1765–1784, 1967.

4. CHANCE, B., Q. H. GIBSON, R. H. EISENHARDT, and K. K. LONBERG-HOLM. *Rapid Mixing and Sampling Techniques in Biochemistry.* New York: Academic, 1964.

5. CHASAN, B., and A. K. SOLOMON. Urea reflection coefficient for the human red cell membrane. *Biochim. Biophys. Acta* 821: 56–62, 1985.

6. COLOMBE, B. W., and R. I. MACEY. Effects of calcium on potassium and water transport in human erythrocyte ghosts. *Biochim. Biophys. Acta* 363: 226–239, 1974.

7. CURRAN, P. F., and A. K. SOLOMON. Ion and water fluxes in the ileum of rats. *J. Gen. Physiol.* 41: 143–168, 1957.

8. DAINTY, J. Water relations of plant cells. *Adv. Bot. Res.* 1: 279–326, 1963.

9. DEKRUIJFF, B., and R. A. DEMEL. Polyene antibiotic-sterol interactions in membranes of *Acholeplasma Laidlawii* cells and lecithin liposomes. *Biochim. Biophys. Acta* 339: 57–70, 1974.

10. DIRKEN, M. N. J., and H. W. MOOK. The rate of gas exchange between blood cells and serum. *J. Physiol. Lond.* 73: 349–360, 1931.

11. DIX, J. A., and A. K. SOLOMON. Role of membrane proteins and lipids in water diffusion across red cell membranes. *Biochim. Biophys. Acta* 773: 219–230, 1984.

12. DURBIN, R. P. Osmotic flow of water across permeable cellulose membranes. *J. Gen. Physiol.* 44: 315–326, 1960.

13. DURBIN, R. P., H. FRANK, and A. K. SOLOMON. Water flow through frog gastric mucosa. *J. Gen. Physiol.* 39: 535–551, 1956.

14. EISENMAN, G., D. O. RUDIN, and J. U. CASBY. Glass electrode for measuring sodium ion. *Science Wash. DC* 126: 831–834, 1957.

15. FRIEDMANN, E., A. K. SOLOMON, and N. T. WERTHESSEN. Radioactive organic bromo-compounds. *Nature Lond.* 143: 472–473, 1939.

16. GILL, T. J., and A. K. SOLOMON. Effect of ouabain on sodium flux in human red cells. *Nature Lond.* 183: 1127–1128, 1959.

17. GLYNN, I. M. The action of cardiac glycosides on sodium and potassium movements in human red cells. *J. Physiol. Lond.* 136: 148–173, 1957.

18. GOLDSTEIN, D. A., and A. K. SOLOMON. Determination of equivalent pore radius for human red cell by osmotic pressure measurement. *J. Gen. Physiol.* 44: 1–17, 1960.

19. HARTRIDGE, H., and F. J. W. ROUGHTON. A method of measuring the velocity of very rapid chemical reactions. *Proc. R. Soc. Lond. A Math. Phys. Sci.* 104: 376–394, 1923.

20. HEVESY, G. v., E. HOFER, and A. KROGH. The permeability of the skin of frogs to water as determined by D_2O and H_2O. *Scand. Arch. Physiol.* 72: 199–214, 1935.

21. HOLZ, R., and A. FINKELSTEIN. The water and nonelectrolyte permeability induced in thin lipid membranes by the polyene antibiotics nystatin and amphotericin B. *J. Gen. Physiol.* 56: 125–145, 1970.

22. INGRAHAM, R. C., H. C. PETERS, and M. B. VISSCHER. On the movement of materials across living membranes against concentration gradients. *J. Phys. Chem.* 42: 141–150, 1938.

23. JACOBS, M. H. Osmotic properties of the erythrocyte. III. Applicability of osmotic laws to the rate of hemolysis in hypotonic solutions of nonelectrolytes. *Biol. Bull.* 62: 178–194, 1932.

24. JACOBS, M. H. Surface properties of the erythrocyte. *Ann. NY Acad. Sci.* 50: 824–834, 1950.

25. KATCHALSKY, A., and P. F. CURRAN. *Nonequilibrium Thermodynamics in Biophysics.* Cambridge, MA: Harvard Univ. Press, 1964.

26. KEDEM, O., and A. KATCHALSKY. Thermodynamic analysis of the permeability of biological membranes to nonelectrolytes. *Biochim. Biophys. Acta* 27: 229–246, 1958.

27. KEDEM, O., and A. KATCHALSKY. A physical interpretation of the phenomenological coefficients of membrane permeability. *J. Gen. Physiol.* 45: 143–179, 1961.

28. KHURI, R. N., D. A. GOLDSTEIN, D. L. MAUDE, C. EDMONDS, and A. K. SOLOMON. Single proximal tubules of Necturus kidney. VIII. Na and K determinations by glass electrodes. *Am. J. Physiol.* 204: 743–748, 1963.

29. KNAUF, P. A., and A. ROTHSTEIN. Chemical modification of membranes. II. Permeation paths for sulfhydryl agents. *J. Gen. Physiol.* 58: 211–223, 1971.

30. KOEFOED-JOHNSEN, V., and H. H. USSING. The contributions of diffusion and flow to the passage of D_2O through living membranes. *Acta Physiol. Scand.* 28: 60–76, 1953.

31. KOPITO, R. R., and H. F. LODISH. Structure of the murine anion exchange protein. *J. Cell. Biochem.* 29: 1–17, 1985.

32. LEVIN, R. L., S. W. LEVIN, and A. K. SOLOMON. Improved stop-flow apparatus to measure permeability of human red cells and ghosts. *J. Biochem. Biophys. Methods* 3: 255–272, 1980.

33. LEVITT, D. G., and H. J. MLEKODAY. Reflection coefficient and permeability of urea and ethylene glycol in the human red cell membrane. *J. Gen. Physiol.* 81: 239–253, 1983.

34. MACEY, R. I. Transport of water and urea in red blood cells. *Am. J. Physiol.* 246 (Cell Physiol. 15): C195–C203, 1984.

35. MACEY, R. I., and R. E. L. FARMER. Inhibition of water and solute permeability in human red cells. *Biochim. Biophys. Acta* 211: 104–106, 1970.

36. ONSAGER, L. Theories and problems of diffusion. *Ann. NY Acad. Sci.* 46: 241–265, 1945.

37. OWEN, J. D., and E. M. EYRING. Reflection coefficients of permeant molecules in human red cell suspensions. *J. Gen. Physiol.* 66: 251–265, 1975.

38. PAGANELLI, C. V., and A. K. SOLOMON. Rate of exchange of tritiated water across the human red cell membrane. *J. Gen. Physiol.* 41: 259–277, 1957.

39. PAPPENHEIMER, J. R. Passage of molecules through capillary walls. *Physiol. Rev.* 33: 387–423, 1953.

40. PAPPENHEIMER, J. R., E. M. RENKIN, and L. M. BORRERO. Filtration, diffusion and molecular sieving through peripheral capillary membranes. *Am. J. Physiol.* 167: 13–46, 1951.

41. POST, R. J., and P. C. JOLLY. The linkage of sodium, potassium and ammonium active transport across the human erythrocyte membrane. *Biochim. Biophys. Acta* 25: 118–128, 1957.

WATER
MOVEMENT
ACROSS
MEMBRANES

42. Prescott, D. M., and E. Zeuthen. Comparison of water diffusion and water filtration across cell surfaces. *Acta Physiol. Scand.* 28: 77–94, 1953.

43. Rao, A., and R. A. F. Reithmeier. Reactive sulfhydryl groups of the band 3 polypeptide from human erythrocyte membranes. Location in the primary structure. *J. Biol. Chem.* 254: 6144–6150, 1979.

44. Renkin, E. M. Filtration, diffusion, and molecular sieving through porous cellulose membranes. *J. Gen. Physiol.* 38: 225–243, 1954.

45. Richards, A. N. Processes of urine formation. *Proc. R. Soc. Lond. B Biol. Sci.* 126: 398–432, 1938.

46. Richards, A. N., and A. M. Walker. Methods of collecting fluid from known regions of the renal tubules of amphibia and of perfusing the lumen of a single tubule. *Am. J. Physiol.* 118: 111–120, 1937.

47. Robinson, C. V. A methane, proportional counting method for the assay of tritium. *Rev. Sci. Instrum.* 22: 353–355, 1951.

48. Ruben, S., and M. D. Kamen. Radioactive carbon of long half-life. *Phys. Rev.* 57: 549, 1940.

49. Schatzmann, H. J., E. E. Windhager, and A. K. Solomon. Single proximal tubules of the Necturus kidney. II. Effect of 2,4-dinitrophenol and ouabain on water reabsorption. *Am. J. Physiol.* 195: 570–574, 1958.

50. Schloerb, P. R., B. J. Friis-Hansen, I. S. Edelman, A. K. Solomon, and F. D. Moore. The measurement of total body water in the human subject by deuterium oxide dilution. *J. Clin. Invest.* 29: 1296–1310, 1950.

51. Schwartz, A., G. E. Lindenmeyer, and J. C. Allen. The sodium, potassium ATPase: Pharmacological, physiological and biochemical aspects. *Pharmacol. Rev.* 27: 3–134, 1975.

52. Sha'afi, R. I., and M. B. Feinstein. Membrane water channels and SH-groups. *Adv. Exp. Med. Biol.* 84: 67–80, 1977.

53. Sha'afi, R. I., C. M. Gary-Bobo, and A. K. Solomon. Permeability of red cell membranes to small hydrophilic and lipophilic solutes. *J. Gen. Physiol.* 58: 238–258, 1971.

54. Sha'afi, R. I., G. T. Rich, V. W. Sidel, W. Bossert, and A. K. Solomon. The effect of the unstirred layer on human red cell water permeability. *J. Gen. Physiol.* 50: 1377–1399, 1967.

55. Shipp, J. C., I. B. Hanenson, E. E. Windhager, H. J. Schatzmann, G. Whittembury, H. Yoshimura, and A. K. Solomon. Single proximal tubules of the Necturus kidney. Methods for micropuncture and microperfusion. *Am. J. Physiol.* 195: 563–569, 1958.

56. Sidel, V. W., and A. K. Solomon. Entrance of water into human red cells under an osmotic pressure gradient. *J. Gen. Physiol.* 41: 243–257, 1957.

57. Solomon, A. K. The half-life of sodium 24. *Phys. Rev.* 79: 403, 1950.

58. Solomon, A. K. Characterization of biological membranes by equivalent pores. *J. Gen. Physiol.* 51: 335s–364s, 1968.

59. Solomon, A. K. On the equivalent pore radius. *J. Membr. Biol.* 94: 227–232, 1986.

60. Solomon, A. K., and D. C. Caton. Modified flame photometer for microdetermination of sodium and potassium. *Anal. Chem.* 27: 1849–1850, 1955.

61. Solomon, A. K., B. Chasan, J. A. Dix, M. F. Lukacovic, M. R. Toon, and

A. S. Verkman. The aqueous pore in the red cell membrane: band 3 as a channel for anions, cations, nonelectrolytes, and water. *Ann. NY Acad. Sci.* 414: 97–124, 1983.

62. Solomon, A. K., and C. M. Gary-Bobo. Aqueous pores in lipid bilayers and red cell membranes. *Biochim. Biophys. Acta* 255: 1019–1021, 1972.

63. Solomon, A. K., T. J. Gill, and G. L. Gold. The kinetics of cardiac glycoside inhibition of potassium transport in human erythrocytes. *J. Gen. Physiol.* 40: 327–350, 1956.

64. Solomon, A. K., B. Vennesland, F. W. Klemperer, J. M. Buchanan, and A. B. Hastings. The participation of carbon dioxide in the carbohydrate cycle. *J. Biol. Chem.* 140: 171–182, 1941.

65. Staverman, A. J. The theory of measurement of osmotic pressure. *Rec. Trav. Chim. Pays-Bas Belg.* 70: 344–352, 1951.

66. Steck, T. L., and J. A. Kant. Preparation of impermeable ghosts and inside-out vesicles from human erythrocyte membranes. *Methods Enzymol.* 31: 172–180, 1974.

67. Terwilliger, T. C., and A. K. Solomon. Osmotic water permeability of human red cells. *J. Gen. Physiol.* 77: 549–570, 1981.

68. Toon, M. R., and A. K. Solomon. Control of red cell urea and water permeability by sulfhydryl reagents. *Biochim. Biophys. Acta* 860: 361–375, 1986.

69. Toon, M. R., and A. K. Solomon. Interrelation of ethylene glycol, urea, and water transport in the red cell. *Biochim. Biophys. Acta* 898: 275–282, 1987.

70. Ussing, H. H. The distinction by means of tracers between active transport and diffusion. *Acta Physiol. Scand.* 19: 43–56, 1950.

71. Vieira, F. L., R. I. Sha'afi, and A. K. Solomon. The state of water in human and dog red cell membranes. *J. Gen. Physiol.* 55: 451–466, 1970.

72. Villegas, R., T. C. Barton, and A. K. Solomon. The entrance of water into beef and dog red cells. *J. Gen. Physiol.* 42: 355–369, 1958.

73. Visscher, M. B., E. S. Fetcher, C. W. Carr, H. P. Gregor, M. S. Bushey, and D. E. Barker. Isotopic tracer studies on the movement of water and ions between intestinal lumen and blood. *Am. J. Physiol.* 142: 550–575, 1944.

74. Walker, A. M., and C. L. Hudson. The reabsorption of glucose from the renal tubule in amphibia and the action of phlorhizin upon it. *Am. J. Physiol.* 118: 130–143, 1937.

75. Wang, J. H. Self-diffusion and structure of liquid water. *J. Am. Chem. Soc.* 73: 510–513, 1951.

76. Williams, J. B., and H. Kutchai. Studies of the diffusion boundary layer around the red blood cell in a stopped-flow apparatus (Abstract). *Biophys. J.* 47: 160a, 1985.

77. Windhager, E. E., G. Whittembury, D. E. Oken, H. J. Schatzmann, and A. K. Solomon. Single proximal tubules of the Necturus kidney. III. Dependence of H_2O movement on NaCl concentration. *Am. J. Physiol.* 197: 313–318, 1959.

78. Wood, H. G., and C. H. Werkman. The relationship of bacterial utilization of CO_2 to succinic acid formation. *Biochem. J.* 34: 129–138, 1940.

V

Sodium-Potassium Pump

JENS C. SKOU

THE idea of a pump in the cell membrane was introduced by R. B. Dean in 1941 in a paper entitled "Theories of Electrolyte Equilibrium in Muscle" (20). Referring to experiments by L. A. Heppel in 1938, by Heppel and C. L. A. Schmidt in 1939, and by H. B. Steinbach in 1940 (see ref. 20), Dean concluded: "*the muscle can actively move potassium and sodium against concentration gradients . . . this requires work. Therefore there must be some sort of a pump possibly located in the fiber membrane, which can pump out the sodium or, what is equivalent, pump in the potassium.*"

In the following decennium the whole concept of active transport of sodium and of a sodium pump responsible for the transport was developed, not least helped by the introduction of isotopic tracers in biology (102).

With the knowledge available in the beginning of the fifties about active transport of Na^+ and K^+ across the cell membrane (see ref. 28) it was possible to foresee that the transport system is a membrane-located enzyme system that has ATP as substrate and is activated by Na^+ on the cytoplasmic side and by K^+ on the extracellular side of the membrane. To demonstrate this it was necessary to take membrane fragments, which give access to both sides of the membrane, and test for ATPase activity as a function of Na^+ and K^+. With this in mind it seems peculiar that the Na^+-K^+ pump was identified as Na^+- and K^+-activated ATPase (Na^+-K^+-ATPase) by someone whose research interest was not active transport, who met it more or less by accident, and who did not even know that a countryman, H. H. Ussing in Copenhagen, had made important contributions to the development of the concept of active transport (102). Had I not identified the pump I am sure that logical reasoning soon would have led someone in the transport field to the discovery; the time was ripe.

MEMBRANE
TRANSPORT

My interest was surgery. In the surgical ward where I worked during the second half of the 1940s there was no anesthetist, and to avoid the unpleasant ether anesthesia, we used whenever possible spinal and local anesthesia. Using these aroused my interest in the mechanism of action of local anesthetics. From pharmacological studies I knew the Meyer-Overton theory: the correlation between solubility of general anesthetics in lipids and anesthetic potency (59, 68). Did this apply also to local anesthetics, and what did this mean from the point of view of the mechanism of anesthesia? I decided to spend some time on the problem and use it as the subject for a thesis before continuing my surgical career. I obtained a position at the Institute for Medical Physiology at the University of Aarhus.

The University of Aarhus was founded in 1928. The Institute for Medical Physiology was opened in 1937, but the Medical Faculty was not complete until 1957. The Institute for Medical Physiology and the Institute for Medical Biochemistry were the only biological departments at the university. There were no PhD students. The scientific milieu in biology was poor with little or no contact with the outside scientific world. With an intake of 140 medical students a year and only four teachers in physiology, the teaching load was heavy.

Local anesthetics did not follow the Meyer-Overton theory but experiments suggested that their solubility in lipids in some way played a role in their potency. In *The Physics and Chemistry of Surfaces* by N. K. Adam (1) I read about I. Langmuir's work with lipid monolayers and of J. H. Schulmann's demonstration that capillary-active drugs penetrate from the water phase into a lipid monolayer. These works and the Danielli-Davson (19) model of the cell membrane with a bilayer of lipids suggested to me that a monolayer of lipids extracted from nerves could be used as a simple model for half of the membrane and that such a monolayer could be used to test the membrane effect of local anesthetics.

The experiments showed that local anesthetics penetrate from the water phase into the monolayer and that at a given area of the monolayer they increase the surface pressure. There was a good correlation between local anesthetic potency and the increase in surface pressure. Also the effect of pH on the local anesthetic potency correlated with the effect of pH on the penetration of the monolayer. Because it seemed likely that the Na$^+$ channels were in proteins in the nerve membrane, the idea that developed from the monolayer experiments was that the change in surface pressure, due to penetration of local anesthetics into the lipid part of the nerve membrane, has an effect on the conformation of proteins in the membrane and thereby blocks the opening of the Na$^+$ channels.

A step on the way to testing this was to determine whether the conformation of a protein in a monolayer is dependent on surface pressure and thereafter to determine whether the increase in surface pressure, due to penetration of local anesthetics into the lipid part of a monolayer (which consists of a mixture of lipids and proteins), has an effect on the conformation of the protein. The plan was to test the effect of surface pressure on the activity of a surface spread enzyme and, if there was an effect, to use this as an indication of a change in the conformation of the protein. Because the amount of protein in a monolayer is very small, it was necessary to have a highly active enzyme. A candidate was acetylcholinesterase (AChe)—not because of David Nachmansohn's view on the role of AChe in nerve conduction (61) but because AChe with a high degree of purity could be obtained. Its use had the further advantage of requiring a visit with professor Nachmansohn in the United States to prepare the enzyme from electric eel tissue. I was introduced to Nachmansohn by his friend Professor E. Lundsgaard from the Institute of Physiology in Copenhagen. I wanted to combine the trip to the United States with participation in the 19th International Congress of Physiology in Montreal. At that time Nachmansohn spent his summers at the Marine Biological Laboratory in Woods Hole, Massachusetts and he suggested that I join him there in July. He also suggested that after participating in the Congress in Montreal I should return to Columbia University in New York and prepare the enzyme.

For a young man from a remote university with a poor scientific milieu, Woods Hole was a new world. I was very impressed and realized that science is a very serious affair and not just a temporary hobby for people who for a period will accept a low salary. I envied the young people brought up in such a milieu. I also realized that science is competitive.

Nachmansohn and Harry Grundfest shared a laboratory, which Robert Post from Nashville visited at the same time as I did. We observed Eric Kao injecting different drugs into squid giant axons. When there were no squid available Post's wife Elizabeth came by with the car and took the other young scientists and me to the beach. Robert preferred to stay in the laboratory.

Professor Nachmansohn took good care of me, and Woods Hole gave me an opportunity to listen to lectures and to meet people, many of whom I knew from textbooks and from the literature. I was a shy young man so I listened and learned.

At other times when there were no squid, I read that B. Libet (56) had shown that there is a Mg^{2+}-activated ATPase in the sheath part of the squid giant axon. I wondered what the function of this ATPase could be. Because it was located in the membrane, it was probably a lipoprotein, and I considered it a candidate for the planned experiments on the effect of local anesthetics on monolayers of lipids and proteins.

August was spent at Columbia with Nachmansohn preparing AChe.

SODIUM-POTASSIUM PUMP

[157]

After my return to Aarhus I continued the monolayer experiments. Acetylcholinesterase loses its activity when unfolded in a monolayer, but the activity can be regained by increasing the surface pressure: the activity depends on the surface pressure. Similar results were obtained with catalase as the enzyme source. This suggested to me that a change in surface pressure could influence the conformation of a protein in a membrane.

MEMBRANE TRANSPORT

Concurrent with the monolayer experiments, I began looking for the nerve membrane ATPase. I had no access to squid giant axons but used crab nerves instead. A membrane fragment of a homogenate of the nerves showed Mg^{2+}-activated ATPase activity, and with Mg^{2+} there was a low stimulation by Na^+. The results varied, however. The Ca^{2+} was considered as a candidate for the variation, but this was experimentally excluded. The K^+ had no effect on the activity in the presence of Mg^{2+}, but the effect of a combination of Na^+ and K^+ was not considered. Additionally the concentration of Na^+ and K^+ in the test medium varied, partly because the nerves in some of the experiments were homogenized in 0.58 M KCl instead of sucrose and partly because the Ba^{2+} salt of ATP, which was used as a source for ATP, sometimes was converted to the Na^+ salt and sometimes to the K^+ salt. It was an experiment in which the activity with Mg^{2+} and Na^+ of the enzyme prepared in sucrose was higher with the K^+ salt of ATP than with the Na^+ salt, which showed that K^+ activates in the presence of Na^+. The next experiments showed that K^+ in low concentrations relative to the concentration of Na^+ produced a pronounced increase in activity, whereas higher concentrations not only reversed this activation but also inhibited the low activity due to Na^+ alone (Fig. 1). The results suggested that there were different sites for the activating effects of Na^+ and K^+ and that the inhibition by K^+ was due to a competition for Na^+. This was in August 1955 and had taken more than half a year of intermittent work during which time the lack of reproducibility nearly had driven me to despair. However, as mentioned earlier, a knowledge of the literature on active transport might have saved time. A link between the transport of Na^+ and K^+ was suggested from experiments on red cells by E. J. Harris and M. Maizels in 1951, from experiments on muscle cells by H. B. Steinbach in 1951 and 1952, and by R. D. Keynes in 1954 (see ref. 28). Had I known I might have looked for a combined effect of Na^+ and K^+.

Later experiments showed that crab nerves were a lucky choice as a source for the enzyme. With mammalian tissue most of the activity is hidden because of vesicle formation of the plasma membrane fragments, and the vesicles must be opened (e.g., with a detergent) in order to see the combined effect of Na^+ and K^+. This is not the case with crab nerve membranes. If I had used mammalian tissue as the enzyme source, I would probably not have seen the combined effect of Na^+ and K^+.

[158]

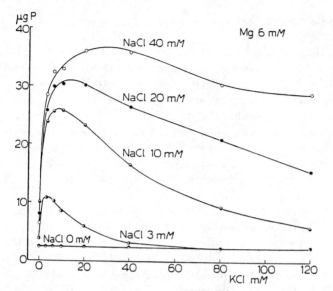

Figure 1. Effects of Na⁺, and of K⁺ in presence of Na⁺, on activity of ATPase located in microsomes isolated from nerves of shore crab *Carcinus maenas*. The 1-ml test solution contained 0.1 ml of microsomal suspension, 30 mM histidine HCl as buffer (pH 7.2), 3 mM ATP, 6 mM Mg^{2+}, and Na⁺ and K⁺ concentrations as shown. Activity is given as μg P_i (inorganic phosphate) hydrolyzed from ATP in 30 min at 36°C. [From Skou (86).]

What was the physiological function of the enzyme? I was interested in the blockage of nerve conduction, so I first thought that it was the Na⁺ channel. However, this was soon rejected; it seemed unlikely from what was known about nerve conduction that the opening and closing of the sodium channel should be ATP dependent. It seemed more likely that the enzyme was involved in the active transport of Na⁺ and that ATP was the energy source for the transport. I found support for this view from a paper by A. L. Hodgkin and R. D. Keynes published in 1955 (39). They showed that poisoning giant axons with dinitrophenol, cyanide, or azide reduced the efflux of sodium, suggesting that high-energy phosphate esters are the energy source for the active transport of Na⁺. They also showed a link between transport of Na⁺ and of K⁺. I was not aware (and apparently neither were Hodgkin and Keynes, since it is not mentioned in the paper) that G. Gardos in a paper published in a Hungarian journal in 1954 (27) had shown that ATP supports the accumulation of K⁺ in red blood cells. However, in 1956 Hodgkin and Keynes (40) reported that there was no dramatic recovery of the sodium efflux in cyanide-poisoned giant axons by a microinjection of ATP. Fortunately I did not see their paper until after I had sent my paper on the crab nerve enzyme for publication. In 1957 P. C. Caldwell and Keynes (12) showed that ATP is the high-energy phosphate ester that supports the active extrusion of Na^+ in giant axons.

My interest was local anesthesia and monolayers; therefore it was a bit disturbing suddenly to have in hand what seemed to be the Na^+-K^+ pump (or at least part of it), but in the end the pump turned me away from monolayers and surgery. The effect of local anesthetics on a mixture of lipid and protein in a monolayer was never investigated; in fact the enzyme could not be used for such an experiment unless it was extracted from the membranes, and no methods were available for this. The results of the monolayer work were used for a thesis that was defended in 1954 and was published in a series of papers (see ref. 88).

In 1956 there was an international neurochemistry meeting in Aarhus and after a paper by R. D. Keynes entitled "Electrolytes and Nerve Activity" (52) I showed in the discussion the results of the experiments with the crab nerve ATPase (85). A. Pope, who participated in the meeting, looked for the enzyme in rat brain slices after his return home and observed the combined activation of Na^+ and K^+ on the Mg^{2+}-activated ATPase activity (38).

The paper on crab nerve enzyme was written up in 1956 and published in 1957 (86). I was not very experienced in writing English. I needed some help and got it from Mogens Schou who had spent a year in the United States. He had spent two years in the Institute of Physiology and had then moved to the Psychiatric University Hospital in Aarhus and had started his important work on the effect of lithium on mania (77). I remember we discussed whether the words "sodium pump" should be in the title, but I found it too provocative. The final title became "The Influence of Some Cations on an Adenosine Triphosphatase From Peripheral Nerves." No wonder that it took some time before it was realized that this enzyme had something to do with transport of Na^+ and K^+.

In 1958 I participated in the 4th International Congress of Biochemistry in Vienna. I presented a paper entitled "The Influence of the Degree of Unfolding and the Orientation of Side Chains on the Activity of a Surface Spread Enzyme." At the conference, Robert Post told me about his experiments on the stoichiometry of the Na^+-K^+ pump in red blood cells. I told him that I had identified what seemed to be the pump and that it was a membrane-bound Na^+-K^+-ATPase. He had not seen the paper and from his reaction I understood that this interested him more than surface-spread enzymes. To his first question, "*Is it inhibited by ouabain?*" my reply was, "*What is ouabain?*" I was not aware that in 1953 H. Schatzmann (76) had shown that cardiac glycosides inhibit the active transport of Na^+ and K^+, and therefore I had not done the crucial experiment that could show Na^+-K^+-ATPase involvement in active transport of the cations. I called Aarhus and arranged for the experiment to be done.

Post came to Aarhus after the conference. The experiment with ouabain showed inhibition (87). I had started to look for the enzyme in red blood cells, but because I had no experience with transport in these cells and Post had, I agreed that he should do the experiments. Instead I turned to other tissues: mammalian brain, kidney, and muscle.

In 1959 I participated in the 21st International Congress of Physiology in Buenos Aires and presented my first paper on the Na^+-K^+-ATPase. Professor Hodgkin invited me for lunch to hear more about the enzyme. His interest impressed me and I began to realize that the identification of the sodium pump was of a certain importance. SODIUM-POTASSIUM PUMP

The experiments on red blood cells by Post et al. (73) showed a good correlation between cation effects on fluxes and on the activity of Na^+-K^+-ATPase. This further supported the view that the enzyme was involved in active transport. Support was also given by the finding that the enzyme could be isolated from mammalian brain, kidney, and muscle (89) as well as many other tissues (9) and that in tissues for which data were available on the active transport of cations there was a correlation between transport activity and enzyme activity (8).

In 1962 Arnold Schwartz came to Aarhus and entertained us for four months while he worked on the Na^+-K^+-ATPase in muscle.

During the early 1960s so much evidence was collected from a number of laboratories that it was possible in a 1965 review (90) to conclude that the enzyme system fulfilled the following requirements for a system responsible for active transport of Na^+ and K^+ across the cell membrane: 1) It is located in the cell membrane; 2) on cytoplasmic sites, it has a higher affinity for Na^+ than for K^+; 3) it has an affinity for K^+ on extracellular sites, which is higher than for Na^+; 4) it has enzymatic activity and catalyzes ATP hydrolysis; 5) the rate of ATP hydrolysis depends on cytoplasmic Na^+ as well as on extracellular K^+; 6) it is found in all cells that have coupled active transport of Na^+ and K^+; 7) the effect of Na^+ and of K^+ on transport in intact cells and on the activity of the isolated enzyme correlates quantitatively; and 8) the enzyme is inhibited by cardiac glycosides, and the inhibitory effect on the active fluxes of the cations correlates with the inhibitory effect on the isolated enzyme system. The enzyme system was named the Na^+- and K^+-activated ATPase or Na^+-K^+-ATPase.

In the test tube Na^+-K^+-ATPase activity is measured, but the transport is not. An unanswered question was whether the Na^+-K^+-ATPase is only a part of the transport system that converts chemical energy from the hydrolysis of ATP to a movement of a carrier for Na^+ and K^+ (the engine, so to speak). Alternatively, are the catalytic activity and the carrier activity part of the same molecule? that is, is Na^+-K^+-ATPase the complete transport system and can the Na^+-K^+-ATPase activity be taken as an indication of transport activity through the membrane fragments in the test tube? The answer could not be given until it became possible to dissolve purified fully active enzyme in detergent, reincorporate the enzyme in the membrane of liposomes with no loss of activity, and measure transport as well as ATP hydrolysis. This was 20 years in the future, at which time studies showed that the Na^+-K^+-ATPase is the transport system (see ref. 17).

ACHIEVEMENT OF THE DISCOVERY

For us at the Institute of Physiology, the importance of the identification of the Na^+-K^+-ATPase as the Na^+-K^+ pump was not just its identification. Had it not been identified at Aarhus it would certainly have been identified elsewhere. However, it provided the Institute with a contact to the international scientific world. In 1961 I met R. W. Berliner in Stockholm at the first international pharmacological meeting. We discussed the possibility of obtaining a grant from the National Institutes of Health. I applied and, I am sure not least because of Berliner's support, I obtained a two-year grant in 1962. The grant was of a great help not only financially, but also because it showed international interest in the work we were doing.

In 1963 when the chairman of the department, S. L. Ørskov, retired, I was appointed chairman. The scientific milieu that developed because of the enzyme work attracted diligent young people. This was important in the 1960s when more money was allocated to the universities in Denmark and it became possible to increase the staff of the institute from 4 to 20–25 scientists. Most important perhaps was the effect this had on the teaching and thereby the education of the medical students in physiology.

Research interest at the Institute was not concentrated on Na^+-K^+-ATPase. The policy was also to attract people whose research interests covered other fields. Scientists who in the following years studied Na^+-K^+-ATPase and who made important contributions included P. L. Jørgensen (purification and structure), I. Klodos (phosphorylation), O. Hansen (effect of cardiac glycosides and vanadate), P. Ottolenghi (effect of lipids), J. Jensen (ligand binding), J. G. Nørby (ligand binding and kinetics), L. Plesner (kinetics). M. Esmann (solubilization and determination of molecular weight), T. Clausen and co-workers (hormonal control), A. Maunsbach and E. Skriver from the Institute of Anatomy in collaboration with Jørgensen (electron microscopy and crystallization of Na^+-K^+-ATPase in the membrane), and I. Plesner from the Department of Physical Chemistry (enzyme kinetics and evaluation of models).

In 1977 I was offered the medical faculty chair in Biophysics. This department was smaller with a staff of seven, but with five of the positions open. I moved with Nørby, Klodos, L. Plesner, and Esmann. As a result of the move Klodos and Esmann, who had fellowships, obtained permanent positions. Department members M. Mulvany, who was interested in hypertension, and F. Cornelius became interested in the connection between pump activity and vasoconstriction and reconstitution of the enzyme, respectively. Other advantages of the move included increased space and common research interest among department members as well as reduced administrative responsibilities; in addition I gained freedom from teaching obligations.

Na$^+$-K$^+$-ATPase

The field developed quickly from research done in many laboratories. Notably, there were five international meetings with Na$^+$-K$^+$-ATPase as the subject held in 1973 at New York, in 1978 at Sandbjerg, Denmark, in 1981 at Yale University in New Haven, CT, in 1984 at Cambridge, UK, and in 1987 at Fuglsø in Denmark. The proceedings from these meetings (4, 29, 41, 98) provide good insight in the development of Na$^+$-K$^+$-ATPase research.

Next I discuss some of the steps in the development in order to illustrate the present situation and some of the current problems. [For more details and for detailed literature references see the proceedings and the recent excellent review by Glynn (28).]

Purification

There was no recipe for purifying a membrane-bound enzyme. Attempts with detergents showed that the specific Na$^+$-K$^+$-ATPase activity of preparations from mammalian brain or kidney could be increased by treatment of the membranes with deoxycholate (DOC), but the preparations obtained were far from pure (89). Because DOC in higher concentrations inactivated the enzyme, it seemed unlikely in 1962 that pure preparations could be obtained by the use of detergents. However, the necessary trick proved to be finding the right tissue as an enzyme source. It seems to be a nearly general rule that somewhere in nature there is always a tissue that is especially suitable for a given purpose, and one was found for this purpose.

In 1965 P. L. Jørgensen, a young M.D., came to the Department. He was interested in the effect of aldosterone on Na$^+$-K$^+$-ATPase in the kidney. The high content of Na$^+$-K$^+$-ATPase in the outer medulla of the kidney led him to suggest this as a source for purification. By using DOC under well-controlled conditions (pH, temperature, time, and detergent to protein ratio) it was possible to remove enough non–Na$^+$-K$^+$-ATPase proteins from the membranes so that the Na$^+$-K$^+$-ATPase in the membrane was ~50% pure, and had a specific activity of ~1,200 μmol P$_i$·mg^{-1} protein·h^{-1} (46) in the best preparations.

Another approach was to dissolve with detergents the membranes from the outer medulla of kidney [J. Kyte (54) and L. K. Lane et al. (55)] or from shark rectal glands [L. E. Hokin et al. (42)], which are another rich source of Na$^+$-K$^+$-ATPase. Thereafter the enzyme was

[163]

purified by gel filtration (54) or ammonium sulfate precipitation (42, 55). This yielded preparations that consisted of two polypeptide chains, α-chain (89,000–97,000 daltons), and β-chain (glycoprotein; 55,000–57,000 daltons) with a mole ratio of 1:1. These filtration and precipitation procedures provided an enzyme that was >90% pure. However, when enzyme purified in the membranes was used as reference, the specific activity did not increase as much as the purity of the protein, suggesting that the filtration and precipitation procedures lead to inactivation of some of the enzyme molecules.

By using sodium dodecyl sulfate (SDS) instead of DOC and protecting the enzyme by ATP, Jørgensen managed to obtain practically pure membrane preparations with specific activities of 2,000 to 2,200 μmol $P_i \cdot mg^{-1}$ protein$\cdot h^{-1}$ (45). Practically all the non–Na^+-K^+-ATPase protein was removed from the membrane by SDS but the Na^+-K^+-ATPase protein remained. The SDS procedure has become the most widely used purification method. An unsolved problem is whether treatment with SDS has an effect on the activity. Ligand-binding studies show that the number of ligands bound is lower than expected from the molecular weights of the α- and β-chains (see ref. 28), which may mean that this procedure also leads to inactivation of a part of the enzyme molecule.

As previously mentioned, the enzyme can be dissolved in detergent. However, loss of activity was a problem. This was overcome by the use of the nonionic detergent octaethyleneglycol dodecyl-monoether ($C_{12}E_8$). The Na^+-K^+-ATPase partially purified with DOC in membranes from shark rectal glands is selectively extracted from the membrane by $C_{12}E_8$ (24). This detergent-dissolved enzyme is pure α- and β-polypeptides with a specific activity as high as in the most active membrane preparation. The detergent can also extract the enzyme from partially purified membranes from the outer medulla of the kidney, but kidney enzyme dissolved in $C_{12}E_8$ is more labile than shark enzyme (10).

The solubilization of the pure active enzyme made possible molecular weight determinations from analytical ultracentrifugation (10, 22, 24, 34). Furthermore, the reconstitution experiments, which showed that purified Na^+-K^+-ATPase has the full capacity of transport of Na^+ and K^+, indicated that the purified enzyme not only contains the catalytic activity but is the intact transport system (17, 43).

Molecular Weight

The experiments by Kyte (54), Lane et al. (55), and Hokin et al. (42) showed that the enzyme consists of two polypeptide chains, α and β with a mole ratio of 1:1. This has since been substantiated (see ref. 45). The molecular weight of the α-chains is ~100,000; the protein part of the β-chain has a molecular weight of ~40,000, and its carbohydrate part has a molecular weight of ~15,000. The α-

chain has catalytic activity, but the function of the β-chain is unknown.

Lipids bound to the proteins are necessary for activity (see ref. 65). However, the reason lipids are necessary, their exact number, and their nature are unknown. Therefore the molecular weight is expressed as that of the protein part of the molecule.

With impure preparations the molecular weight can be determined by radiation inactivation, which was first attempted by G. Kepner and R. Macey (51) in 1968. They estimated molecular weights of 300,000 for Na^+-K^+-ATPase in red blood cell membranes and 190,000 in kidney membranes.

Results from radiation inactivation experiments with impure preparations, from analytical ultracentrifugation, and from low-angle neutron scattering of dissolved purified enzyme vary (see ref. 28), but the values center around 250,000–300,000, suggesting that the enzyme is an $(\alpha,\beta)_2$ structure. However, dissociation of $(\alpha,\beta)_2$ into (α,β) by treatment with the nonionic detergent $C_{12}E_8$ does not lead to inactivation of the enzyme; (α,β) has retained at least part of its Na^+-K^+-ATPase activity (10, 21). It has not been establish yet whether catalytic activity of (α,β) also means transport activity or this is confined to $(\alpha,\beta)_2$.

Crystallization

With enzyme purified in membrane pieces it became possible to crystallize the enzyme in two dimensions (35). Dependent on the conditions used for crystallization, the enzyme crystals exist in a form that suggests an (α,β) and an $(\alpha,\beta)_2$ structure, respectively.

From a Fourier analysis of tilted membrane preparations of negatively stained $(\alpha,\beta)_2$ crystals a three-dimensional model of the Na^+-K^+-ATPase has been constructed (36, 66). The resolution is down to 20 Å. The molecule consists of two symmetrical rodlike structures that each seem to consist of an α-β protomer. The height perpendicular to the membrane is $\sim$100 Å. There is a cleft $\sim$20 Å wide between the rods and they are connected by an area $\sim$20 Å in that half of the molecule that faces the cytoplasmic side of the membrane, with a part of the connection inside the lipid bilayer. The molecule seems to protrude $\sim$40 Å on the cytoplasmic side of the lipid bilayer and $\sim$20 Å on the extracellular side. The intramembraneous part of each protomer is $\sim$25% of the mass of the α- as well as of the β-subunit. Practically all the rest—75%—of the β-subunit is on the extracellular side, while the cytoplasmic part of the α subunit is $\sim$3 times larger than the extracellular part.

Sequence of Amino Acids

New possibilities for characterization of the system has been opened by the isolation of the complementary DNA of the α- (50,

[165]

67, 80) as well as of the β-subunit (11, 62, 67, 81). The amino acid sequence has been deduced from the cDNA. The α-chain consists of 1,016 amino acids and has eight hydrophobic regions suggesting eight transmembrane segments. The N-terminal hydrophilic region is on the cytoplasmic side.

In 1979, K. Sweadner showed that there are two isoforms of the α-chain, α and $\alpha(+)$, in brain tissue (100). Enzyme with $\alpha(+)$ has a higher sensitivity toward inhibition by cardiac glycosides. Also, $\alpha(+)$ has been found in muscle cells and in adipocytes, and is the form that is responsive to insulin (57). The cDNA of the two isoforms has been isolated and characterized from rat brain tissue and a third isoform, αIII, has been identified (82). The $\alpha(+)$ and αIII consist of 1,015 and 1,013 amino acids, respectively. The main difference in the amino acid sequence of the three isoforms is in the N-terminal region and in a region between the aspartic acid, nr. 369, which becomes phosphorylated from ATP, and a lysine, nr. 501, which seems to be part of the ATP-binding site.

The protein part of the β-chain consists of 302 amino acids and seems to have one transmembrane segment located near the cytoplasmic N terminal with the hydrophilic part of the molecule on the extracellular side.

Reactions with ATP and Conformational Transitions

Early experiments on the ADP-ATP exchange catalyzed by Na^+-K^+-ATPase suggested that the reaction with ATP involves a phosphorylation-dephosphorylation (87). More direct evidence came from experiments by Albers and co-workers (3, 25, 83) and by Post and co-workers (71, 72, 74) who demonstrated an Na^+-dependent phosphorylation and a K^+-dependent dephosphorylation of the phosphoenzyme. The experiments showed that the first step in the reaction is formation of an ADP-sensitive phosphoenzyme that is converted to a K^+-sensitive phosphoenzyme, suggesting two different molecular conformations, denoted $E_1 \sim P$, which has a high-energy phosphate bond and E_2-P, which has a low-energy phosphate bond. More recently, J. G. Nørby et al. (64) have shown that besides the phosphoenzyme bound with ADP, at least three consecutive acid-stable phosphoenzymes are formed when the enzyme reacts with ATP in the presence of Na^+. One is ADP sensitive, another has a fast rate of conversion to the ADP-sensitive phosphoenzyme, that is, both disappears fast when ADP is added ($E_1' \sim P$ and $E_1'' \sim P$ respectively in the following). The third phosphoenzyme is the K^+-sensitive E_2-P.

In 1972 Post et al. (70) made the important observation that dephosphorylation of E_2-P by K^+ leads to a conformation of the enzyme from which K^+ has a low rate of release; the experiments suggested that ATP increases the rate of K^+ release. This led to the

[166]

concept of occluded K^+ (30). It was later shown by I. Glynn et al. (31) that phosphorylation of the enzyme leads to occlusion of Na^+.

These reactions can be described by the following schemes, in which the parentheses show occlusion

$$E_1 \cdot ATP \cdot Na_n \rightleftharpoons E_1 \sim P \cdot ADP \cdot Na_n \rightleftharpoons E_1' \tag{1}$$

$$\sim P \cdot Na_m \rightleftharpoons E_1'' \sim P \cdot Na_n \rightleftharpoons E_2\text{-}P \cdot Na_n$$

$$E_2\text{-}P \cdot Na_n \rightleftharpoons E_2\text{-}P \rightleftharpoons E_2\text{-}P \cdot K_m \rightleftharpoons E_2 \cdot K_m \rightleftharpoons E_2'(K_m) \tag{2}$$

where n and m are numbers. The affinity of E_2-P for K^+ is high and comparable with the affinity for K^+ for activation of the hydrolysis, suggesting that the dephosphorylation is due to an effect of K^+ on extracellular sites. This probably means that it is K^+ on the extracellular sites that becomes occluded by the dephosphorylation.

In 1971 it was shown independently by Nørby and J. Jensen (63) in Aarhus and by C. Hegyvary and Post (37) in Nashville that Na^+-K^+-ATPase in the presence of Na^+ has a high affinity for ATP with a dissociation constant, K_d, of 0.1–0.2 μM whereas in the presence of K^+, K_d is higher. This indicates that the nonphosphorylated enzyme also exists in two different conformations and that the nature of the cation bound to the enzyme determines the conformation, an Na^+ conformation, E_1, and a K^+ conformation, E_2. This was subsequently confirmed by differences in the reactivities of the Na^+ and of the K^+ conformations toward tryptic digestion (44). Additional confirmation came from studies of the differences in fluorescence signals in the presence of Na^+ and of K^+, which is observed with the use of the intrinsic fluorescence of tryptophan (48) and the extrinsic fluorescence of probes like FTP, an ATP analogue (49); of eosin, which binds noncovalently to what seems to be the ATP site (96); and of fluorescein isothiocyanate, which binds covalently to a lysine group near the ATP site (47).

With no Na^+ or K^+ in the medium, the enzyme is in the E_2 conformation (97). As shown by Glynn and D. E. Richards (30), the binding of K^+ to E_2, like the dephosphorylation of E_2-P by K^+, leads to an occlusion of K^+, that is, to $E_2(K_m)$; with the nonphosphorylated enzyme, it seems to be K^+ on the cytoplasmic sites, which become occluded. Binding of Na^+ to E_2 leads to a transition to the E_1 conformation of the enzyme, that is, to E_1Na_n

$$E_2'(K_m) \rightleftharpoons E_2 \cdot K_m \rightleftharpoons E_2 \rightleftharpoons E_2 \cdot Na_n \rightleftharpoons E_1 \cdot Na_n \tag{3}$$

Because of the low rate of deocclusion of K^+, the rate of the transition from $E_2(K_m)$ to E_1Na_n is low (at 22°C, $t_{1/2}$ is several hundred milliseconds), whereas the rate of the reverse reaction is high (a few milliseconds). As previously mentioned, experiments by Post et al. (70) suggested that ATP increases the rate of deocclusion of K^+ from $E_2(K_m)$. This was confirmed by S. J. D. Karlish and D. Yates (48), who showed that ATP increases the rate of the transition from $E_2'(K_m)$

[167]

to E_1Na_n; the affinity for ATP for this effect is low ($K_d = 0.45$ mM). This means that at a given Na^+ and K^+ concentration addition of ATP shifts the distribution between the two conformations toward E_1Na_n, and in steady state this is observed as an increase in apparent affinity for Na^+ for activation of ATP hydrolysis in the presence of $Na^+ + K^+$ (91).

$$E_2'(K_m) \rightleftharpoons E_2' \cdot ATP(K_m) \rightleftharpoons E_2 \cdot ATP \cdot K_m \rightleftharpoons E_2$$
$$\cdot ATP \rightleftharpoons E_2 \cdot ATP \cdot Na_n \rightleftharpoons E_1 \cdot ATP \cdot Na_n \quad (4)$$

A combination of Equations 1, 2, and 4 gives a scheme for an overall reaction with Na^+, K^+, and ATP (Mg^{2+}, although necessary for reaction, is omitted). There are three main features of the reaction: 1) the enzyme undergoes a cyclic change from a K^+ conformation, $E_2'(K_m)$, to an Na^+ conformation, $E_1 \cdot ATP \cdot Na_n$ facilitated by ATP, and via a phosphorylation-dephosphorylation the Na^+ conformation is reversed to the K^+ conformation; 2) K^+ as well as Na^+ in the course of the reaction becomes occluded and deoccluded; and 3) these reactions are governed by the reaction with ATP. Phosphorylation gives occlusion of Na^+, and the following molecular rearrangements of the phosphoenzyme leads to deocclusion of Na^+ and transition from the E_1 to the E_2 conformation (E_2-P, Eq. 1). Dephosphorylation of E_2-P leads to occlusion of K^+ (Eq. 2) and ATP facilitates the deocclusion and the transition from the E_2 to the E_1 conformation (Eq. 4).

TRANSPORT OF Na^+ AND K^+

In parallel with the work on the isolated system, extensive research was done on transport in intact cells. A major contribution was made by Glynn and co-workers in Cambridge, UK, from work with red blood cells, but many others have added to the information (see ref. 28). The results show six different cation-exchange reactions, all of which are inhibited by cardiac glycosides, suggesting that they are due to the same transport system, the Na^+-K^+-ATPase.

1. In Na^+-K^+ exchange, three intracellular Na^+ are exchanged for two extracellular K^+ for each ATP hydrolyzed. The exchange is electrogenic.

In the reverse reaction, besides an Na^+ gradient into the cell and K^+ gradient out of the cell this requires a low intracellular ATP and high intracellular ADP and P_i concentration. The backward running of the pump leads to a synthesis of ATP from ADP and P_i.

2. With no extracellular K^+ but extracellular Na^+ there are two different reactions in which intracellular Na^+ is exchanged for extracellular Na^+.

One exchange requires ADP, besides ATP, and is not accompanied by a net hydrolysis of ATP. The reaction seems to be a phosphorylation from ATP, with an outward transport of Na^+, and a reversal of

[168]

the phosphorylation due to a reaction with ADP with an inward transport of Na⁺. The exchange is electroneutral.

Another exchange is accompanied by a net hydrolysis of ATP and is inhibited by ADP. Experiments with reconstituted Na⁺-K⁺-ATPase in liposomes show that the exchange is not one for one (26); about three Na⁺ are transported from the cytoplasmic side to the extracellular side and one or two Na⁺ are transported in the opposite direction for each ATP hydrolyzed. The exchange is electrogenic like the Na⁺-K⁺ exchange (18).

3. An uncoupled Na⁺ efflux is seen in the absence of Na⁺ and K⁺ in the extracellular medium. It is accompanied by a net hydrolysis of ATP.

4. A K⁺-K⁺ exchange occurs in the absence of extracellular Na⁺ but in the presence of extracellular K⁺. It requires intracellular ATP and P_i, but is not accompanied by a hydrolysis of ATP; ATP can be replaced by a nonhydrolyzable ATP analogue (84).

Coupling Between Chemical Reactions, Conformational Transitions, and Transport

Sodium ions from the cytoplasmic medium, which has a high K⁺ and a low Na⁺ concentration, are exchanged for K⁺ from the extracellular medium, which has a low K⁺ and a high Na⁺ concentration. For this to proceed the transport system must be able to undergo a change in affinity from a high K⁺ and low Na⁺ affinity to a high Na⁺ and low K⁺ affinity on the cytoplasmic side and vice versa on the extracellular side. Furthermore the cation sites must in some way change their exposure from one side of the membrane to the other— a gating reaction. The shift in affinity as well as the gating reaction must be governed by a reaction with the substrate that delivers the energy for the transport.

As discussed in the previous section, the Na⁺-K⁺-ATPase can undergo a conformational change with a change in apparent affinity for the cations from a K⁺ affinity (E_2) to an Na⁺ affinity (E_1), which is facilitated by an effect of ATP, and from an Na⁺ to a K⁺ affinity, which is due to the phosphorylation from ATP (E_2-P).

The occlusion of the cations can be viewed as a transfer of the cations to the membrane phase, a closing of a gate. Assuming this, the occlusion shows that the translocation of the cations is not a single step in which the cation sites change their exposure from the medium on the one side of the membrane to the medium on the other side but rather that it involves steps in which the cations are inside gates closed toward the cytoplasmic as well as the extracellular medium. As discussed above these steps are also governed by the reaction with ATP.

A combination of the reactions described in Equations 1, 2, and 4 gives a scheme for an overall reaction of the enzyme with ATP and

[169]

the cations. Adding sidedness to the reactions with the cations converts the reactions into a scheme for a coupling between the reaction with ATP, the conformational transitions, and the translocation of the cations (Fig. 2). Basically it is the so-called Albers-Post scheme (2, 72) with modifications by Karlish et al. (49), inclusion of the scheme for phosphorylation in the presence of Na^+ by Nørby et al. (64), and some further modifications. The exchange of 3 Na^+ for 2 K^+ suggests that in Equations 1 and 4 the values of n and m are 3 and 2, respectively.

Na^+-K^+ Exchange

ATP increases the rate of deocclusion of K^+ from $E_2'(K_2)$ and in the presence of Na^+ shifts the distribution toward $E_1 \cdot ATP \cdot Na_3$; this is followed by a phosphorylation from ATP with an occlusion of Na^+. These reactions may describe the transfer of K^+ from the membrane phase to the cytoplasmic medium, followed by the exchange of the inwardly translocated K^+ for outgoing Na^+, and by the transfer of Na^+ to the membrane phase (Fig. 2).

The phosphorylation leads not only to an occlusion of Na^+ but,

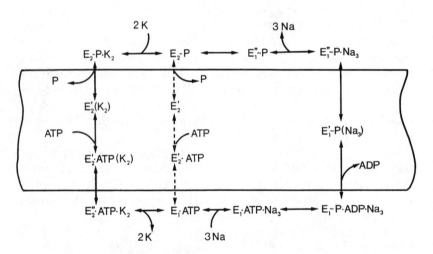

Figure 2. Na^+-K^+ exchange by Na^+-K^+-ATPase. Reaction with the cations is consecutive, and the scheme is based on the Albers-Post scheme (2, 72), the modifications of this scheme by Karlish et al. (49), and on the scheme by Nørby et al. (64) for formation of phosphoenzymes, and with some further modifications. Symbols E_1, E_1', E_1'', E_2, E_2', and E_2'' refer to different enzyme conformations. Parentheses indicate cations are occluded: that is, in the membrane phase inside gates, which are closed toward cytoplasmic and extracellular media.

[170]

through a step with deocclusion of Na^+, to the formation of the ADP-insensitive, K^+-sensitive phosphoenzyme E_2-P (Eq. 1). This phosphoenzyme has a high affinity for K^+ and is dephosphorylated by K^+ in concentrations that suggest it faces the extracellular side of the system. The phosphorylation from ATP in the presence of cytoplasmic Na^+ thus seems to lead to the translocation of Na^+ from the cytoplasmic side to the extracellular side of the system.

Binding of K^+ to E_2-P leads to a dephosphorylation, which leads to an occlusion of K^+ (Eq. 2). This suggests that the reaction leads to a transfer of K^+ from the extracellular medium to the membrane phase. The subsequent deocclusion of K^+ to the cytoplasmic side by a low-affinity effect of ATP terminates the cycle (Fig. 2; Eq. 4).

The Michaelis constant (K_m) for ATP for the phosphorylation of the enzyme in the presence of Na^+ is a fraction of a micromole per liter, whereas K_m for the overall hydrolysis in the presence of Na^+ and K^+ is a fraction of a millimole per liter. The higher K_m for ATP in the presence of Na^+ and K^+ is explained by the low-affinity effect of ATP necessary to increase the rate of deocclusion of K^+ from E_2' (K_2) to the cytoplasmic medium [Fig. 2; (48)]. At a low ATP concentration, this step is rate limiting in the Na^+-K^+ exchange reaction.

ADP-Dependent Na^+-Na^+ Exchange

In the ADP-dependent Na^+-Na^+ exchange, the conformation that delivers Na^+ to the extracellular medium must be sensitive to the addition of ADP and therefore cannot be the ADP-insensitive E_2-P. It must therefore be the phosphoenzyme that precedes formation of E_2-P and that follows the phosphoenzyme with Na^+ occluded. In Figures 2 and 3 this phosphoenzyme is denoted $E_1'' \sim P \cdot Na_3$. In the ADP-sensitive Na^+-Na^+ exchange the enzyme shuttles between $E_1 \cdot ATP \cdot Na_3$ and $E_1'' \sim P \cdot Na_3$ (Fig. 3A).

Oligomycin inhibits the Na^+-Na^+ exchange but not the ADP-ATP exchange that accompanies the reaction (see ref. 28). Oligomycin occludes Na^+ (23); thus oligomycin prevents the transition from $E_1' \sim P(Na_3)$ to $E_1'' \sim P \cdot Na_3$ and thereby the exchange of intracellular Na^+ for extracellular Na^+, but it does not prevent the phosphorylation with formation of $E_1' \sim P(Na_3)$ and the reversal of this reaction by ADP.

ATP-Hydrolysis–Dependent Na^+-Na^+ Exchange

The conformation that delivers Na^+ to the extracellular medium must be the same in both the ATP-hydrolysis–dependent and in the ADP-stimulated Na^+-Na^+ exchange. However, the stoichiometry with 3 Na^+ transported outward for 1–2 Na^+ transported inward (18) suggests that the conformation of the enzyme that accepts Na^+ for inward transport in the ATP-hydrolysis–dependent Na^+-Na^+ exchange is not the same as the conformation that delivers Na^+ to the

[171]

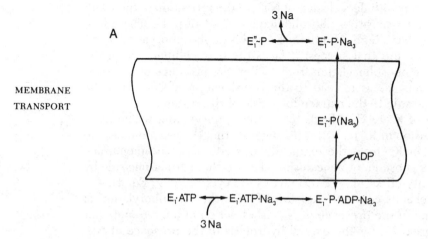

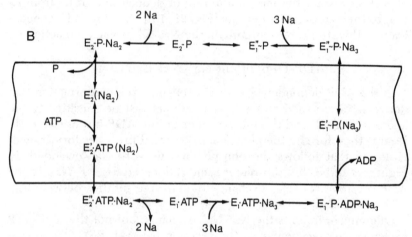

Figure 3. A: ADP-dependent Na$^+$-Na$^+$ exchange; *B*: ATP-hydrolysis–dependent Na$^+$-Na$^+$ exchange.

extracellular medium. The K$^+$ behavior of extracellular Na$^+$ in the ATP-hydrolysis-dependent Na$^+$-Na$^+$ exchange (stoichiometry and electrogenic effect) suggests that the phosphoenzyme that accepts Na$^+$ for the inward transport is the same as the phosphoenzyme that accepts K$^+$ for the inward transport in the Na$^+$-Na$^+$–exchange reaction, that is, E$_2$-P (Fig. 3*B*; cf. Fig. 2).

The steady-state distribution between the ADP-sensitive phosphoenzyme E$_1''$ ~ P·Na$_3$, and the K$^+$-sensitive phosphoenzyme E$_2$-P determines whether the enzyme shows ADP-stimulated Na$^+$-Na$^+$ exchange or ATP-hydrolysis–dependent Na$^+$-Na$^+$ exchange. This steady-state distribution varies for enzymes from different tissues.

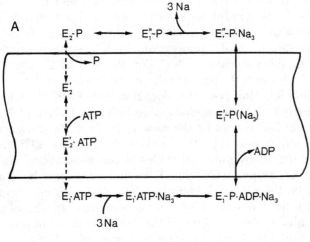

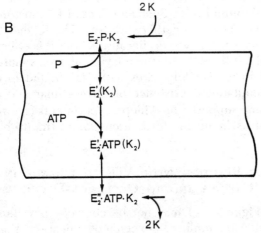

Figure 4. A: uncoupled Na⁺ efflux; B: K⁺-K⁺ exchange.

Uncoupled Na⁺ Efflux and K⁺-K⁺ Exchange

The uncoupled Na⁺ efflux is described in Figure 4A; a scheme for the ATP- and P_i-dependent K⁺-K⁺ exchange is shown in Figure 4B.

CONSECUTIVE VERSUS SIMULTANEOUS REACTION

The schemes in Figures 2–4 seem to give a reasonable description of the sequence of steps in the coupling reaction. There are, however, observations that are not explained by the schemes. Two examples follow.

First, in Figure 2 the enzyme reacts with one ATP molecule. According to ligand-binding experiments this denotes an (α,β) structure. However, molecular-weight determinations suggest an $(\alpha,\beta)_2$ structure and thus a reaction in which two (α,β) interact in some way.

[173]

Second, in a consecutive reaction with Na^+ and K^+, as illustrated in Figure 2, the phosphoenzymes formed in the reaction with Na^+ alone are also part of the reaction in the presence of Na^+ and K^+. Thus the rate of dephosphorylation by K^+ of the phosphoenzyme formed in the presence of Na^+ but absence of K^+ must be high enough to account for the overall rate of hydrolysis in the presence of Na^+ and K^+. However, this does not seem to be the case (64). If this is correct, the intermediary steps in the hydrolysis of ATP in the presence of Na^+ cannot be the same as those in the presence of Na^+ and K^+, suggesting that the enzyme hydrolyzes ATP along two different but interconnected cycles in the presence of Na^+ and of Na^+ and K^+, respectively (69). Potassium ions must have an effect not only on the rate of dephosphorylation but also on the formation of the phosphoenzyme. This view is supported by the observation that K^+ inhibits the transition of the ADP-sensitive phosphoenzymes (64); in Equation 1 it is the transition from $E_1' \sim P(Na_n)$ to $E_1'' \sim P \cdot Na_n$ that is inhibited by K^+. Such an effect of K^+ must mean that K^+ is bound to the enzyme together with Na^+. The K^+ may be bound to the cytoplasmic sites; however, because K^+ has this effect in concentrations at which K^+ activates the enzyme, it seems more likely that K^+ is bound to extracellular sites with Na^+ bound to cytoplasmic sites. The simultaneous existence of extracellular and cytoplasmic sites finds some support from kinetic experiments (see ref. 28). For a simultaneous scheme and for a discussion of the problem see ref. 94.

COUPLING OF REACTION WITH ATP TO CHANGE IN AFFINITIES FOR CATIONS AND TO OCCLUSION-DEOCCLUSION

Although Figures 2–4 may not be correct, they illustrate some basic principles in coupling between a chemical reaction and transport of the cations, namely the change in affinities and the occlusion-deocclusion of the cations governed by the reaction with ATP. However, the schemes say nothing about the molecular events. What determines the specificity for the cations? What is the nature of the molecular events that lead to occlusion-deocclusion? How is the effect of ATP conveyed to a change in affinities and to occlusion-deocclusion? How are the cations translocated? Our knowledge about these questions is sparse.

The enzyme E_1Na_3 has a high affinity for ATP, whereas $E_2'(K_2)$ has a low ATP affinity (37, 63). The nature of the cations on the cation-binding sites thus has an effect on the structure of the ATP-binding site. The transition from the Na^+ conformation to the K^+ conformation is accompanied by an increase in pK of pyridoxal 5-phosphate–reactive amino groups (probably lysine) and diethyl pyrocarbonate–reactive amino groups, as well as a decrease in pK of carbodiimide-reactive carboxyl groups of the enzyme (93, 95). Apparently the cation-binding sites adapt to the nature of the cations bound, and the molecular rearrangement of the cation-binding sites leads to a change in the tertiary/quaternary structure of the system and thereby to a change in the structure of the ATP-binding site; the tertiary/quaternary structure seems to link the structure of the cation-binding sites to the structure of the ATP site. This probably means that the effect not only of ATP but also of phosphorylation-dephosphorylation on the affinities for the cations and on occlusion-deocclusion is conveyed via an effect on the tertiary/quaternary structure of the system.

The transition from the Na^+ conformation to the K^+-occluded conformation leads to a net uptake of H^+; 4 Na^+ are replaced by 2 K^+ and 1 H^+ (95). It is unknown whether the proton occupies one of the cation binding sites or whether the H^+ participates in forming a salt bridge or hydrogen bond and thereby influences the tertiary/quaternary structure.

Modification of amino groups of the enzyme with pyridoxal 5-phosphate (93) or with diethyl pyrocarbonate or modification of carboxyl groups with carbodiimide (95) has little effect on the rate of transition between E_2Na_n and E_1Na_n but gives a pronounced increase in the rate of deocclusion of K^+ from $E_2'(K_m)$ and a decrease in rate of occlusion. The molecular rearrangement of the cation binding sites that is due to the binding of K^+ and leads to the occlusion of K^+ is thus much more sensitive to a change in the tertiary/quaternary structure, than the molecular rearrangement that is due to the binding of Na^+, and that leads to the transition from E_2 to E_1. It is unknown what the effect of the modification is on the conformational transition that leads to occlusion of Na^+ when the enzyme is phosphorylated.

Inhibitors

Cardiac Glycosides

As mentioned previously Schatzmann (76) showed that the active transport of Na^+ and K^+ is inhibited by cardiac glycosides, which have since become important tools in the identification of the cation transport system and in experiments with the isolated system. The most water-soluble and thus the most widely used cardiac glycoside is ouabain (G-strophanthin). Cardiac glycosides bind to the extracellular side of the enzyme system, and phosphorylation is necessary for binding (see ref. 79).

Cardiac glycosides differ in their binding affinities. With a given cardiac glycoside and a given combination of ligands, the inhibitory effect varies for different tissues. The $K_{0.5}$ for ouabain inhibition of Na^+ and K^+-dependent ATPase activity is for most tissues of the order of 10^{-7} to 10^{-5} M, while for example Na^+-K^+-ATPase from rat heart and from crab nerves is much less sensitive.

Based on the observation that there is a correlation between the

[175]

effects of several variables on the inotropic action of cardiotonic steroids and on their inhibition of Na^+-K^+-ATPase, K. Repke in 1961 (75) suggested that Na^+-K^+-ATPase is the receptor for the cardiotonic effect. This has since been substantiated. However, a small inotropic effect has been observed with low concentrations of ouabain (10^{-9} to 10^{-8} M), which does not inhibit but in some of the experiments stimulates Na^+-K^+-ATPase in the myocardium (see refs. 78, 99). This has raised the question of whether there are two different inotropic effects of cardiac glycosides. The pump stimulation is inhibited by propanolol, indicating that it is due to a catecholamine effect (see ref. 99). Cardiac glycoside inhibition of Na^+-K^+-ATPase in nerve endings inhibits reuptake and enhances release of catecholamines (see ref. 78). An explanation of the pump stimulation in the myocardium and the inotropic effect by the low concentrations of cardiac glycosides is that the Na^+-K^+-ATPase of nerve endings has a higher affinity for cardiac glycosides than the Na^+-K^+-ATPase in the myocardium; the release of catecholamines that is due to inhibition of the nerve-ending ATPase stimulates the pump of the noninhibited myocardium ATPase, and the released catecholamines induce the inotropic effect.

In 1969 a link between pump inhibition and a positive inotropic effect was suggested by P. F. Baker et al. (5) from studies on an Na^+-Ca^{2+}–exchange carrier in squid axons. Increased intracellular Na^+ resulting from a partial inhibition of the pump leads to enhanced Ca^{2+} influx and/or reduced Ca^{2+} efflux through the Na^+-Ca^{2+}–exchange carrier; increased intracellular Ca^{2+} concentration leads to increased contractility. Support has been given for this view, but other explanations have been offered (see ref. 78, 99).

Cardiac glycoside inhibition of the pump in vascular smooth muscle causes vasoconstriction (see ref. 60). Consequently a connection between pump activity and hypertension has been proposed (see ref. 7). This view has been stimulated by the observation that plasma partially inhibits the isolated Na^+-K^+-ATPase and that it has been reported that the inhibitory effect of plasma from human beings increases with the mean blood pressure of the plasma donors (see ref. 7). The vasoconstrictory effect is explained by the increase in the concentration of intracellular Ca^{2+} that follows from partial inhibition of the Na^+-K^+-ATPase. It has been proposed that plasma contains an endogenous factor similar to cardiac glycosides (see ref. 7), but so far the agent in plasma that inhibits the Na^+-K^+-ATPase is unknown.

Vanadate

Vanadate, which was found as a contaminant in certain commercial preparations of ATP, inhibits Na^+-K^+-ATPase in nanomolar concentrations (14). In contrast to the cardiac glycosides, vanadate binds to the cytoplasmic side of the system; Mg^+ is necessary for the binding (13, 15).

Vanadate also inhibits Ca^{2+}-ATPase and other ATPases and is thus not as specific an inhibitor as the cardiac glycosides, which inhibits only the Na^+-K^+-ATPase.

Why the Interest in Na^+-K^+-ATPase?

It may seem awkward to end this chapter by asking this question. Inasmuch as interest was not there when my transport research started but grew with the work, it may be justified.

The Na^+-K^+-ATPase is an energy transducer that converts energy from the hydrolysis of ATP into an electrochemical gradient for the cations. The gradient is a free energy source that is used for a number of physiological reactions.

It leads to the membrane potential. It is the driving force for the transient Na^+ influx and K^+ efflux during the action potential; besides, the Na^+ influx and K^+ efflux is compensated for by an increased activity of the Na^+-K^+-ATPase.

Furthermore the gradient for Na^+ is used to drive a number of carrier mediated uphill fluxes across cell membranes. An inward, cotransport of amino acids, glucose (see ref. 101), Cl^-, $Cl^- + K^+$ (33), as well as an outward, countertransport of Ca^{2+} (5) and of H^+. These fluxes are of importance for regulation of the cell interior and for cell volume regulation (see ref. 58). They play an important role in transepithelial transport in intestine, kidney, and secretory glands.

Top left: Robert Lichely Post; *top right*: Hans Jürgen Schatzmann; *bottom left*: Jens Christian Skou; *bottom right*: Ian Michael Glynn.

[177]

By creating and sustaining the gradients for Na^+ and for K^+, the Na^+-K^+-ATPase thus indirectly makes the energy from the hydrolysis of ATP available for a large number of reactions, and has thereby a key function in relation to transfer of substances across cell membranes. It is therefore of great interest to know the mechanism of the exchange reaction and what regulates the system.

The pump is activated by a combined effect of intracellular Na^+ and of extracellular K^+ and there is an intracellular as well as extracellular competition between Na^+ and K^+ for binding to the enzyme. This means that a change in the intracellular as well as in the extracellular Na^+-K^+ concentration ratios changes the activity. This couples a change in the passive flux of Na^+ and K^+ across the cell membrane to a change in the oppositely directed active flux. With the normal intracellular and extracellular Na^+-K^+ concentration ratios, the Na^+-K^+ pump operates at a low fraction of the maximal activity, 5%–20%. The large reserve power is important, for example, during exercise, when the muscles gain Na^+ and lose K^+. During maximal exercise, the K^+ concentration in arterial plasma may increase by 3 mM within 1 min, producing a net loss of K^+ of at least 40 mmol/min to the extracellular space (see ref. 16). Such a loss will very quickly give a deleterious K^+ concentration in plasma and extracellular fluid unless it is counteracted by an increase in the active Na^+-K^+ exchange. The maximal capacity for K^+ reabsorption in the total muscle pool is about 125 mmol/min (see ref. 16). In addition to the stimulation of the active transport by the change in Na^+-K^+ concentration ratios on the two sides of the membrane, the active transport is stimulated by insulin, epinephrine, and norepinephrine (see ref. 16) and this response may also play a role during exercise. These hormones have no direct effect on the Na^+-K^+-ATPase, and the mediator of the effect from the hormone receptors to the Na^+-K^+-ATPase in the intact membrane is unknown. Thyroid hormone increases the activity of Na^+-K^+-ATPase in various tissues and this appears to be due to a de novo synthesis of the enzyme (32). Corticosteroids also seem to have an effect on the synthesis of the enzyme molecules (see ref. 53).

Have the Na^+-K^+-ATPase dependent Na^+-Na^+ and K^+-K^+ exchange reactions a physiological function? The enzyme has an ATP-stimulated $Na^+ + K^+$ dependent phosphatase activity (see ref. 28). Does this have a physiological function?

The cardiac glycoside effect on the Na^+-K^+-ATPase prompts another question. The Na^+-K^+-ATPase is the receptor not only for the toxic effect of cardiac glycosides but also for the therapeutic effect on the heart. Is there an endogenous factor similar to cardiac glycosides in plasma of importance in regulating pump activity? The vasoconstriction that is observed when pump activity is decreased by cardiac glycosides or by low concentrations of extracellular K^+ may be due partly to pump inhibition in nerve terminals with an increased transmitter release and/or decreased transmitter reuptake and partly to pump inhibition in vascular smooth muscle (see ref. 60). Does the vasoconstriction by pump inhibition denote a connection between pump activity and blood pressure? Does the factor in plasma that inhibits the pump play a role in this connection?

Conclusion

It may seem disappointing that despite thirty years extensive work by a large number of investigators from many countries, it has not yet been possible to come to a better understanding of the transport process. On the other hand, at the beginning of the 1960s it was difficult to see how to handle a membrane-bound enzyme, and much more has been achieved than seemed possible at that time. When I listened to papers at the Vienna conference in 1958 on DNA and on biosynthesis it did not occur to me that thirty years later it should be possible to isolate and characterize the complementary DNA of the α and the β chains of the Na^+-K^+-ATPase, and from that deduce the amino acid sequence. However, there is still a long way to go before we understand the molecular events behind the transport process.

BIBLIOGRAPHY

1. ADAM, N. K. *The Physics and Chemistry of Surfaces.* London: Oxford Univ. Press, 1941.
2. ALBERS, R. W. Biochemical aspects of active transport. *Annu. Rev. Biochem.* 36: 727–756, 1967.
3. ALBERS, R. W., S. FAHN, and G. J. KOVAL. The role of sodium ions in the activation of *Electrophorus* electric organ adenosine triphosphatase. *Proc. Natl. Acad. Sci. USA* 50: 474–481, 1963.
4. ASKARI, A. (editor). *Properties and Function of (Na⁺ + K⁺)-Activated Adenosinetriphosphatase.* New York: NY Acad. Sci., 1974, vol. 242.
5. BAKER, P. F., M. P. BLAUSTEIN, A. L. HODGKIN, and R. A. STEINHARDT. The influence of calcium on sodium efflux in squid axons. *J. Physiol. Lond.* 200: 431–458, 1969.
7. BLAUSTEIN, M. P., and J. M. HAMLYN. Sodium transport inhibition, cell calcium, and hypertension. The natriuretic hormone/Na^+-Ca^{2+}-exchange/hypertension hypothesis. *Am. J. Med.* 77, Suppl. 4A: 45–59, 1984.
8. BONTING, S. L., and L. L. CARAVAGGIO. Studies on sodium-potassium-activated ATPase. V. Correlation of enzyme activity with cation flux in six tissues. *Arch. Biochem. Biophys.* 101: 37–46, 1963.
9. BONTING, S. L., K. A. SIMON, and N. M. HAWKINS. Studies on sodium-potassium-activated adenosine triphosphatase. I. Quantitative distribution in several tissues of the cat. *Arch. Biochem. Biophys.* 95: 416–423, 1961.
10. BROTHERUS, J. R., J. V. MØLLER, and P. L. JØRGENSEN. Soluble and active renal Na,K-ATPase with maximum protein molecular mass 170,000 ±

9,000 daltons; formation of larger units by secondary aggregation. *Biochem. Biophys. Res. Commun.* 100: 146–154, 1981.

11. BROWN, T. A., B. HOROWITZ, R. P. MILLER, A. A. McDONOUGH, and R. A. FARLEY. Molecular cloning and sequence analysis of the $(Na^+ + K^+)$-ATPase β subunit from dog kidney. *Biochim. Biophys. Acta* 912: 244–253, 1987.

12. CALDWELL, P. C., and R. D. KEYNES. The utilization of phosphate bond energy for sodium extrusion from giant axons (Abstract). *J. Physiol. Lond.* 137: 12P, 1957.

13. CANTLEY, L. C., JR., L. G. CANTLEY, and L. JOSEPHSON. A characterization of vanadate interactions with the (Na,K)-ATPase. Mechanistic and regulatory implications. *J. Biol. Chem.* 253: 7361–7368, 1978.

14. CANTLEY, L. C., JR., L. JOSEPHSON, R. WARNER, M. YANAGISAWA, C. LECHENE, and G. GUIDOTTI. Vanadate is a potent (Na,K)-ATPase inhibitor found in ATP derived from muscle. *J. Biol. Chem.* 252: 7421–7423, 1977.

15. CANTLEY, L. C., JR., M. D. RESH, and G. GUIDOTTI. Vanadate inhibits the red cell (Na^+,K^+)ATPase from the cytoplasmic side. *Nature Lond.* 272: 552–554, 1978.

16. CLAUSEN, T. Regulation of active Na,K-transport in skeletal muscle. *Physiol. Rev.* 66: 542–580, 1986.

17. CORNELIUS, F., and J. C. SKOU. Reconstitution of $(Na^+ + K^+)$-ATPase into phospholipid vesicles with full recovery of its specific activity. *Biochim. Biophys. Acta* 772: 357–373, 1984.

18. CORNELIUS, F., and J. C. SKOU. Na^+-Na^+ exchange mediated by $(Na^+ + K^+)$-ATPase reconstituted into liposomes: Evaluation of pump stoichiometry and response to ATP and ADP. *Biochim. Biophys. Acta* 818: 211–221, 1985.

19. DANIELLI, J. F., and H. DAVSON. A contribution to the theory of permeability of thin films. *J. Cell. Comp. Physiol.* 5: 495–508, 1935.

20. DEAN, R. B. Theories of electrolyte equilibrium in muscle. *Biol. Symp.* 3: 331–348, 1941.

21. ESMANN, M. The distribution of $C_{12}E_8$-solubilized oligomers of the $(Na^+ + K^+)$-ATPase. *Biochim. Biophys. Acta* 787: 81–89, 1984.

22. ESMANN, M., C. CHRISTIANSEN, K. A. KARLSSON, G. C. HANSSON, and J. C. SKOU. Hydrodynamic properties of solubilized $(Na^+ + K^+)$-ATPase from rectal glands of *Squalus acanthias*. *Biochim. Biophys. Acta* 603: 1–12, 1980.

23. ESMANN, M., and J. C. SKOU. Occlusion of Na^+ by the Na,K-ATPase in the presence of oligomycin. *Biochem. Biophys. Res. Commun.* 127: 857–863, 1985.

24. ESMANN, M., J. C. SKOU, and C. CHRISTIANSEN. Solubilization and molecular weight determination of Na,K-ATPase from rectal glands of *Squalus acanthias*. *Biochim. Biophys. Acta* 567: 410–420, 1979.

25. FAHN, S., G. J. KOVAL, and R. W. ALBERS. Sodium-potassium-activated adenosine triphosphatase of *Electrophorus* electric organ. V. Phosphorylation by adenosine triphosphate-^{32}P. *J. Biol. Chem.* 243: 1993–2002, 1968.

26. FORGAC, M., and G. CHIN. K^+-independent active transport of Na^+ by the $(Na^+ + K^+)$-stimulated adenosine triphosphatase. *J. Biol. Chem.* 256: 3645–3646, 1981.

[180]

27. Gardos, G. Akkumulation der Kaliumionen durch Menschliche Blut-körperchen. *Acta Physiol. Scient. Hung.* 6: 191–199, 1954.

28. Glynn, I. M. The Na$^+$,K$^+$-transporting adenosine triphosphatase. In: *The Enzymes of Biological Membranes*, edited by A. N. Martonosi. New York: Plenum, 1985, vol. 3, 28–114.

29. Glynn, I. M., and J. C. Ellory (editors). *The Sodium Pump. The 4th International Conference on Na,K-ATPase.* Cambridge, UK: Company of Biologists, 1985.

30. Glynn, I. M., and D. E. Richards. Occlusion of rubidium ions by the sodium-potassium pump: its implications for the mechanism of potassium transport. *J. Physiol. Lond.* 330: 17–43, 1982.

31. Glynn, I. M., Y. Hara, and D. E. Richards. Trapping of sodium ions by a phosphorylated form of the sodium-potassium pump (Na,K-ATPase). *J. Physiol. Lond.* 351: 531–547, 1984.

32. Guernsey, D. L., and I. S. Edelman. Regulation of thermogenesis by thyroid hormones. In: *Molecular Basis of Thyroid Hormone Action*, edited by J. N. Oppenheimer and H. H. Samuels. New York: Academic, 1983, p. 293–320.

33. Haas, M., W. F. Schmidt, and F. J. McManus. Catecholamine-stimulated ion transport in duck red cells. *J. Gen. Physiol.* 80: 125–147, 1985.

34. Hastings, D. F., and J. A. Reynolds. Molecular weight of (Na$^+$,K$^+$)ATPase from shark rectal gland. *Biochemistry* 8: 817–821, 1979.

35. Hebert, H., P. L. Jørgensen, E. Skriver, and A. B. Maunsbach. Crystallization patterns of membrane-bound (Na$^+$ + K$^+$)-ATPase. *Biochim. Biophys. Acta* 689: 571–574, 1982.

36. Hebert, H., E. Skriver, and A. B. Maunsbach. Three-dimensional structure of renal N,K-ATPase determined by electron microscopy of membrane crystals. *FEBS Lett.* 187: 182–186, 1985.

37. Hegyvary, C., and R. L. Post. Binding of adenosine triphosphate to sodium and potassium ion-stimulated adenosine triphosphatase. *J. Biol. Chem.* 246: 5234–5240, 1971.

38. Hess, H. H., and A. Pope. Effect of metal cations on adenosinetriphosphate activity of rat brain (Abstract). *Federation Proc.* 16: 196, 1957.

39. Hodgkin, A. L., and R. D. Keynes. Active transport of cations in giant axons from Sepia and Loligo. *J. Physiol. Lond.* 128: 28–60, 1955.

40. Hodgkin, A. L., and R. D. Keynes. Experiments on the injection of substances into squid giant axons by means of a microsyringe. *J. Physiol. Lond.* 131: 592–616, 1956.

41. Hoffman, J. F., and B. Forbush, III (editors). *Current Topics in Membranes and Transport. Structure, Mechanism and Function of the Na/K Pump.* London, Academic, 1982, vol. 19.

42. Hokin, L. E., J. L. Dahl, J. D. Deupree, J. F. Dixon, J. F. Hackney, and J. F. Perdue. Studies on the characterization of the sodium-potassium transport adenosine triphosphatase. *J. Biol. Chem.* 7: 2593–2603, 1973.

43. Hokin, L. E., and J. F. Dixon. Parameters of reconstituted Na$^+$ and K$^+$ transport in liposomes in which purified Na,K-ATPase is incorporated by "freeze-thaw-sonication. In: *Na,K-ATPase. Structure and Kinetics*, edited by J. C. Skou and J. G. Nørby. London: Academic, 1979, p. 47–67.

SODIUM-
POTASSIUM
PUMP

[181]

44. JØRGENSEN, P. L. Purification and characterization of (Na$^+$ + K$^+$)-ATPase. V. Conformational changes in the enzyme. Transitions between the Na-form and the K-form studied with tryptic digestion as a tool. *Biochim. Biophys. Acta* 401: 399–415, 1975.

45. JØRGENSEN, P. L. Isolation and characterization of the components of the sodium pump. *Q. Rev. Biophys.* 7: 239–274, 1975.

46. JØRGENSEN, P. L., J. C. SKOU, and L. P. SOLOMONSEN. II. Preparation by zonal centrifugation of highly active (Na$^+$ + K$^+$)-ATPase from outer medulla of rabbit kidneys. *Biochim. Biophys. Acta* 233: 381–394, 1971.

47. KARLISH, S. J. D. Characterization of conformational changes in (Na,K)ATPase labeled with fluorescein at the active site. *J. Bioenerg. Biomembr.* 12: 111–136, 1980.

48. KARLISH, S. J. D., and D. W. YATES. Tryptophan fluorescence of (Na$^+$ + K$^+$)-ATPase as a tool for study of the enzyme mechanism. *Biochim. Biophys. Acta* 527: 115–130, 1978.

49. KARLISH, S. J. D., D. W. YATES, and I. M. GLYNN. Conformational transitions between Na$^+$-bound and K$^+$-bound forms of (Na$^+$ + K$^+$)-ATPase, studied with formycin nucleotides. *Biochim. Biophys. Acta* 525: 252–264, 1978.

50. KAWAKAMI, K., S. NOGUCHI, M. NODA, H. TAKAHASHI, T. OHTA, M. KAWAMURA, H. NOJIMA, K. NAGANO, T. HIROSE, S. INAYAMA, H. HAYASHIDA, T. MIYATA, and S. NUMA. Primary structure of the α-subunit of *Torpedo californica* (Na$^+$ + K$^+$)ATPase deduced from cDNA sequence. *Nature Lond.* 316: 733–736, 1985.

51. KEPNER, G. R., and R. J. MACEY. Membrane enzyme systems. Molecular size determinations by radiation inactivation. *Biochim. Biophys. Acta* 163: 188–203, 1968.

52. KEYNES, R. D. Electrolytes and nerve activity. In: *Metabolism of the Nervous System*, edited by D. Richter. London: Pergamon, 1957, p. 159–173.

53. KLEIN, L. E., M. S. BARTOLOMEI, and C. S. LO. Corticosterone and triiodothyronine control of myocardial Na$^+$-K$^+$-ATPase activity in rats. *Am. J. Physiol.* 247 (*Heart Circ. Physiol.* 16): H570–H575, 1984.

54. KYTE, J. Purification of the sodium- and potassium-dependent adenosine triphosphatase from canine renal medulla. *J. Biol. Chem.* 246: 4157–4165, 1971.

55. LANE, L. K., J. H. COPENHAVER, JR., G. E. LINDENMAYER, and A. SCHWARTZ. Purification and characterization of and (^{3}H)ouabain binding to the transport adenosine triphosphatase from outer medulla of canine kidney. *J. Biol. Chem.* 20: 7197–7200, 1973.

56. LIBET, B. Adenosine triphosphatase (ATP-ase) in nerve (Abstract). *Federation Proc.* 7: 72, 1948.

57. LYTTON, J., J. C. LIN, and G. GUIDOTTI. Identification of two molecular forms of (Na$^+$ + K$^+$)-ATPase in rat adipocytes. Relation to insulin stimulation of the enzyme. *J. Biol. Chem.* 260: 1177–1184, 1985.

58. MACKNIGHT, A. D. C., and A. LEAF. Regulation of cellular volume. *Physiol. Rev.* 57: 510–573, 1977.

59. MEYER, H. H. Zur Theorie der Alkoholnarkose. Erste Mitteilung. *Arch. Exp. Path. Pharmak.* 42: 109–118, 1899.

60. MULVANY, M. J. Changes in sodium pump activity and vascular contraction. *J. Hypertension* 3: 429–436, 1985.

61. Nachmansohn, D. Chemical mechanism of nerve activity. In: *Modern Trends of Physiology and Biochemistry*, edited by E. E. G. Barron. New York: Academic, 1952, p. 229–276.

62. Noguchi, S., M. Noda, H. Takahashi, K. Kawakami, T. Ohta, K. Nagano, T. Hirose, S. Inayama, M. Kawamura, and S. Numa. Primary structure of the β-subunit of *Torpedo californica* $(Na^+ + K^+)$-ATPase deduced from the cDNA sequence. *FEBS Lett.* 196: 315–320, 1986.

63. Nørby, J. G., and J. Jensen. Binding of ATP to brain microsomal ATPase. Determination of the ATP-binding capacity and the dissociation constant of the enzyme-ATP complex as a function of K^+ concentration. *Biochim. Biophys. Acta* 233: 104–116, 1971.

64. Nørby, J. G., I. Klodos, and N. O. Christiansen. Kinetics of Na-ATPase activity by the Na,K-pump. Interactions of the phosphorylated intermediates with Na^+, $Tris^+$, and K^+. *J. Gen. Physiol.* 82: 725–759, 1983.

65. Ottolenghi, P. The relipidation of delipidated Na,K-ATPase. An analysis of complex formation with dioleoylphosphatidylcholine and with dioleoylphosphatidylethanolamine. *Eur. J. Biochem.* 99: 113–131, 1979.

66. Ovchinnikov, Y. A., V. V. Demin, A. N. Barnakov, A. P. Kuzin, A. V. Lunev, N. N. Modyanov, and K. N. Dzhandzhugazyan. Three-dimensional structure of $(Na^+ + K^+)$-ATPase revealed by electron microscopy of two-dimensional crystals. *FEBS Lett.* 190: 73–76, 1985.

67. Ovchinnikov, Y. A., N. N. Modyanov, N. E. Broude, K. E. Petrukhin, A. V. Grishin, N. M. Arzamazova, N. A. Aldanova, G. S. Monastyrskaya, and E. D. Sverlov. Pig kidney Na^+,K^+-ATPase. Primary structure and spatial organization. *FEBS Lett.* 201: 237–245, 1986.

68. Overton, E. *Studien über die Narkose zugleich ein Beitrag zut allgemeinen Pharmakologie.* Jena: Fischer, 1901.

69. Plesner, I. W., L. Plesner, J. G. Nørby, and I. Klodos. The steady-state kinetic mechanism of ATP hydrolysis catalyzed by membrane-bound $(Na^+ + K^+)$-ATPase from ox brain. III. A minimal model. *Biochim. Biophys. Acta* 643: 483–494, 1981.

70. Post, R. L., C. Hegyvary, and S. Kume. Activation by adenosine triphosphate in the phosphorylation kinetics of sodium and potassium ion transport adenosine triphosphatase. *J. Biol. Chem.* 247: 6530–6540, 1972.

71. Post, R. L., and S. Kume. Evidence for an aspartyl phosphate residue at the active site of sodium and potassium ion transport adenosine triphosphatase. *J. Biol. Chem.* 248: 6993–7000, 1973.

72. Post, R. L., S. Kume, T. Tobin, B. Orcutt, and A. K. Sen. Flexibility of an active centre in sodium-plus-potassium adenosine triphosphatase. *J. Gen. Physiol.* 54: 306S–326S, 1969.

73. Post, R. L., C. R. Merritt, C. R. Kinsolving, and C. D. Albright. Membrane adenosine triphosphatase as a participant in the active transport of sodium and potassium in the human erythrocyte. *J. Biol. Chem.* 235: 1796–1802, 1960.

74. Post, R. L., A. K. Sen, and A. S. Rosenthal. A phosphorylated intermediate in adenosine triphosphate-dependent sodium and potassium transport across kidney membranes. *J. Biol. Chem.* 240: 1437–1445, 1965.

75. Repke, K. Metabolism of cardiac glycosides. In: *New Aspects of Cardiac Glycosides*, edited by W. Wilbrandt. New York: Pergamon, 1963, vol.

SODIUM-
POTASSIUM
PUMP

3, p. 47–73. (Proc. Internat. Pharmacol. Meet. Stockholm 1961.)

76. SCHATZMANN, H. J. Herzglykoside als Hemmstoffe fur der aktiven Kalium und Natrium Transport durch die Erythrocytenmembran. *Helv. Physiol. Pharmacol. Acta* 11: 346–354, 1953.

77. SCHOU, M., N. JUEL-NIELSEN, E. STRÖMGREN, and H. VOLDBY. The treatment of manic psychoses by the administration of lithium salts. *J. Neurol. Neurosurg. Psychiatry* 17: 250–260, 1954.

78. SCHWARTZ, A. Positive inotropic action of digitalis and endogenous factors: Na,K-ATPase and positive inotropy; "Endogenous Glycosides." In: *Current Topics in Membranes and Transport. Structure, Mechanism and Function of the Na/K Pump*, edited by J. F. Hoffman and B. Forbush, III. London: Academic, 1983, vol. 19, p. 825–855.

79. SCHWARTZ, A., D. E. LINDENMAYER, and J. C. ALLEN. The sodium-potassium adenosine triphosphatase: pharmacological, physiological and biochemical aspects. *Pharmacol. Rev.* 27: 3–134, 1975.

80. SHULL, G. E., A. SCHWARTZ, and J. B. LINGREL. Amino-acid sequence of the catalytic subunit of the $(Na^+ + K^+)$ATPase deduced from a complementary DNA. *Nature Lond.* 316: 691–695, 1985.

81. SHULL, G. E., L. K. LANE, and J. B. LINGREL. Amino-acid sequence of the β-subunit deduced from cDNA. *Nature Lond.* 321: 429–431, 1986.

82. SHULL, G. E., J. GREEB, and J. B. LINGREL. Molecular cloning of three distinct forms of the Na^+,K^+-ATPase α-subunit from rat brain. *Biochemistry* 25: 8125–8132, 1986.

83. SIEGEL, G. J., and R. W. ALBERS. Sodium-potassium activated adenosine triphosphatase of *Electrophorus* electric organ. IV. Modification of responses to sodium and potassium by arsenite plus 2,3-dimercaptopropanol. *J. Biol. Chem.* 242: 4972–4975, 1967.

84. SIMONS, T. J. B. The interaction of ATP-analogues possessing a blocked γ-phosphate group with the sodium pump in human red cells. *J. Physiol. Lond.* 244: 731–739, 1975.

85. SKOU, J. C. Discussion in ref. 52. In: *Metabolism of the Nervous System*, edited by D. Richter. London: Pergamon, 1957, p. 173.

86. SKOU, J. C. The influence of some cations on an adenosine triphosphatase from peripheral nerves. *Biochim. Biophys. Acta* 23: 394–401, 1957.

87. SKOU, J. C. Further investigations on a $Mg^{++} + Na^+$-activated adenosine triphosphatase, possibly related to the active linked transport of Na^+ and K^+ across the nerve membrane. *Biochim. Biophys. Acta* 42: 6–23, 1960.

88. SKOU, J. C. The effect of drugs on cell membranes with special reference to local anaesthetics. *J. Pharm. Pharmacol.* 13: 204–217, 1961.

89. SKOU, J. C. Preparation from mammalian brain and kidney of the enzyme system involved in active transport of Na^+ and K^+. *Biochim. Biophys. Acta* 58: 314–325, 1962.

90. SKOU, J. C. Enzymatic basis for active transport of Na^+ and K^+ across cell membrane. *Physiol. Rev.* 45: 596–617, 1965.

91. SKOU, J. C. Effects of ATP on the intermediary steps of the reaction of the $(Na^+ + K^+)$-ATPase. IV. Effect of ATP on $K_{0.5}$ for Na^+ and on hydrolysis at different pH and temperature. *Biochim. Biophys. Acta* 567: 421–435, 1979.

92. SKOU, J. C. The effect of pH, of ATP and of modification with pyridoxal 5-phosphate on the conformational transition between the Na^+-form

and the K^+-form of the $(Na^+ + K^+)$-ATPase. *Biochim. Biophys. Acta* 688: 369–380, 1982.

93. Skou, J. C. Effect on the equilibrium between the Na^+-form of the $(Na^+ + K^+)$-ATPase of modification of the enzyme with pyridoxal 5-phosphate. *Biochim. Biophys. Acta* 789: 44–50, 1984.

94. Skou, J. C. Considerations on the reaction mechanism of the Na,K-ATPase. In: *The Sodium Pump. The 4th International Conference on Na,K-ATPase,* edited by I. M. Glynn and J. C. Ellory. Cambridge, UK: Company of Biologists, 1985, p. 575–588.

95. Skou, J. C. Modification of the $(Na^+ + K^+)$-ATPase with 1-ethyl-3-(3-dimethylaminopropyl)carbodiimide and with diethylpyrocarbonate. Effect on the conformational transition. *Biochim. Biophys. Acta* 819: 119–130, 1985.

96. Skou, J. C., and M. Esmann. Eosin, a fluorescent probe of ATP binding to the $(Na^+ + K^+)$-ATPase. *Biochim. Biophys. Acta* 647: 232–240, 1981.

97. Skou, J. C., and M. Esmann. The effects of Na^+ and K^+ on the conformational transitions of $(Na^+ + K^+)$-ATPase. *Biochim. Biophys. Acta* 746: 101–113, 1983.

98. Skou, J. C., and J. G. Nørby (editors). *Na,K-ATPase. Structure and Kinetics.* London: Academic, 1979.

99. Smith, T. W., and W. H. Barry. Monovalent cation transport and mechanisms of digitalis-induced inotropy. In: *Current Topics in Membranes and Transport. Structure, Mechanism and Function of the Na/K Pump,* edited by J. F. Hoffman and B. Forbush, III. London: Academic, 1983, vol. 19, p. 857–884.

100. Sweadner, K. J. Two molecular forms of $(Na^+ + K^+)$-stimulated ATPase in brain. Separation and difference in affinity for strophanthidin. *J. Biol. Chem.* 254: 6060–6067, 1979.

101. Ullrich, K. J. A. Sugar, amino acid and Na^+ co-transport in proximal tubule. *Annu. Rev. Physiol.* 41: 181–195, 1979.

102. Ussing, H. H. Transport of ions across cellular membranes. *Physiol. Rev.* 29: 127–155, 1949.

SODIUM-
POTASSIUM
PUMP

VI

From Frog Lung to Calcium Pump

WILHELM HASSELBACH

W HEN I recall decisive events in my scientific career starting around 1946, I must admit that most things occurred more or less spontaneously. My entry into the field of physiology, to which I had little prerogative, and also most of my scientific achievements were not purposefully designed but developed incidentally.

The person who acted as a signpost at the beginning of my scientific journey was Hans Lullies, the late professor of physiology at the University of Kiel. He had settled there after a long odyssey that led him from Königsberg via Cologne, Strassburg, and Homburg/Saar to Kiel. He was a well-esteemed electrophysiologist in Germany (29). I met Lullies at Marburg, a small university town north of Frankfurt, where he lived as a refugee. Marburg had suffered comparatively little in the war, and the university, which was founded in the Reformation, was one of the first that was permitted to readmit students.

During the war I initially studied natural science and later medicine at various universities, although interrupted by military service. After being released as a prisoner of war by the US Army, 1st Infantry Division, quite early in 1945, I applied for admission at the University of Marburg and resumed my medical studies in November 1945. With the hopeful spirit of the returnee, I overcame the first term and soon attempted a medical thesis. More or less by chance, I started working at the Department of Physiology. It was housed in a neo-Gothic building opposite the famous Gothic cathedral dedicated to Elizabeth of Hungary. Except for the small but well-kept library and an electric pendulum designed by von Helmholtz, the department appeared to be deprived of nearly everything: no glassware, no chemicals, no distilled water, and little or no heating. In wintertime the frogs in the cellar had to be hacked out from under the ice covering their basins. My studies in the library led me to consider, as a worthwhile task, the determination of the quantity of acetylcholine liberated from one end plate of a muscle fiber during a single twitch. However, I remained stuck with the reevaluation of a method to measure small quantities of acetylcholine using the frog lung as the sensitive tissue. One day, while I was standing in front of my self-made, self-paid tissue bath in which a frog lung was mounted, a tiny agile man peered into the room and asked for the head of the department. With a swift glance he evidently had detected the reddish-brown frog lung in the glass tube. *"Ah, if you are working with frog lungs, I can perhaps help you,"* he said and introduced himself as Professor Lullies. I recalled his name from the literature and knew that together with Karl Brecht (4, 30) he had introduced the isolated frog lung as an apparently highly acetylcholine-sensitive contractile tissue. On talking further, he became enthusiastic and asked me to let him read my thesis. Half a year later I did so, and he invited me to his small apartment where he lived with his large family and his parents-in-law. He engaged me in a long discussion during which he painstakingly analyzed the problems. It was the first time that I had the opportunity to discuss scientific questions with an expert and devoted scientist. Finally, Lullies asked me if I would be interested in working in physiology, and he recommended the department of Hans Hermann Weber at Tübingen, who I had not heard of before. Spontaneously, Lullies wrote a letter of recommendation and, by return mail, Weber asked me to send him a copy of my thesis. We met at the first postwar meeting of the German Physiological Society in Frankfurt in 1948. I had prepared myself for this interview by reading Weber's 1934 review in the *Ergebnisse der Physiologie* (45), which Lullies had mentioned to me. I was greatly impressed by the way Weber applied physicochemical principles and methods to biological problems and how he discussed the results. While I was studying Weber's article, I found in a new issue of the *Acta Physiologica Scandinavica* a review by Albert Szent-Györgyi (41) in which he had summarized the work done by his group in Budapest between 1940 and 1942. I was overwhelmed by the abundance of new findings, and I got the feeling that the problem of energy transduction in biological structures appeared to be within the reach of experimental approach. In Frankfurt, I met Weber after a presentation on biological rhythms by Albrecht Bethe, the father of the physicist Hans Bethe, from Ithaca, NY. Weber clearly expressed his dislike of Bethe's speculative approach. On the way to lunch Weber asked me if I had any problem I would like to work on in the range of his interest. He distractedly sucked at his cigarette when I suggested pursuing Szent-Györgyi's work on the interaction of ATP with isolated muscle proteins. *"At first we have to accurately determine the molecular weight of myosin, and we also have to look for globulin X in the muscle,"* was his short answer. On the 1st of May, 1949, I moved to Tübingen with a promised salary of 100 deutsche marks per month. Weber introduced me (often as Herr Hagenbach) to Hildegard Portzehl, his daughter Annemarie Weber, and Gerhard Schneider. In his studies on the molecular properties of myosin, Weber often applied viscosity measurements that were cumbersomely corrected "according to Hagenbach." Weber lived with his family on the first floor of the Institute; the lower floor housed his office and three larger laboratories. They were somewhat better equipped than the Department of Physiology at Marburg. Weber was proud of the new equipment that consisted of a Stock low-speed centrifuge, a butcher's refrigerator, a highly explosive hydrogen-pH meter, as well as a Kjeldahl apparatus. Together with Schneider, who had some experimental experience, I tried to look for globulin X, which Weber had described in 1934 as a major protein constituent of the muscle. However, by using neutral instead of alkaline high–ionic-strength salt solutions for extracting the muscle, we could not find this protein. We had a hard time convincing Weber that we were most likely searching for a degradation product of Straub's actin formed in the alkaline Weber-Edsall solution. Actin was discovered as a new muscle protein by F. B. Straub in Szent-Györgyi's laboratory in 1942. During these experiments we also observed that myosin was selectively extracted from the muscle such that its band structure disappeared (20). The experiments began to furnish reproducible results after we used salt solutions containing pyrophosphate as a substitute for the nonavailable ATP. These solutions were later called Hasselbach-Schneider solutions by Jean Hanson and Hugh E. Huxley. In long and intense discussions, Weber often challenged our results and arguments. He sharply questioned not only our reasoning but also every experimental detail. With unlimited cigarette sharing, the discussions often lasted until midnight. Weber often deviated from science and would start discussing general problems such as the political development in Germany before and after 1933. In his opinion scientists were as easily misled in politics as other people. He often expressed his admiration for Otto Meyerhof, who left Germany from Heidelberg at the last moment in 1938.

With the work on the selective extraction of myosin from striated muscle and the subsequent electron-microscopic studies, which I performed with the support of Gerhard Schramm and Friedrich Freksa, members of the Kaiser-Wilhelm-(later Max-Planck-)Institut für Virusforschung, I first earned international recognition (14). It was Hanson who took the initiative and visited me shortly after I had moved from Tübingen to Heidelberg in 1954. Her visit started my most stimulating and close relations with muscle research groups in London and Cambridge.

One of the most exciting events during our time at Tübingen was the second visit of the Unitarian Service Committee, which was sent by the Unitarian Church of America to Europe to help to restore scientific relations. Linus Pauling presented results of his recent work on the structure of proteins, and I had to write an article about it in the local newspaper. F. Karsh reported on the ion-binding prop-

CALCIUM PUMP

erties of proteins. At the end of the meeting, the Commission handed Weber a collection of books, a check, and, as a personal present, a fountain pen. Weber, visibly moved, expressed our thanks. The money was used to buy an air-driven analytical ultracentrifuge. I also remember with gratitude the first visits to Weber and the Institute by German-born emigrants who had left Germany during Hitler's regime. Ernst Fischer from Richmond, Virginia, who had to leave Bethe's institute in Frankfurt, and David Nachmannsohn from Columbia University in New York, who emigrated from Berlin, were among the first to come to Tübingen. Another especially memorable event was the spontaneous visit of Manuel Morales, who was at that time a member of the Navy Medical Research Institute in Bethesda, Maryland. I recall heated discussions with him and his three young co-workers. We did not agree with Morales' complicated concept for the ATP-induced viscosity drop of actomyosin solutions later directly disproved by A. Weber (43).

After the structural work my interest turned to the enzymatic properties of the contractile proteins. I was fascinated by the remarkable change that occurs in the enzymatic and physical behavior of the proteins when the ionic strength of the medium is altered. These transitions were suspected of playing an important role in the regulation of muscle activity. With this work I entered the mainstream of a new scientific development. It was initiated by a study by B. B. Marsh (33) in which he described the effects of a factor present in aqueous muscle extracts that could prevent ATP from causing the contractile proteins in myofibrils to contract. Marsh's study originated from an observation made by many muscle biochemists: the volumes of muscle homogenates pelleted at low centrifugal force were sometimes small and dense and sometimes large and loose. The possible significance of this observation had evidently been recognized by Kenneth Bailey from the biochemistry department at the University of Cambridge. His student Marsh showed that the unpredictable volume changes depended on the presence of a labile factor that could be separated from the myofibrils and that its effectiveness depended on the presence of ATP. Furthermore Marsh found that calcium ions at astonishingly low concentrations abolish the effect of the factor; i.e., on addition of calcium the myofibrils start to shrink. J. R. Bendall (1), at that time a member of the Low Temperature Station at Cambridge, added the factor to glycerinated muscle fibers and demonstrated that it could cause relaxation, and that the addition of calcium ions caused contraction of the relaxed fibers. Fascinated by these findings we tried to integrate them into the framework of Weber's concept of activity regulation (46). Weber assumed that the interaction between myosin, actin, and ATP leading to ATP hydrolysis causes contraction and that the inhibition of hydrolysis results in relaxation. Accordingly ATP was supposed to have a dual function, acting both as an energy donor and a relaxing agent. Switching between the two functions was accomplished in vitro by the addition of enzyme inhibitors that did not interfere with ATP binding or by high concentrations of ATP alone.

I found that both the relaxing effect and the inhibition of enzyme activity, which ATP alone could exert, was tremendously augmented in the presence of the factor (21). However, these studies were greatly hampered by the lability and the unpredictable effectiveness of the factor. Calcium contaminations were suspected to be the main cause of the troublesome variability. Recalling my basic inorganic chemistry, I added calcium binding agents (citrate, phosphate, and oxalate) to the isolation media and found that the factor stability and its effectiveness was considerably enhanced by the addition of relatively low concentrations of oxalate, whereas the other agents were comparatively ineffective. In the presence of 5 mM oxalate, the assay became so sensitive that we often had difficulty removing the residual factor activity in the myofibrillar preparations. These findings later proved to be quite important.

In 1954 Weber became the successor of Otto Meyerhof and Hermann Rein as the director of the Institute of Physiology in the Institute for Medical Research of the Max Planck Society for the Advancement of Science. After our move from Tübingen to Heidelberg, the relaxing-factor activity remained one aspect of our interest. We did not trust the explanation of Bendall (2) and L. Lorand (28) that the ubiquitous enzymes adenylate kinase or creatine transphosphorylase might be the active principles of the factor. We also knew of the work of S. Ebashi (9), who attempted to analyze the effect of calcium on the activity of the isolated contractile protein. However, at that time information still flowed quite weakly. The gap resulting from the disconnection of German science from the scientific development in the United States during the war had not yet closed. Thus we did not know that W. W. Keilley and O. Meyerhof (27) had shown that muscle extracts contain a factor of particulate material, muscle microsomes, that exhibits a most characteristic ATPase activity. Meyerhof was interested in this enzyme because, in contrast to the ATPase of myosin, the activity of the enzyme was strongly inhibited by calcium ions. Knowledge of this fact was brought to us by A. Weber, who had spent some months in J. Edsall's department in Boston. She suggested that we check whether the removal of the microsomes by centrifugation would affect the relaxing activity of the aqueous muscle extract. This experiment could successfully be done only after we bought our first American-manufactured preparative ultracentrifuge. The relaxing activity was found present in the microsomal pellet, and thus the idea that the relaxing effect was brought about by proteins of low molecular weight, adenylate kinase, or creatine phosphokinase could be excluded (37). These findings were in agreement with Ebashi's (9) observation that the active principle of the relaxing factor could be precipitated by ammonium sulfate, a procedure applied by Keilley and Meyerhof to remove the microsomes from the aqueous muscle extract.

In Heidelberg we found much better working conditions, and H. H. Weber could increase his staff. Although A. Weber moved permanently to the United States, H. Hoffman-Berling, and M. and G. Ulbrecht joined us. Weber wanted the Ulbrechts to study the question of whether contractile proteins are phosphorylated during chemomechanical energy transformation.

Hoffmann-Berling continued his studies on the mechanism of cellular motility. His findings that in some motile structures complete contraction-relaxation cycles can be induced merely by the addition and removal of calcium ions greatly excited us (25). At that time we were working on the concept that in the presence of ATP the microsomes might produce a soluble calcium-sensitive relaxing factor. We attributed calcium sensitivity to a soluble factor because we had good evidence that the properties of the purified contractile proteins were not affected by physiological calcium concentrations. It seemed logical to assume that the production of a soluble factor should finally exhaust the microsomes. When Madoka Makinose joined us in 1957 on a postdoctoral fellowship, while accompanying Professor T. Nagai, we tried to solve this problem. Despite working day and night we could not obtain an unequivocal answer (35). After our Japanese guests' departure, I continued to explore the factor's properties by studying its inactivation by calcium ions. I tried to inactivate the factor in a stepwise manner by the addition of increasing amounts of calcium ions and by measuring the factor's effect on the activity of the actomyosin ATPase of isolated myofibrils. Unexpectedly, the factor was only transiently inactivated when small amounts of calcium ions were added; i.e., calcium addition initiated bursts of ATPase activity of the actomyosin ATPase. On adding larger amounts, the activity transients became longer and finally, inactivation persisted. Now there was a new and interesting problem ahead of us. I thought it would be a good idea to apply radioactive calcium to find out which component of the system was the target for calcium. Radioactive calcium had just become available at sufficiently high specific activities. When I asked Weber's permission to use radioactive calcium, he at first tried to convince me to use some classic analytical procedure and he referred to Warburg's aversion to tracer experiments. However, when I found that the radioactive calcium (bought without permission) was neither in the myofibrils nor in the particle-free medium but had been completely concentrated in the microsomes, Weber gave up his previous misgivings. The ability of the microsomes to remove calcium proved to be dependent on the presence of ATP and magnesium ions in the system. Furthermore the calcium-removing ability of the microsomes and their relaxing effect were affected in parallel by various interventions. In media of appropriate composition, 0.1 mg microsomes/ml were able to reduce the calcium concentration in the solution to <0.1 μM, which is far below the concentration of the calcium contaminants in the system. If sufficient amounts of calcium were added to the system, the total calcium concentration in the microsomes even reached 10 μmol/mg. This enormous calcium load strongly ruled out the possibility that calcium removal was a result of the binding of calcium ions to the microsomal proteins. Because excessive calcium storage only occurred in the presence of oxalate, it was necessary to assume that calcium is stored with oxalate as insoluble calcium oxalate by the microsomes. Under these assumptions, the concentration of ionized calcium inside the microsomes could be delineated from the solubility product of calcium oxalate and the oxalate concentration in the medium. Thus calculated, the free-calcium concentration inside the microsomes turned out to be much higher than the calcium concentration in the solution. Consequently the microsomes should function as a calcium-concentrating system, i.e., a calcium pump, an expression at first used with hesitation (17). The energy needed for calcium accumulation could only stem from ATP because the media did not contain any other energy-yielding substrates. At that time ion pumps were not considered to be ubiquitous biological instruments. An active extrusion of sodium ions out of the giant nerve fiber of the squid had just been demonstrated (5). Its energy-yielding reactions had been studied by showing that sodium transport could be supported by the injection of many energy-rich phosphate compounds. The red cell ghosts were not yet invented (24), and the demonstration of a sodium-potassium ATPase in nerve membrane fragments merely allowed us to suspect a relationship between enzyme activity and ion transport (40). The intention to study the correlation between calcium uptake and the consumption of ATP had to overcome the objection that the ATPase activity of the microsomes was not activated but rather inhibited by calcium ions. Nevertheless we tried to simultaneously measure calcium uptake and ATP splitting. Makinose, who did these experiments after returning from Japan in 1960, and I were very disappointed when we at first found that the enzymatic activity did not change under calcium-uptake conditions. However, we soon realized that we had missed the main event by measuring enzymatic activity in the time interval between 2 and 7 min after the addition of calcium ions, a mistake that is sometimes still made today. When we screened our protocols in which we had not only written down the 2- and 7-min values but also the zero blank values for each experiment, we discovered that the ATPase of the microsomes was activated only transiently after the addition of calcium. In retrospect this finding appears to be a major breakthrough relieving us from the aforementioned bias of calcium inhibition. An important experimental problem in these studies was the determination of the time course of calcium uptake. We terminated

CALCIUM
PUMP

[193]

the reaction by cooling and subsequent centrifugation. This naturally did not yield precise values. We found that the uptake of ~1.2 calcium ions was accompanied by the splitting of 1 ATP molecule. A causal relationship between calcium transport and ATP hydrolysis could be further substantiated by interventions that inhibited both processes simultaneously, like blockage of the microsome's thiol groups with mersalyl. We also tried to estimate the osmotic work done by the pump. From the calculated free-calcium concentration inside the microsomes, the measured calcium concentration in the medium, and a few reasonable assumptions, we ended up with a pump efficiency of 17%–35% (17).

At that time our view of the structure of the microsomes had slowly changed. When I first visited England in 1957, Hanson, Huxley, and J. Lowey proposed that we should look at the relaxing-factor granules, as we called the microsomes. We prepared microsomes and processed them for the electron microscope in a Soho restaurant. The fixation took place in test tubes that Huxley carried in his coat pocket. The pictures made it clear that the microsomes were in fact membrane vesicles (35). Vesicles of a similar appearance had been observed by Huxley scattered in between the myofibrils of glycerinated muscle fiber. The relation of these vesicles to the sarcoplasmic reticulum membrane, which had just been rediscovered by Porter and Palade (36), appeared evident. At that time an effective cooperation to pursue the problem of the origin and structure of the relaxing-factor membranes was quite difficult to accomplish. During a visit in late 1959 we searched for calcium oxalate in actively loaded membrane vesicles and obtained the first electron-microscopic pictures of calcium oxalate precipitates inside the vesicles (16) (Figs. 1 and 2). Thus the ability of the membranes to concentrate calcium was pictorially demonstrated. However, for a long time when I showed these pictures I met with considerable reservation. The critics mainly referred to the findings of Rossi and Lehninger (38), who at the same time had shown that mitochondria could also store calcium ions. Our calcium-storing membranes were considered to be of mitochondrial origin despite the fact that R. M. Berne (3), who had cooperated with us for some time, had shown that the preparations were only slightly contaminated by mitochondrial enzymes. It was even disregarded that mitochondria proved to be unable to store calcium in the presence of oxalate. We received final proof that the sarcoplasmic reticulum membranes were the calcium storing structures we studied in 1964. The electron-microscopic pictures showing numerous terminal cisternae of the sarcoplasmic reticulum filled with calcium oxalate were taken by Heidi Ganzler-Nemecek from the Department of Pathology at the University of Heidelberg, with whom we occasionally cooperated since 1961. The picture was included in the manuscript of a presentation given at the 1964 meeting of the Federation of American Societies for Experimental

MEMBRANE TRANSPORT

[194]

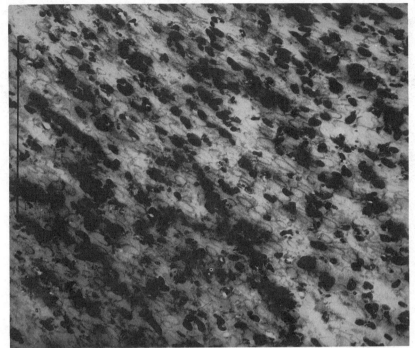

Figure 1. First electron micrograph of sarcoplasmic reticulum vesicles loaded with calcium oxalate in ATP- and magnesium-containing solutions. Vesicles were pelleted, fixed, and stained. Thin sections of the embedded material were examined with the electron microscope. *Bar,* 1,000 nm. [Obtained in cooperation with H. E. Huxley, J. Lowy, and J. Hanson (16).]

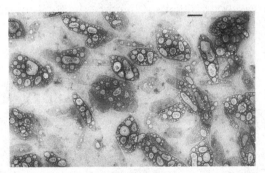

Figure 2. Unstained preparation of calcium oxalate–filled sarcoplasmic reticulum microsomes isolated from smooth muscle preparations. *Bar,* 100 nm. (Courtesy of B. Agostini.)

Biology in Chicago to which I was invited (15). Evidently Costantin, Podolsky, and Francini-Armstrong had done corresponding experiments and nearly simultaneously published a similar picture in *Science* (6).

In the meantime Makinose had developed a procedure to rapidly

[195]

remove calcium-loaded vesicles from the suspension by precipitating them with a solution of $HgCl_2$, which allowed us to terminate the calcium uptake rapidly and thus measure calcium-uptake kinetics more precisely (19). We established anew the stoichiometry of the pump. Now the transport of 2 calcium ions was found to be coupled to the hydrolysis of 1 molecule of ATP. Furthermore we were able to correlate calcium uptake and ATP hydrolysis with a rapidly proceeding transfer of the terminal phosphate residue of ATP to ADP (18) (Fig. 3). In these studies we made use of M. Ulbrecht's studies on the phosphorylation of actomyosin by ATP and her attempt to eliminate interference with the exchange activity of the microsomes (13). The activation of the exchange activity of the transport system by calcium ions was the first indication of the occurrence of phosphorylated intermediates in a transport reaction cycle. We precipitously assumed that the enzyme gained high calcium affinity by this reaction and that the release of calcium at the luminal surface was connected hydrolytically with the phosphate residue. That instead phosphorylation has to reduce an inborn high affinity of the calcium transport system was, at last, realized when we found that ATPase is equipped with an appropriate number of high-affinity calcium binding sites, which are occupied by calcium at concentrations needed to activate the pump (12).

In the spring of 1962, I had my first opportunity to visit the United States. John Gergely, who we knew from the work he had done with Szent-Györgyi, had invited H. H. Weber, Caspar Rüegg, and me to a meeting on the occasion of the opening of the Retina Foundation Institute in Boston, Massachusetts. Now when I enjoy the center of Boston on occasional visits, I still picture the fields of rubble where

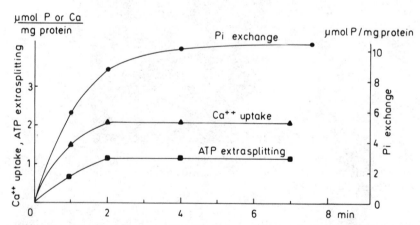

Figure 3. First simultaneous kinetic measurements of calcium uptake, ATP extrasplitting, and calcium-dependent phosphate exchange between ATP and ADP. Calcium uptake was not terminated by filtration but rather by precipitation with $HgCl_2$ (19). A calcium-transport–ATP-splitting stoichiometry of 2 was measured.

the Institute was built. We met near Boston at Dedham in a luxurious New England mansion and were kindly received, together with other visitors from Japan and Poland, by young members of the American science community. As a representative of the older generation, Hermann Kalckar participated. We had long unconstrained discussions and enjoyed a most wonderful springtime. After the meeting I was invited to give two seminars at the Medical School, one in the pharmacology department and the other in the biochemistry department. On Divinity Avenue I was rewarded by great interest, long discussions, and careful listening. On my way home I paid a visit to the Muscle Research Institute, where I met A. Weber and Alexander Sandow, and to Fritz Lipmann's department at the Rockefeller Institute. Although my presentation at the Muscle Research Institute received friendly praise, I encountered fierce criticism and a hardly understandable aversion in Lipmann's department. Above all, it was the use of oxalate as a calcium-trapping agent to which Lipmann and most of his co-workers objected. They did not want to have calcium accumulated by an ATP-supported pump but preferred an ATP-dependent calcium-binding process. Such a mechanism was originally proposed by Ebashi, who had observed in Lipmann's lab that in the presence of ATP a small amount of calcium became associated with the membranes. I was not aware of these findings, which were still in one of Lipmann's drawers (11). Peter Siekewicz was the only person in the audience from whom I received support.

At home we slowly started publishing the results reported in Boston, including the subfractionation of the vesicular membranes, the establishment of the 2:1 stoichiometry of the pump, and an improved procedure for measuring the calcium concentration ratio more accurately (19). Simultaneously we had started to change our opinion about the function of the sarcoplasmic reticulum in the living muscle. We originally tried to incorporate the calcium pump into our previous concept by assuming that the microsomes not only pump calcium but also produce a calcium-sensitive relaxing factor. We abandoned this concept after A. Weber (44) showed that quite a number of well-known phenomena concerning the inhibition of the actomyosin ATPase and of the contraction could be explained by the removal of calcium ions from the contractile proteins as they are present in myofibrils and in crude preparations of actomyosin. The soluble relaxing factor finally became obsolete when Ebashi (10) showed that the contractile proteins themselves contain a complex protein system that in the absence of calcium ions prevents the interactions between ATP, myosin, and actin.

The calcium sensitivity of the contractile protein and the intracellular calcium transport system of the sarcoplasmic reticulum have now become the main elements of the calcium concept of excitation-contraction coupling. This concept has quite a long history. It was envisaged by L. V. Heilbrunn (22), criticized by A. V. Hill (23), and experimentally supported by D. L. Gilbert and W. O. Fenn (13) as well as by the previous findings of Else Weise (47). In recent years the calcium concept has been extended to all kinds of cellular activities. A calcium-sensitive muscle protein was in fact the first isolated high-affinity calcium target protein. It has now become just one member of a large family of proteins involved in diverse calcium-dependent cellular functions. Membrane-bound calcium transport systems have been detected in other intracellular membranes and also are essential constituents of the plasma membrane (39), where it is responsible for the cells' long-term calcium homeostasis. Because of the low calcium permeability of most cell membranes, calcium extrusion can evidently be accomplished by very few pump molecules in the cell surface membrane. This is in striking contrast to the densely packed calcium transport molecules in the sarcoplasmic reticulum membranes. The system has to eliminate large quantities of calcium over very short time intervals; thus it needs a large fraction of the muscle's metabolism.

The abundance of the sarcoplasmic reticulum membranes in many muscles and the exceptional properties of their fragments, which allow simultaneous monitoring of calcium transport and the related energy-yielding or energy-consuming reactions as well as structural changes, have made them highly preferred structures for analyzing energy transformation. In the United States, research activity on sarcoplasmic reticulum membranes was initiated by A. Martonosi's report (34) in which he introduced Millipore filtration techniques

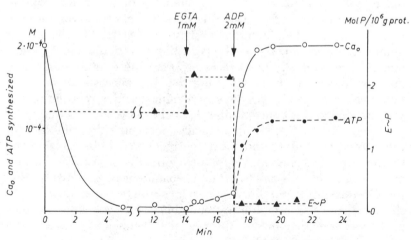

Figure 4. Reverse operation mode of sarcoplasmic reticulum calcium pump. Efflux of calcium initiated by addition of 2 mM ADP results in ATP formation. Release of 2 calcium ions gives rise to formation of 1 molecule of ATP. Phosphoprotein formed by incorporation of inorganic phosphate during calcium loading and calcium gradient largely disappear when ATP synthesis proceeds (32).

[198]

for measuring calcium-uptake kinetics, a more convenient procedure than our HgCl$_2$ method. In the meantime the boom of reports dealing with the sarcoplasmic calcium transport has slowly leveled off. It has provided us with the knowledge that the large affinity changes needed for calcium transport result from the mutual interaction of the membrane with calcium and phosphate originating from ATP when the pump runs forward and from inorganic phosphate when the pump runs backward (7, 26) (Fig. 4). However, the discovery period of new phenomena appears to be over. Hard and continuous work will be required to combine our knowledge of the function of the active transport systems with the events occurring in the transport molecule itself. After the enzyme has been crystallized (8) and its amino acid sequence unraveled (31), this problem can be approached.

CALCIUM
PUMP

BIBLIOGRAPHY

1. BENDALL, J. R. Further observations on a factor (the Marsh-factor) effecting relaxation of ATP shortened muscle fibre models and the effect of Ca and Mg upon it. *J. Physiol. Lond.* 121: 247–260, 1953.
2. BENDALL, J. R. The relaxing effect of myokinase on muscle fibres, its identity with the "Marsh"-factor. *Proc. R. Soc. Lond. B Biol. Sci.* 142: 409–426, 1954.
3. BERNE, R. M. Intracellular localization of skeletal muscle relaxing factor. *J. Biochem. Tokyo* 83: 364–368, 1962.
4. BRECHT, K. Über die Wirkung des Acetylcholins auf die Froschlunge, ihre Beeinflussung und ihre theoretischen Grundlagen. *Pfluegers Arch. Gesamte Physiol. Menschen Tiere* 246: 553–576, 1943.
5. CALDWELL, P. C., A. C. HODGKIN, R. P. KEYNES, and T. T. SHAW. The effect of injecting "energy-rich" phosphate compounds on the active transport of ions in the giant axons of *Loligo. J. Physiol. Lond.* 152: 561–590, 1960.
6. COSTANTIN, L. L., C. FRANZINI-ARMSTRONG, and R. J. PODOLSKY. Localization of calcium-accumulating structures in striated muscle fibers. *Science Wash. DC* 147: 158–160, 1965.
7. DEMEIS, L. The sarcoplasmic reticulum transport and energy transduction transport. In: *Life Sciences*, edited by E. E. Bittar. New York: Wiley, 1984, 1–163.
8. DUX, L., and A. MARTONOSI. Membrane crystals of Ca^{2+}-ATPase in sarcoplasmic reticulum of fast and slow skeletal and cardiac muscles. *Eur. J. Biochem.* 141: 43–49, 1984.
9. EBASHI, S. A granule-bound relaxation factor in skeletal muscle. *Arch. Biochem. Biophys.* 76: 410–413, 1958.
10. EBASHI, S., F. EBASHI, and A. KODAMA. Troponin as the Ca^{++}-receptor protein in the contractile system. *J. Biochem. Tokyo* 62: 137–138, 1967.
11. EBASHI, S., and F. LIPMANN. Adenosine triphosphate-linked concentration of calcium ions in a particulate fraction of rabbit muscle. *J. Cell Biol.* 14: 389–400, 1962.

12. FIEHN, W., and A. MIGALA. Calcium binding to sarcoplasmic membranes. *Eur. J. Biochem.* 20: 245–248, 1971.

13. GILBERT, D. L., and W. O. FENN. Calcium equilibrium in muscle. *J. Gen. Physiol.* 40: 393–408, 1957.

14. HASSELBACH, W. Elektronenmikroskopische Untersuchungen an Muskelmyofibrillen bei totaler und partieller Extraktion des L-Myosins. *Z. Naturforsch. Teil B. Anorg. Chem. Org. Chem. Biochem. Biophys. Biol.* 8b: 449–454, 1953.

15. HASSELBACH, W. Relaxation and the sarcotubular calcium pump. *Federation Proc.* 32: 909–912, 1964.

16. HASSELBACH, W. Relaxing factor and the relaxation of muscle. *Prog. Biophys. Mol. Biol.* 24: 167–222, 1964.

17. HASSELBACH, W., and M. MAKINOSE. Die Calciumpumpe der "Erschlaffungsgrana" des Muskels und ihre Abhängigkeit von der ATP-Spaltung. *Biochem. Z.* 333: 518–528, 1961.

18. HASSELBACH, W., and M. MAKINOSE. ATP and active transport. *Biochem. Biophys. Res. Commun.* 7: 132–136, 1962.

19. HASSELBACH, W., and M. MAKINOSE. Über den Mechanismus des Calciumtransportes durch die Membranen des sarkoplasmatischen Retikulums. *Biochem. Z.* 339: 94–111, 1963.

20. HASSELBACH, W., and G. SCHNEIDER. Der L-Myosin- und Aktingehalt des Kaninchenmuskels. *Biochem. Z.* 321: 461–476, 1951.

21. HASSELBACH, W., and H. H. WEBER. Der Einfluss des M-B-Faktors auf die Kontraktion des Fasermodells. *Biochim. Biophys. Acta* 11: 160–161, 1953.

22. HEILBRUNN, L. V., and F. J. WIERCINSKI. The action of various cations on muscle protoplasma. *J. Cell. Comp. Physiol.* 29: 15–32, 1947.

23. HILL, A. V. On the time required for diffusion and its relation processes in muscle. *Proc. R. Soc. Lond. B Biol. Sci.* 135: 446–453, 1948.

24. HOFFMAN, J. F., D. C. TOSTESON, and R. WHITTAM. Retention of potassium by human erythrocyte ghosts. *Nature Lond.* 185: 186–187, 1960.

25. HOFFMANN-BERLING, H. Die Beweglichkeit der Zellen. *Mikrokosmos* 46: 73–77, 1957.

26. INESI, G. Mechanism of calcium transport. *Annu. Rev. Physiol.* 47: 573–601, 1985.

27. KIELLEY, W. W., and O. MEYERHOF. Studies on adenosine-triphosphatase of muscle. *J. Biol. Chem.* 176: 591–601, 1948.

28. LORAND, L. Adenosine triphosphate-creatine transphosphorylase as relaxing factor of muscle. *Nature Lond.* 172: 1181–1183, 1953.

29. LULLIES, H. Reiz- und Erregungsbedingungen vegetativer Nerven. *Ergeb. Physiol. Biol. Chem. Exp. Pharmakol.* 38: 621–673, 1936.

30. LULLIES, H., and S. MEINERS. Die Wirkung von Acetylcholin und Methylacetylcholin auf die Blutgefässe des Frosches bei Durchströmung mit kontinuierlich ansteigenden Konzentrationen. *Pfluegers Arch. Gesamte Physiol. Menschen Tiere* 246: 525–542, 1943.

31. MACLENNAN, D. H., C. J. BRANDL, B. KORCZAK, and N. M. GREEN. Aminoacid sequence of a Ca^{2+}-Mg^{2+}-dependent ATPase from rabbit muscle sarcoplasmic reticulum, deduced from its complementary DNA sequence. *Nature Lond.* 316: 696–700, 1985.

32. MAKINOSE, M. Phosphoprotein formation during osmochemical energy conversion in the membrane of the sarcoplasmic reticulum. *FEBS Lett.*

25: 113–115, 1972.

33. MARSH, B. B. The effect of adenosinetriphosphatase on the fiber volume of muscle homogenate. *Biochim. Biophys. Acta* 9: 247–260, 1952.

34. MARTONOSI, A., and R. FERRETOS. Sarcoplasmic reticulum. II. Correlation between adenosine triphosphatase activity and Ca^{++} uptake. *J. Biol. Chem.* 239: 659–668, 1964.

35. NAGAI, T., M. MAKINOSE, and W. HASSELBACH. Der physiologische Erschlaffungsfaktor und die Muskelgrana. *Biochim. Biophys. Acta* 44: 334–340, 1960.

36. PORTER, K. R., and G. E. PALADE. Studies on the endoplasmatic reticulum. III. Its form and distribution in striated muscle cells. *J. Biophys. Biochem. Cytol.* 3: 269–300, 1957.

37. PORTZEHL, H. Bewirkt das System Phosphokreatin-Phosphokreatinkinase die Erschlaffung des lebendigen Muskels? *Biochim. Biophys. Acta* 24: 477–482, 1957.

38. ROSSI, C. S., and A. L. LEHNINGER. Stoichiometry of respiratory stimulation, accumulation of Ca^{++} and phosphate, and oxidative phosphorylation in rat liver mitochondria. *J. Biol. Chem.* 239: 3971–3980, 1964.

39. SCHATZMANN, H. J., and F. F. VINCENZI. Calcium movements across the membrane of human red cells. *J. Physiol. Lond.* 201: 369–395, 1969.

40. SKOU, J. C. Further investigations on a $(Mg^{++} + Na^+)$-activated adenosine triphosphatase possibly related to the active, linear transport of Na^+ and K^+ across the nerve membrane. *Biochim. Biophys. Acta* 42: 6–23, 1960.

41. SZENT-GYÖRGYI, A. Studies on muscle. *Acta Physiol. Scand.* 9, Suppl. 25: 7–116, 1945.

42. ULBRECHT, M. Beruht der Phophate-Austausch zwischen Adenosin-Triphosphat und Adenosin ^{32}P diphosphat in gereinigten Fibrillen- und Actomyosin-Präparaten auf einer Verunreinigung durch Muskelgrana? *Biochim. Biophys. Acta* 57: 438–454, 1962.

43. WEBER, A. The ultracentrifugal separation of L-myosin and actin in an actomyosin gel under the influence of ATP. *Biochim. Biophys. Acta* 19: 345–351, 1956.

44. WEBER, A., and R. HERZ. The binding of calcium to actomyosin systems in relation to their biological function. *J. Biol. Chem.* 238: 599–605, 1963.

45. WEBER, H. H. Die Muskeleiweisskörper und der Feinbau des Skelettmuskels. *Ergeb. Physiol. Biol. Chem. Exp. Pharmakol.* 36: 109–150, 1934.

46. WEBER, H. H., and H. PORTZEHL. The transference of the muscle energy in the contraction cycle. *Prog. Biophys. Biophys. Chem.* 4: 60–111, 1954.

47. WEISE, E. Untersuchung zur Frage der Verteilung des Calciums in Muskel. *Arch. Exp. Pathol. Pharmakol.* 176: 367–372, 1934.

VII

Anion Exchanges and Band 3 Protein

ASER ROTHSTEIN

THE anion-exchange system of the red blood cell, mediated by the transport protein band 3, has been the subject of intensive investigation for the past decade. Consequently numerous reviews have appeared, and no shortage of source material exists for those who are interested. The bibliography in this chapter includes thirty-six reviews or quasi-reviews (symposium articles) and there are others that are not cited. Clearly band 3–anion transport is an adequately if not overreviewed topic. Why then undertake yet another? Most reviews incorporate a catalogue of relevant research and/or an assembly of information supporting particular models, hypotheses, or points of view in various mixtures. The impression is usually given that a logical progression of ideas propelled by intelligent analysis provides increasing insight into particular biological mysteries. In real life, of course, research is not quite like that. There is a lot of stumbling and fumbling, unexpected results, and chance events that provide considerable impact. In this chapter I tell the story of how and why research on anion exchange came to be done, rather than to simply summarize what has been done and what conclusions can be reached. I attempt to place the development of knowledge of the anion-exchange system in some historical perspective and to describe events and people that substantially influenced the early directions of the research and its ultimate outcome. In doing so I present a highly personal view of the research developments and how they came about. I cannot claim to be a completely objective historian because I was a participant as well as an observer, so this effort is also something of a personal history. Undoubtedly I was unaware of certain influences that shaped the research effort; thus my history may be somewhat flawed and incomplete. I hope,

BAND 3–
ANION
EXCHANGE

however, that it is at least entertaining. Much of the paper is concerned with earlier events that in retrospect proved to be important. The mainstream of current research is also considered, but largely in the context of its historical origins. (For current status of the field see refs. 35, 38, 39, 56, 66.)

Although I examine research on anion exchange in broad perspective, it is of course difficult to be completely evenhanded. I undoubtedly have overemphasized the work of my colleagues and myself. Assigning priority and credit is a little slippery because ideas generated in one lab usually flowed rapidly to others. By the mid-1970s this subject achieved considerable popularity. Indeed for a period of time organizers of major meetings felt impelled to include in their programs at least one or two presentations on anion exchange. The situation resulted in rapid communication. Consequently parallel ideas often evolved in two or three laboratories. In this connection it is interesting that the network of interactions occurring during initial phases of band 3 research was the subject of a *Current Contents* article (27) on the nature of knowledge transfer.

The effort was truly international. I cannot mention everyone and I apologize for omissions. The laboratories that were involved early, maintained some continuing interest, and made major contributions included Z. I. Cabantchik (Canada and Israel), B. Deuticke (Germany), G. Guidotti (USA), R. B. Gunn (USA), P. A. Knauf (Canada and USA), R. Motais (France), H. Passow (Germany), A. Rothstein (Canada), K. F. Schnell (Germany), T. L. Steck (USA), M. J. A. Tanner (England), and J. O. Wieth and co-workers (Denmark). I estimate that at least 400 papers have now been published concerning anion transport and/or band 3; thus it is obvious that many other investigators, some trained in these laboratories, joined the chase and contributed extensively. Consequently I avoid the necessity of compiling an extended bibliography by referring in many cases to reviews rather than to the original literature.

The concept of Cl^--HCO_3^- exchange as an important physiological process is relatively ancient. Even when I was a student, more years ago than I like to remember, it was an old precept devised to explain the role of red blood cells in the transport of CO_2 from tissues to lungs. The exchange, first described over 100 years ago by H. Nasse for horse blood, preceded the theory of electrolytic dissociation by nine years. As first noted, I believe by Passow, it is somewhat startling to know that "*anion exchange was discovered before anions.*" (Nasse demonstrated the exchange of the corresponding acids.) Over a period of time the physiological importance of the phenomenon became evident, and it became the basis for explaining the "chloride shift" described in chapters on respiration in all physiology textbooks. Briefly, as metabolic CO_2 levels increase in plasma, the dissolved gas rapidly equilibrates across the red blood cell membrane. Within the cell, the CO_2 reaction with water catalyzed by carbonic anhydrase rapidly produces HCO_3^- and H^+. The HCO_3^- leaves the cell in exchange for Cl^- (the "chloride shift"), so that the flow of CO_2 into the cell continues. By this clever and simple mechanism, the carrying capacity of the blood for CO_2 is substantially enhanced. The solubility of CO_2 is relatively low, but large amounts can be carried in the form of HCO_3^-. The protons produced are largely buffered by hemoglobin, so pH changes are minimal. Hemoglobin thus serves double duty, with essential involvement in both O_2 and CO_2 transport. In the lungs, as the CO_2 tension in the blood is reduced, the whole process is reversed. The HCO_3^- moves into the cell in exchange for Cl^-, is protonated to H_2CO_3, and is converted to CO_2 and H_2O (catalyzed by carbonic anhydrase), with the CO_2 moving out of the cell and out of the body. The process has to be very rapid because the passage time of red blood cells in lung capillaries is less than one second (16, 100).

BAND 3— ANION EXCHANGE

Although the CO_2 cycle has been well understood for many years, the mechanism of Cl^--HCO_3^- exchange across the membrane was quite a mystery until relatively recently and in some respects remains one today. My own involvement in trying to unravel the mystery is also relatively recent, but as I look back it is clear that fortuitous earlier events had much to do with both my involvement and my contributions. I became specifically interested in anion transport in about 1969. At that time my associates and I were primarily interested in studying K^+ permeability of red blood cells through the use of mercurials and other sulfhydryl agents to modify the membrane transport, presumably by modification of membrane proteins (for review see ref. 76). I had been interested in cation permeability for a long time. In an unanticipated manner this interest was to lead to studies of anion permeability.

In retrospect the first important event was the appearance in my laboratory at the University of Rochester of a German postdoctoral fellow, Hermann Passow, who was to become a key figure in the development of anion-exchange knowledge. He was interested at the time in our use of chemical agents to probe membrane functions. In addition to publishing original papers Passow and I, together with the young English toxicologist Tom Clarkson, a postdoctoral fellow in my laboratory, undertook writing a review on the pharmacology of heavy metals (70). In it we expounded a number of ideas that represented our working hypothesis on the uses of impermeant or slowly penetrating chemical toxic agents to explore membrane function, to identify functional ligands, and to determine their location in the membrane. These ideas influenced my own and Passow's future research considerably.

When Passow's year in my laboratory came to an end, the review was not yet finished. I arranged to spend some time during the following year at the University of Hamburg (to which Passow had returned) to finish it. While there I was introduced by Passow to his mentor, Professor R. Mond, head of the Physiology Department. I had no idea at the time that Mond's seminal ideas concerning the anion-exchange system (59), amplified and extended by Passow in the years to come, were to have a great impact on my own future research. Passow developed a kinetic model that suggested a population of fixed positive charges in the membrane controlling both anion and cation permeabilities (62, 71). The model effectively accounted for much that was known at the time about relative rates of cation and anion permeation as a function of ion concentrations and pH. This concept stimulated research in a relatively quiescent field to follow two interconnected themes: a further exploration of transport kinetics and the application of chemical agents that could modulate anion transport. I deal first with the latter technology because it led in a few years to identification of band 3 as the transport protein.

EARLY EVENTS USING CHEMICAL AGENTS TO PROBE ANION TRANSPORT SYSTEM

While Passow was exploring the fixed-charge concept, I became advisor at the University of Rochester to Phillip Knauf, a biophysics PhD student who is now a professor at the University of Rochester. Knauf became familiar with the fixed-charge theory early in its evolution when Passow revisited Rochester. In fact he helped Passow with some of the calculations through access to our departmental computer. The following summer in Passow's laboratory in Germany, Knauf continued the computer-aided calculations. Knauf has reminded me that on his return to Rochester I, as his advisor, steered him away from studies on anion permeability *"because anions were not very interesting."* We all make mistakes. Fortunately my advice was soon to become irrelevant and ignored. A paper by H. C. Berg, J. M. Diamond, and P. S. Marfey (3) published in 1965 demonstrated that the amino-reactive reagent, fluoro-2,4-dinitrobenzene (FDNB) could substantially increase cation permeability. Passow and Knauf independently found that FDNB also inhibited anion fluxes. The reciprocal effects of FDNB on cation and anion permeability provided considerable support for the fixed-charge theory, and Passow used this agent effectively (62, 71). For his PhD thesis Knauf continued studying the effects of sulfhydryl and amino-reactive agents on cation and anion permeability. Some of his (and Passow's) results were not entirely consistent with Passow's initial concept of a single population of positively charged ligands controlling both cation and anion permeabilities. It appeared that the cation-controlling sites might be deeper within the membrane and therefore less accessible to slowly penetrating modifying agents such as p-chloromercuribenzenesulfonate. Knauf began looking for amino-reactive agents that had a limited permeability. One day he came across a paper describing a nonpermeating fluorescent compound proposed for use as a marker for plasma membrane fractions during their preparation and separation (57). It was 4-acetamido-4'-isothiocyanatostilbene-2,2'-disulfonic acid (SITS), a double-ring disulfonate (disulfonic stilbene). It had an isothiocyano group that should react covalently with membrane amino groups. Fortunately SITS was available. Knauf tried it. Great excitement! It turned out to be a very potent, impermeant inhibitor of anion transport with no effect on K^+ permeability—the first agent to clearly differentiate anion and cation permeability. We were therefore able to determine that different populations of sites control the cation and anion pathways and that the anion-controlling sites are more superficially located (52). More important, we now had in our hands a highly specific, impermeant, anion-transport inhibitor; the disulfonic stilbene, SITS and related compounds were destined to become widely used to study anion transport in many cell types. They turned the emphasis of my laboratory away from cation permeability toward anion permeability and soon provided the means for definitive identification of the anion-transport protein.

With this goal in mind Knauf initiated experiments using the fluorescence enhancement associated with SITS binding to try quantifying its interaction with membrane components. This approach proved to be technically difficult and SITS-binding studies could not be done effectively until radioactive forms were synthesized.

Fortunately for me, Knauf's departure for a postdoctoral fellowship coincided with the appearance in 1970 of another student, Ioav Cabantchik from Israel. I suggested that he try to prepare radioactive SITS to covalently label and identify the anion transporter. In preliminary studies he found SITS to be somewhat unsatisfactory because its covalent reaction was slow and incomplete. For his PhD thesis he therefore set out to test the inhibitory potency of a series of disulfonic stilbene derivatives. A few he was able to buy, but many he made himself. Fortunately he was an accomplished organic chemist, which I was not. In all, he tested 16 derivatives, of which about half were capable of forming covalent bonds (11). They all inhibited, but potency varied over a thousandfold. From the data published in 1972, function-structure relationships that provided information concerning the binding site were evident. The substantial fluorescence enhancement that accompanied stilbene binding was quantified. This characteristic became important many years later in energy-transfer experiments related to the stilbene-binding site (56). One of the most important aspects of Cabantchik's thesis turned out to be the synthesis of a new disulfonic stilbene, 4,4'-diisothiocyanostilbene-2,2'-disulfonic acid (DIDS), which proved to be the most potent analogue tested. This compound could react rapidly and quantitatively to form covalent bonds. It turned out to be the key not only to identification of band 3 as the transport protein but also to further understanding of the nature of the transport reaction, to the arrangement of the transport protein in the bilayer, to quantification of transport sites, and to other good things. It was to become a standard reagent for study anion transport. Had we anticipated all these developments, we might have patented DIDS and made some pocket money. Our immediate concern, however, was to place a radioactive label in the DIDS. Cabantchik succeeded in labeling the rings with ^{125}I, and with this probe he demonstrated for the first time that DIDS is highly localized in an abundant peptide of 95,000 daltons, which had been assigned the prosaic name band 3 just one year earlier in 1971. It was the third major stained peptide from the top of sodium dodecyl sulfate (SDS)–acrylamide gels after electrophoresis (19). The specific activity was not high, however, and we were also worried that the ^{125}I label might escape the DIDS and directly label other proteins. We therefore arranged to custom tritiate a batch of precursor 4,4'-diamino-2,2'-stilbenedisulfonic acid (DADS), which Cabantchik converted to a tritiated reduced form called dihydro-DIDS ([^{3}H]DIDS). We now had a specific impermeant labeled covalent inhibitor of the anion-exchange system (Fig. 1).

With [^{3}H]DIDS we were able to unequivocally demonstrate that almost all of the label, at inhibitory concentrations, was found covalently linked to band 3 [Fig. 2; (12)]. Passow et al. (68), also on the trail of the transporter, soon after published parallel observations pointing to band 3 by demonstrating that disulfonic stilbenes specifically prevented the binding to that peptide of labeled FDNB, and Ho and Guidotti (36) demonstrated the binding of another somewhat less specific inhibitor (a sulfanilic acid derivative) to band 3. The evidence that band 3 was the anion transporter was becoming firm and was ultimately to become an accepted fact. Additional support was derived from several other studies. First, Cabantchik, now a postdoctoral fellow in my lab, identified other covalent-anion–transport inhibitors that were highly localized in band 3, such as pyridoxal phosphate (8), and with Knauf, N-(4-azido-2-nitrophenyl)2-amino-

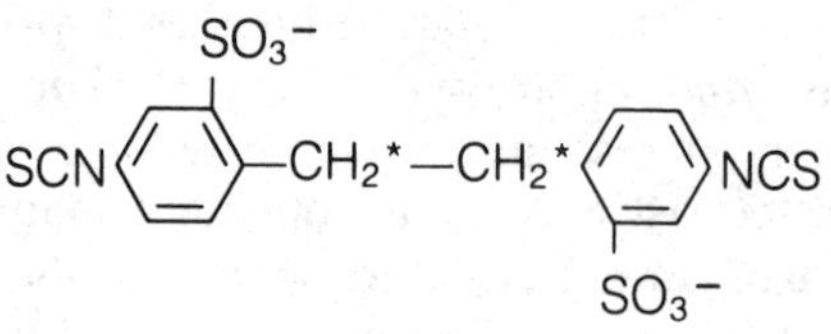

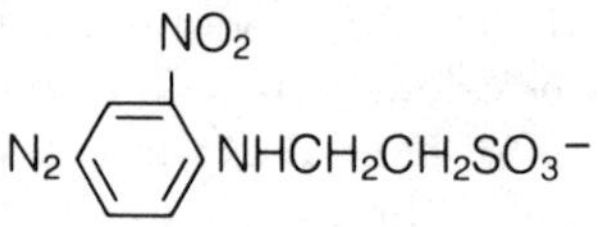

Figure 1. Two biomodal covalent probes used to study anion transport.

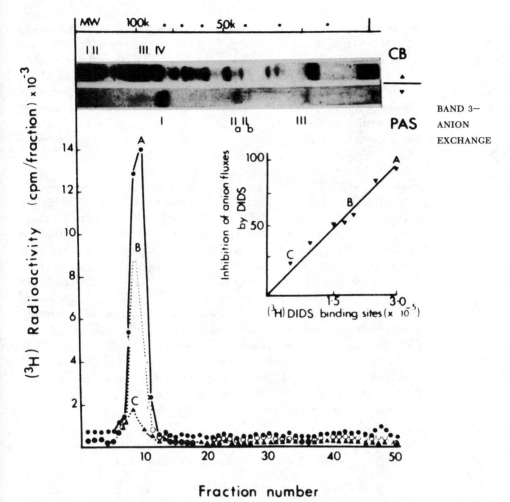

Figure 2. Labeling of band 3 by DIDS. [From Cabantchik et al. (12).]

ethanesulfonic acid (NAP-taurine) (9). The latter compound (see Fig. 1), first reported by J. V. Staros and F. M. Richards (92), became particularly useful as a photoaffinity probe. Second, we were able to reconstitute anion-transport activity using nonionic-detergent–solubilized fractions enriched in band 3 (81). As better purification procedures were developed, it became clear that the reconstituted activity was due to band 3, and Cabantchik demonstrated retention of anion transport in vesicles from which most other peptides were removed (negative purification) (for review see ref. 14). Third, Passow demonstrated (55) and we later confirmed (90) that the number of inhibitory sites (measured by DIDS binding) was equal to the number of monomers of band 3.

Initially there was some confusion concerning the stoichiometry of DIDS binding to band 3. In our first study using [³H]DIDS (12) we reported that complete inhibition was associated with the binding of 300,000 molecules of agent per cell. About a year later I received a call from Passow. He had also prepared [^{3}H]DIDS and was finding about 1 million binding sites per cell. By this time a value of about 1 million monomers of band 3 per cell had been established; thus Passow's value, indicating a 1:1 stoichiometry, made sense. Rather than publish a contradictory value, he visited us in Toronto (where I had moved in 1972) to compare results. This was very considerate of him. Knauf, who had joined me in Toronto, had also been concerned because he could not repeat the original numbers with a new batch of [^{3}H]DIDS. It was Knauf who finally solved the mystery. He contacted New England Nuclear and obtained their protocol for the tritiation used in preparing the world's first batch of [^{3}H]DIDS. From the data it became clear that they had used insufficient tritium to fully react with the amount of DIDS precursor that we had sent them. Consequently the preparation ended up as a mixture of unlabeled DIDS and [^{3}H]DIDS. Because the covalent reaction of DIDS is more rapid than that of the reduced [^{3}H]DIDS, the amount of labeling associated with full inhibition was substantially reduced. When we performed experiments with a new fully tritiated batch of [^{3}H]DIDS and also with labeled but unreduced DIDS, our results (90) were in full agreement with those of Passow et al. (55). The 1:1 stoichiometry provided additional evidence for band 3 as the transporter and also indicated that each monomer was a functional unit with one transport site.

Technical Innovations Allowing Identification and Study of Anion Transporter

I believe that many advances or breakthroughs in membrane research have resulted from technical rather than conceptual advances (79). The identification and characterization of the anion transporter is a good example. Synthesis of DIDS and identification of other inhibitory agents capable of covalent interactions was the first step (10, 45), but other recent technical developments were also essential. These were the use of detergents to solubilize red blood cell–membrane proteins and the development of SDS–acrylamide-gel electrophoresis to separate the membrane proteins into molecular-weight classes. If DIDS had been developed prior to 1971, it could not have been used to identify the transporter because there was no way to separate the DIDS-labeled peptide. In fact, we had, just a year earlier, tried dissolving the membrane in alcohols, labeling with less specific probes than DIDS, and separating by the then-available electrophoretic procedures. This attempt was frustrating and entirely abortive. The proteins could be dissolved but they couldn't be effectively separated (43). Thus, we were mentally prepared to label the transporter and to separate and identify it.

[210]

When an effective probe was identified (DIDS) and a separation procedure (SDS–acrylamide-gel electrophoresis) became available, we were ready to exploit these techniques.

The first application of SDS–acrylamide-gel electrophoresis to the red blood cell membrane was the now-classic study by G. Fairbanks, T. L. Steck, and D. F. H. Wallach (19). As already indicated the third major stained band later turned out to be the anion transporter. Cabantchik and I met Ted Steck for the first time in 1973 at a meeting in Montreal, where we reported some of our early studies using DIDS. Steck was to have considerable influence on our research direction because he made us aware of the growing body of biochemical knowledge concerning red blood cell membrane proteins including band 3. This information was compiled in his review (93) in 1974—the same year we published our first paper using [^{3}H]DIDS. I believe that we also influenced him because in his continuing research on the biochemistry of band 3 (a field in which he made major contributions) he was sufficiently converted from biochemistry to carry out occasional transport studies.

In his review Steck noted the following characteristics of band 3: 1) it is an intrinsic membrane protein, only extractable by detergents; 2) it can be cleaved in the intact cell by proteolytic enzymes; 3) it is glycosylated; 4) it is exposed at both sides of the bilayer, i.e., it "spans the membrane"; 5) it is the most abundant membrane peptide (~1 million copies per cell); and 6) it exists largely as a dimer in the membrane. Within the context of this background information, we began to try combining transport studies with research on membrane protein biochemistry. Passow's laboratory moved in the same direction. Other laboratories, particularly those of Steck (95), Tanner (97), and Guidotti (18) continued to examine primarily biochemical properties.

Additional techniques were important. To determine the arrangement of band 3 in the membrane it was important not only to use impermeant probes but also to have experimental access to both sides of the membrane. In earlier studies intact cells were compared with leaky ghosts (on the basis that any additional interactions of probes in leaky ghosts indicated binding to cytoplasmic sides). An improvement was the systematic development by Passow's laboratory of resealed ghost preparations (4) allowing incorporation of probes within the cytoplasmic compartment. Another important membrane preparation that became widely used was the vesicle system, "inside out" and "right-side out," developed by Steck (94). Finally, a variety of proteolytic enzymes were applied first to the outside of the intact cells and later to the cytoplasmic side of the membrane (using ghosts or vesicles) followed by SDS–acrylamide-gel electrophoresis to determine the sites of cleavage with respect to side of the membrane as well as with respect to the primary structure of band 3 (using covalent chemical probes as markers) (10,

18, 38, 39, 45, 56, 66, 67, 97). Correlations could then be made with the effects of specific cleavages on transport. This approach was initiated in 1971 when Passow (63) demonstrated that a protease (pronase) applied to the intact cell could result in inhibition of transport and W. W. Bender et al. (1) demonstrated that treatment of red blood cells with proteases resulted in cleavage of band 3.

What We Knew About Anion Transport

While new technologies were allowing and stimulating biochemical studies of the transporter, the more classic physiological and biophysical approaches continued to produce valuable information. Two important features of the transport system were clarified in the early 1970s. One related to its electrical and the other to its kinetic characteristics.

Anion permeation was known to be very rapid compared to cation permeation. In fact, Cl^- fluxes could not be accurately measured except by special rapid techniques not yet developed. For this reason the earlier studies used sulfate, which permeates $\sim 10^4$ times as slowly, as a substitute anion that passes through the membranes by the same mechanism as that of Cl^- (68, 71). To digress briefly, I might note that the use of SO_4^{2-} as a measuring anion rather than Cl^- probably led to the fixed-charge theory. The pH dependence of SO_4^{2-} transport shows a sharp diminution (~ 7.0) that can be explained by the theory. If Cl^- had been the test anion (pH plateau between ~ 7.0 and 11.0) the fixed-charge theory might not have emerged.

As early as 1959 D. C. Tosteson (98) pointed out that with rapid anion and slow cation fluxes, anion movement must behave as a one-for-one exchange in order to preserve electroneutrality. There were two possible mechanisms. 1) The anions, coupled via the membrane potential, might move independently, in which case the rapid flow of negatively charged anions would carry large currents and the electrical resistance of the membrane would have to be very low; or 2) the transport mechanism might involve an obligatory one-for-one anion exchange that was electrically silent, having no net-current flow. Because of technical limitations it took many years to sort out these options.

The earliest evidence came from observations with cation ionophores such as valinomycin or gramicidin. If anion permeability were conductive, the addition of valinomycin to substantially increase the K^+ permeability should result in a massive outflow of KCl (and rapid cell shrinkage) driven by the large outward K^+ gradient. Surprisingly the ionophores did not have the expected effects (for discussions see refs. 10, 45, 46). By implication, conductive-anion fluxes could not account for the observed anion permeability. In 1972 Wieth (99) directly measured Cl^- fluxes at 0°C and concluded, by analogy with the liquid anion-exchange model of Sollner, that most of the flux must represent an electroneutral exchange.

Conductive and electrical permeation can also be distinguished from each other by determining their electrical parameters. Measurements with microelectrodes were impractical in human red blood cells, because of their small size. This deficiency was not present, however, in the case of the giant red blood cells of the amphibian *Amphiuma*, and Lassen et al. (54) succeeded in directly measuring the electrical properties of the membranes of these cells. The electrical resistance of the membrane turned out to be very high, a finding inconsistent with the large conductive-anion flows.

It was evident from the studies cited above that the large anion flux in red blood cells involved an electroneutral exchange with only a very small conductive component.

With clarification of the electrical characteristics of the anion permeability, the kinetics of transport were delineated in great detail by a number of investigators (for discussions, see refs. 10, 16, 31, 32, 34, 35, 38, 45, 66, 87, 100). To account for the effects of pH, R. B. Gunn (31, 32) proposed a titratable carrier model. M. Dalmark (16) noted saturation behavior for Cl^- transport, competition between halides, and the presence of self-inhibition of higher anion concentrations, accounting for the latter phenomenon by proposing a modifier site that modulated the transport rate. The specificity was broad so that many organic anions could be transported and could act as competitive inhibitors. A variety of other agents could also act as inhibitors (10, 17, 45, 60). Mechanisms, based on the classic concept of carriers, were consistent with the information available at that time. [See our simplified model in Fig. 3; (10)]. In this scheme the anion must bind to the transport site, accounting for the saturation behavior. It is then translocated across the bilayer by a seemingly magical process and is released. The unloaded carrier is not permit-

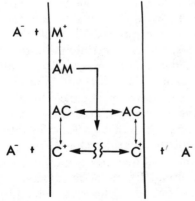

Figure 3. An early carrier-type model for anion transport. [From Cabantchik et al. (10).]

ted to return; therefore the process can only continue when an internal anion binds to the site at the internal surface, allowing the anion's return to the outside. Thus the system is obliged to exchange anions (anions move alternatively in and out). This mechanism acquired the colorful name *Ping-Pong* (a term used by chemical kineticists). Another possibility is a mechanism whereby anion binding sites are exposed simultaneously at both sides. Only after both sites are loaded with anions would translocation occur. This mechanism is called *simultaneous*. Much research effort would be expended to determine whether a Ping-Pong or simultaneous mechanism was involved in anion exchange. Many models would be proposed. The history of transport studies is built on ever more complicated kinetic models, and anion transport is no exception.

Transport Models Using Bimodal Probes That Bind to Band 3

By the mid 1970s it seemed only a matter of time until we would really understand how the anion exchange system worked. As noted previously, a great deal was known about transport behavior; we also knew that band 3 was the mediating protein. It did not turn out to be so easy. Although much data was generated, our understanding of the molecular nature of the transport developed slowly. The first necessity was to try to rationalize transport behavior with the structure and properties of band 3. How could this be done? Studies of transport reactions require the measurement of fluxes (unidirectional and net) and potentials while manipulating parameters such as concentrations and gradients of various ions and inhibitors; the data is then subjected to kinetic analysis. An intact membrane is essential. The characterization of band 3, on the other hand, involves the application of impermeant covalent probes and proteolytic enzymes to the two sides of the membrane, dissolution of the membrane in detergent, and separation of protein fragments (which are then subjected to further proteolytic or chemical cleavages). The primary links between these two disparate sets of information have been agents that act as bimodel probes. These agents can, under appropriate circumstances, act as reversible inhibitors of transport, with sites of action defined by kinetic analysis. Under other circumstances they can react covalently at the sites of inhibition and can therefore be used to link the inhibition to particular sites located in the primary and tertiary structure of the peptide (for review see refs. 10, 38, 39, 45, 56, 64, 66, 67, 69, 78, 84). Thus functionality can be attributed to particular parts of the peptide. One of the most useful bimodal probes is DIDS—the disulfonic stilbene used to first identify band 3 as the transport protein. Its initial reaction is rapid, specific, and reversible, but once bound, a slower covalent reaction locks it in place. Thus it is an affinity probe; its high affinity for the transport site is the determining factor in the specificity of its covalent interaction. In practical terms, at low temperature it is a reversible inhibitor (covalent reaction is slow), useful for determination of inhibition kinetics; whereas at higher temperatures it reacts covalently, allowing studies of its binding site in chemical terms after isolation and cleavage of the peptide (10, 45). It is of considerable advantage that DIDS interacts quantitatively with a single externally exposed functional site in each monomer of band 3 (55, 90).

Y. Shami, in my laboratory, using nuclear magnetic resonance (NMR) technique with a Cl^- probe, first established that the DIDS-binding site was also a Cl^--binding site (88), and also established that reversibly bound DIDS behaved as a competitive inhibitor with respect to Cl^- (89). Thus we could conclude that DIDS binds to the transport site. Competitive inhibition has also been noted with noncovalent disulfonic stilbenes (20, 24, 39, 56).

Another valuable bimodel affinity probe is NAP-taurine (see Fig. 1). This anionic compound contains an azido group that can be activated by light (92). We demonstrated that it was a specific bimodal inhibitor of anion transport (9) and later, defined the nature of its inhibitory effects and the sites in band 3 to which it was bound (see refs. 10, 45, 78, 85). In the dark, NAP-taurine interacts reversibly as a slow substrate and inhibitor from either side of the membrane. By kinetic analysis it behaves as though it competes with Cl^- for binding to the transport site (competitive inhibition) from the cytoplasmic side but as though it binds to a different site of higher affinity from the outside (noncompetitive inhibition). At the time, we thought that the outside site was the modifier site responsible for self-inhibition of Cl^-, reported by M. Dalmark (16). Later studies by Knauf and N. A. Mann (51) indicated that the latter site was located on the cytoplasmic side and that external NAP-taurine was therefore binding to an additional modifier site.

From the studies with DIDS and NAP-taurine, it was evident that the transport site is exposed to both the outside and inside (as in the kinetic models) but that other inhibitory sites are exposed only at the outside (see Fig. 3). Thus the system is asymmetric. It is also asymmetric with respect to DIDS, inasmuch as this agent is a potent inhibitor from the outside but has no effect from the inside. Other forms of asymmetry have also been reported (33, 49, 50, 51, 83). It is only in kinetic models that symmetrical arrangements are proposed. In the real world of proteins asymmetry is to be expected.

With the knowledge that the transport site involved specific protein ligands that bind inhibitors reversibly or covalently, it became necessary to redefine the translocation step in the carrier model (Fig. 3). Earlier concepts of transport had been greatly influenced by the necessity for lipid-insoluble ions to cross a lipid bilayer. Ions were presumed to interact with a membrane-bound carrier to form neutral lipid-soluble complexes that could diffuse across the bilayer to release the ion at the other side (mobile carrier model). These concepts were strongly reinforced by the discovery of ionophores, such as valinomycin, that allow rapid permeation of K^+ by such a mechanism. In the case of inorganic anions, such as Cl^- $(C_2H_5)_3Sn$ (triethyl-tin) can act as an effective mobile carrier (102). How then could a large protein of 90,000 daltons behave kinetically like a

mobile carrier? Certain possibilities were clearly improbable. It was unlikely that the protein was rotating around in the bilayer or even that a large segment of the protein was flipping back and forth across the bilayer. The reasoning was as follows: first, band 3 had been demonstrated to have a fixed arrangement with respect to the bilayer, with its carbohydrate always outside and a large N-terminal segment exposed to the cytoplasm; and second, the extremely rapid turnover was inconsistent with such possibilities; the energy barriers (both hydrophobic and hydrophilic) would be excessive. We had an opportunity to summarize our views on band 3 at a Federation meeting in 1976 (82). By that time, sufficient information was already available to assemble a somewhat simplistic general model. We proposed that 1) band 3 was a fixed structure across the bilayer; 2) it (rather than the lipid bilayer) formed the pathway for anion flow through the membrane; 3) free diffusion of anions was restricted by a local barrier in the protein pathway (explaining the high electrical resistance); 4) the barrier could only be crossed after the anion was bound to specific protein ligands, the transport site; 5) the unloaded ligand could not cross the barrier, opting by this assumption for a Ping-Pong model; 6) a conformational change was substituted for the translocation step of the carrier model (Fig. 3), allowing the transport site to shift from a topologically-out to a topologically-in position and vice versa, with the change occurring only if an anion was bound to the site; 7) a local conformational change occurs to accommodate the rapid turnover (postulating translocation was not across the total bilayer but rather across a localized-diffusion barrier within the peptide structure); and 8) another asymmetry exists inasmuch as the barrier was located at the outer face of the membrane, accommodating the finding that DIDS, an impermeant probe, could inhibit from the outside but had no effect from the inside. The proposed model is illustrated in Figure 4A (82), with the proteinaceous pathway through the bilayer represented by an ovoid structure with an aqueous core and the anion diffusion barrier represented by a barrier near the outside face of the membrane. In later versions the anion-binding site is represented as a gate that, when loaded with an anion, can swing across the barrier (conformational change resulting in translocation of anion) and the peptide as a hollow cylinder representing the aqueous core through which transport occurs; a hydrophobic surface is presented to the lipid continuum of the membrane (Fig. 4B). Other models have since been presented that are more sophisticated, more realistic, and perhaps more artistic.

None of the assumptions or conclusions were particularly astound-

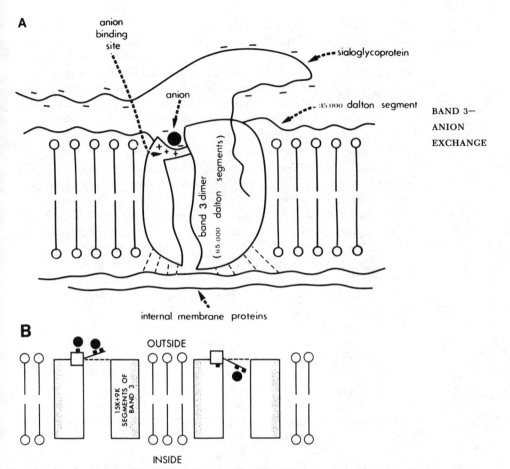

A

anion
binding
site

anion

sialoglycoprotein

:15,000 dalton segment

band 3 dimer
(65,000 dalton segments)

internal membrane proteins

B

OUTSIDE

15K+9K
SEGMENTS OF
BAND 3

INSIDE

Figure 4. Early models for band 3 involvement in anion transport. [Fig. 4*A* from Rothstein et al. (82); Fig. 4*B* from Cabantchik et al. (10).]

ing or even new or unique. They were based on information available at the time, some provided by ourselves but much by others. We opted, at that time, for a Ping-Pong model (we called it *sequential*) because we had preliminary data implying that transport sites could be recruited to one side while they disappeared at the other side. This is a characteristic of a Ping-Pong but not a simultaneous system. The concept is simple. If the Cl$^-$-binding site of band 3 is alternatively exposed to outside and inside by means of conformational change, then reaction at either side with an impermeant-covalent probe should result in locking sites at that side and ultimately in recruitment of all the sites to that side. Technically, however, the experiments are complicated because a recruiting probe must interact at one side, and then the disappearance of sites must be determined at the other side by reaction with a measuring probe. In our first attempt we loaded cells with pyridoxal 5-phosphate (PDP), a slow substrate. Sodium borohydride was then added, inducing covalent binding of the pyridoxal to band 3 by Schiff-base reaction (8). We found a reduction in DIDS binding to the outside (82). Passow, following similar reasoning, used an organic anion, 2-(aminophenyl)-6-methylbenzenethiazol-3-7'-disulfonic acid (APMB), as a recruiting agent (72). Both experiments were successful but suffered some deficiency inasmuch as binding of PDP and aspartame to the transport site had not been unequivocally established. Later S. Grinstein, L. McCulloch, and I (28) used DIDS and NAP-taurine applied to the outside and inside of the membrane, respectively, as recruiting and measuring probes. We were pleased to find that for each DIDS bound to the outside, approximately one binding site became unavailable for NAP-taurine at the cytoplasmic side. Because inhibition kinetics indicate that both probes, under noncovalent-binding conditions, appear to bind to the same site (transport site), this reciprocal relationship of probe to binding provided strong support for the Ping-Pong mode of transport in which the sites can be either in the topologically in or out position but not in both simultaneously.

The Ping-Pong model has been strongly supported by kinetic experiments. It is based on the model that the proportion of sites in the in or out conformations depends on the concentrations of ions at the two sides of the membrane, their affinities for the binding sites, and their relative rates of translocation. For example, after imposition of a large outward gradient of Cl^- or by placement of a fast anion (Cl^-) inside and a slow one (SO_4^{2-}) outside, most of the binding sites should be in the outside-facing conformation. Different distributions of the sites of the two sides of the membrane should lead to unique, testable, kinetic behaviors. In a series of studies Gunn and O. Fröhlich (33, 34, 35) and Knauf and co-workers (26, 49) have concluded that under equilibrium conditions (equal Cl^-) on the two sides of the membrane, the distribution of sites is asymmetric with most of the sites inward facing. These findings impose the requirement that the rate of translocation be considerably greater in the inward rather than outward direction so that the products of concentration time rates are equal in the two directions. The kinetic behavior of transport by manipulation of ion composition and gradients is entirely consistent with the Ping-Pong model and inconsistent with a simultaneous model. Recently M. L. Jennings (37) arranged the experimental conditions so that if the model were correct, all the sites would start outward facing and would transport for one cycle until all the sites were inward facing. The experiment worked. The amount of Cl^- transported in one cycle was equal to the number of monomers of band 3 (each containing one transport site). Recent studies using Cl^--NMR also supported this model (20).

BAND 3 IN CONDUCTIVE ANION FLUXES

The exchanger, as I have pointed out, is an especially abundant system with 10^6 elements per cell, each of which can turn over with great rapidity ($>10^5$/s). Thus the exceptionally rapid flux required to meet the requirements of the physiological function (Cl^--HCO_3^- exchange) can be accommodated. The electrically silent exchanges in the red blood cell are many orders of magnitude greater than those observed in other types of cells. A small fraction of the total anion flux ($\sim$1 in 10,000) is conductive. This is about the same magnitude as conductive-anion fluxes in other cell types. It can be quantitatively measured by cell shrinkage (using optical methods) after addition of valinomycin (46). Normally the loss of K^+ from cells down its large outward gradient is minimal because the K^+ permeability is so low. In the presence of valinomycin, however, the K^+ permeability increases substantially, leading to loss of KCl and osmotic shrinkage. With increasing valinomycin concentrations, the rate of cell shrinkage increases to a maximal value that represents a condition in which the valinomycin-induced K^+ permeability exceeds the conductive-anion permeability. Thus the latter process becomes rate limiting, and the conductive-anion permeabilities can be calculated from these maximal values. We found that almost all of the conductive SO_4^{2-} flux and a major fraction of the conductive Cl^- flux were inhibited by DIDS. Furthermore with a given proportion of band 3 binding sites occupied by DIDS, the inhibitions of exchange and conductive fluxes were in proportion, indicating that a major fraction of the conductive flux involved the DIDS-binding site (transport site) of band 3. Models for band 3 involvement in anion transport must therefore accommodate the presence of a small but finite fraction of conductive Cl^- flow.

An obvious explanation for the conductive flux via band 3 would be that an unloaded–anion binding site occasionally slipped from the outside to the inside position or vice versa (Fig. 4). However, this is not the case. Instead, occasionally an anion bound to the transport site slips across the diffusion barrier without the conformational shift of the transport site (25, 44, 47, 48).

TERTIARY STRUCTURE OF BAND 3

Parallel to the characterization of transport described previously, other investigations aimed at elucidating the structure and arrangement of band 3 in the bilayer were also proceeding. The strategy involved the use of impermeant chemical probes to mark ligands exposed at the outside or inside faces of the membrane, to then cleave band 3 by proteolytic enzymes applied to either face of the membrane, to separate the products by SDS–acrylamide-gel electrophoresis, and to identify those that are labeled. The proteolytic products can then be further split by chemical reactions. Even before band 3 was identified as the anion-exchange protein, Passow (63) had noted that application of pronase to red blood cells results in inhibition of anion transport. It was evident, therefore, that a protein exposed to the outside was involved. At about the same time, as I

[219]

have already noted, cleavage of band 3 in intact cells was reported by Bender et al. The recovered product was a peptide of 60,000 daltons (1), and M. S. Bretscher, using an impermeant probe, concluded that band 3 was exposed to both sides of the membrane (6).

When Cabantchik and I reported studies with DIDS in 1974 identifying band 3 as the transporter, we also applied proteolytic enzymes to DIDS-labeled cells (13). In control cells, as noted in Figure 2, almost all of the DIDS was located in intact band 3 (95,000 daltons), but in the proteolyzed cells most of the DIDS migrated with the 60,000 peptide product noted by Bender et al. (1). Because it bound a small amount of DIDS, we also detected a second poorly stained peptide of 35,000 daltons not present in control cells. These observations indicated that the proteolytic treatment of cells results in a single cleavage in band 3 and that the products (60,000- and 35,000-dalton segments) remain in the membrane. The latter segment is a broad, poorly stained band in the gels because it contains the carbohydrate of band 3. Furthermore we demonstrated that this particular cleavage does not result in inhibition of transport. Additional cleavages at the outside face of the membrane were demonstrated later by Jennings, Passow, and co-workers (39, 40, 42, 66) and by H. Matsuyama et al. (58) with papain, and by our laboratory with chymotrypsin (78, 85) and pepsin (74). These additional cleavages are inhibitory.

Application of enzymes to the cytoplasmic face of the membrane was technically more difficult. Both sides could be exposed in leaky ghosts, but effects on transport could not be evaluated. Passow and co-workers incorporated trypsin into resealed ghosts. They found that as a result of cleavages at the cytoplasmic face, a peptide of 55,000 daltons remains in the membrane that retained the DIDS-binding site (67). The transport measurements were complicated in this case because trypsinization resulted in vesiculation, but it was evident that transport activity still persisted. Steck approached proteolysis first using ghosts and then using a procedure his laboratory developed for making inside-out and right-side-out vesicles. He demonstrated that as a result of cleavages at the external and internal faces of the membrane, band 3 could be divided into 3 domains: the 35,000-dalton C-terminal segment and a 17,000-transmembrane segment (both membrane bound), and a 42,000-dalton N-terminal soluble segment that was on the cytoplasmic side (95). Grinstein and I followed up this information by demonstrating that the DIDS-binding site is in the 17,000-dalton segment and that transport is fully functional in vesicles containing no 42,000-dalton segment and in which cleavage to 17,000 daltons has occurred (29). The 42,000-dalton cytoplasmic segment is not required. It has other functions related to its binding sites for cytoskeletal proteins (2) and for glycolytic enzymes (93, 95) that will not be considered here.

The dissection of the intrinsic parts of band 3 has proceeded in several laboratories in addition to our own (84, 85), particularly those of Tanner (97), K. Drickamer (18), and Jennings (39, 40, 42), allowing construction of increasingly detailed maps. Various binding sites for covalent probes and cleavage sites of proteolytic enzymes have been located in the primary structure of the peptide. Such maps are imprecise because they depend on relatively inaccurate molecular-weight determinations based on migration of peptide fragments in SDS gels. Furthermore the intrinsic sections of band 3 that are functionally important are large (~55,000 daltons) whereas the number of markers are small. Even now, no markers have been found for large portions of the peptide.

The mapping studies have led to the very important concept that band 3 crosses the bilayer several times, as first proposed by Tanner and co-workers (97) and by Drickamer (18). More crossings of the bilayer have been identified: the number of proposed crossings climbed from three to five to seven (39, 40, 42, 84, 85), and recent sequence studies may raise the number to twelve (53).

In essence, the intrinsic part of band 3 appears to wind back and forth across the bilayer with crossing strands alternating with loops exposed outside of the bilayer on one side and then the other. The sites of proteolytic cleavage or binding of covalent probes are presumably located on the loops exposed on one side of the membrane or the other. This structure is substantiated by a recent study we carried out with pepsin, which appears to be one of the most powerful of the proteolytic enzymes tested (74). In ghosts, it appears to cleave all of the loops of peptide exposed on both sides of the membrane. With the loops cleaved, band 3 is reduced to a number of small peptide fragments, all approximately 4,000 daltons long. If hydrolysis is incomplete, peptides are found with sizes in approximate multiples of 4,000. Presumably the 4,000-dalton peptides that are resistant to pepsin are the membrane-crossing strands. A similar result has recently been reported with papain (21). Available sequence data provides strong support for such a structure based on alternate runs of relatively hydrophobic and hydrophilic residues (53).

Some time ago, Guidotti suggested that a stable peptide structure within a bilayer would consist of a hydrophobic sequence of amino acid residues in an α-helical conformation, about 3,200 daltons in length (~32 residues) (30). Glycophorin, an intrinsic glycoprotein of the red blood cell membrane with a single crossing strand, follows that pattern. Band 3, however, must be considerably more complicated; it crosses the bilayer many times and exists in the bilayer as a dimer. Furthermore, the intrinsic segments, which are largely the crossing strands, appeared to contain a proportion of hydrophilic residues. On this basis, we proposed several years ago that the crossing strands are in the form of an α-helical assembly with its outside face in contact with membrane-lipid components that are

[221]

hydrophobic in nature and with its hydrophilic groups internalized to provide an aqueous interior through which transport occurs (78, 83, 84). Sequence data indicate that some of the crossing segments involve primarily hydrophobic residues, but others include polar and charged residues, supporting the idea of a protein structure with an aqueous core that provides a transport pathway through the bilayer (53).

Steck was the first to assess the amount of helical structure in band 3 based on measurements of circular dichroism (93). The percent of α-helical content was not particularly high, but this is not surprising because less than half of the peptide may be within the bilayer. More recently, R. A. F. Reithmeier has reexamined the question using proteolytic enzymes to cleave away those parts of the peptide exposed on the outside of the bilayer. With maximum cleavage using pepsin (74), the α-helical content was increased substantially (to ~85%), indicating that the membrane-spanning segments are indeed largely α-helical (61).

Steck was also the first to directly demonstrate, by cross-linking techniques, that band 3 exists largely as a dimer (93, 95). This finding has been confirmed and extended by a variety of techniques (for review see ref. 38). Of importance in this regard is the finding that the dimer structure occurs between the intrinsic (membrane-bound) parts of band 3 (75), as well as between the cytoplasmic parts (originally cross-linked by Steck); this dimer structure may be important in a functional sense. Initially the functional role of the dimer was not apparent. Each monomer contains one DIDS-binding site and each site seems to be associated with an equal degree of inhibition. Furthermore the kinetics for Cl$^-$ transport (carefully measured) show no evidence of cooperativity. Thus each monomer appears to act independently as a functional entity. On the other hand, studies with inhibitors indicate some interactions between the two monomers in terms of binding constants or interaction rates (56). Thus for Cl$^-$, the transport sites are independent, but for large organic anions they are not: perhaps reflecting the distance between the binding sites on the two monomers of a single dimer. The situation has been potentially clarified by recent studies indicating that the high-affinity disulfonic-stilbene–binding site is only present in the dimer form (4). Thus dimer interaction must influence the conformation of each monomer so that its stilbene-binding site is evident. If the presence of this site is taken as evidence for functional state (a reasonable assumption because stilbenes are competitive inhibitors of transport), then the dimer form may be essential for each monomer to function in transport.

If each monomer of band 3 crosses the membrane as many as twelve times, then each dimer must cross as many as twenty-four times. As previously noted, dimer formation substantially changes the conformation of each monomer with respect to the DIDS-binding site. This suggests that the crossing strands of the two monomers may be substantially interdigitated. It is clear, in any case, that it will be a formidable task to sort out and understand the structure of the band 3 assembly as it exists in the bilayer.

BAND 3 AS A TRANSPORTER

For those of us who were attracted to band 3 studies because of an abiding curiosity about transport, the structural information about the peptide is interesting but not an end in itself. Our goal was to understand the molecular nature of the transport. We still have a long way to go, but progress has been considerable. For example, it was clear from our earliest studies with the disulfonic stilbenes that inhibition resulted from their binding to a rather complex site, involving several ligands (11). Cabantchik was particularly interested in structure-function relationships and he was able to elucidate many properties of the site (10). Analogues without sulfonic acid groups showed no interaction, and related compounds with one sulfonic acid group showed much lower affinities than those with two sulfonic acid groups. It seemed evident therefore that the exceptionally high affinity of the disulfonic stilbenes resulted at least in part from chelation of two positively charged groups in the peptide by the two negatively charged sulfonic acid groups. This conclusion was reinforced by the observed reduction in affinity when positively charged ligands such as amino groups are inserted in the disulfonic stilbene. Another group rapidly interacts covalently with one isothiocyano group of DIDS; it was identified later as a lysine anion group (73). Jennings and Passow (38, 42) demonstrated that a fourth group can react with the second isothiocyano group of DIDS. The latter reaction is slow but can be speeded up by alkaline conditions. The ligand involved is also a lysine amino group (38, 39, 65, 66). Consequently DIDS can, under appropriate conditions, act to cross-link two elements of the peptide. The elements in this case are at a substantial distance in the primary structure, supporting the concept that band 3 is folded within the membrane structure. In addition nucleophilic and hydrophobic interactions of the stilbenes with the peptide are evident. From this it is clear that the domain of the stilbene-binding site is complex, containing a number of ligands, several of which are positively charged. Complex structure-function relationships are also evident from inhibition studies of a variety of other organic anions (60).

The functional roles of the various ligands within the binding site are not entirely clear. When bound covalently, DIDS substantially reduces Cl^- binding to band 3 (88); when bound reversibly, it behaves as a competitive inhibitor of Cl^- transport (89), as does an analogue, 4,4'-dinitro-2-2'-stilbene-disulfonic acid (DNDS) (24). Presumably therefore at least one of the positive groups is a Cl^--binding site, defined kinetically as the transport site. DIDS also prevents the binding of NAP-taurine, inhibiting transport by a noncompetitive mechanism (45). Thus another group, the one involved in interaction with NAP-taurine, may be a close neighbor to the transport site, giving rise to steric hindrance between the two probes. On the other hand the two sites may be at a distance but linked through allosteric relationships. The functional roles of externally exposed lysyl residues have recently been explored by Jennings et al. (41).

The ligand with which DIDS rapidly interacts covalently is not essential for inhibition. This conclusion was evident from the early findings that analogues without covalent reacting groups are also potent inhibitors (11) and reversibly-bound DIDS was equally effective with covalently-bound DIDS as an inhibitor (55, 90). When we first realized that the covalent binding site for DIDS is not itself the transport site we were very disappointed. We had expended considerable effort in localizing a site in the primary structure of band 3 that was only a neighbor of the transport site and not necessarily even a close neighbor in the primary structure.

Information about the nature of the positively charged ligands involved in transport has also been derived from transport studies. The behavior toward pH suggests that lysyl and arginyl residues are both important (31, 32, 34, 100, 101). A role for the latter group is reinforced by the recent evidence that arginyl-specific reagents are transport inhibitors (100, 103). It can be concluded that the transport and inhibitory sites involve a cluster of lysyl and arginyl residues.

Although it is evident that clusters of positively charged ligands of band 3 are exposed to the outside, the surface of the red blood cell is negatively charged because of its high content of sialic acid. The positive clusters associated with the DIDS-binding sites of band 3 may be located within clefts or pits presumably under discontinuities in the negative field. Evidence for clefts or pits is provided by measurements of energy transfer between DIDS and sulfhydryl groups known to be located on the cytoplasmic surface of band 3. The calculated distance is significantly less than the width of the bilayer (for discussion see ref. 56). Location of the DIDS-binding sites with respect to the plane of the surface has been visualized by Cabantchik and myself (80) using electron microscopy after cross-linking ferritin to band 3 using DIDS. The covalently-linked ferritin particles are homogeneously scattered over the surface, as are intralipid particles (of which band 3 is the main constituent) seen by freeze-fracture electron microscopy. These too are homogeneously distributed.

An essential component of the transport models under discussion is a conformational change that allows the transport site and its associated anion to alternate in its topology from inside to outside exposure and vice versa (10, 33, 39, 45, 56, 65). No one has yet been able to directly tune in on this proposed oscillation; a variety of models have been proposed.

[224]

Chemical evidence of transport-related conformational changes is indirect based on measurable changes in accessibility of specific protein ligands to impermeant inhibitors applied to a particular side of the membrane. I have mentioned recruiting experiments in which the anion PDP or APMB bound to the cytoplasmic face of the membrane substantially reduces the binding of DIDS to the outward-facing sites of band 3 or vice versa (72, 82). I also pointed out that the interaction of DIDS at the outside face of band 3 reduces on about a one-to-one basis the sites available for interaction with NAP-taurine at the cytoplasmic face (28). More recently Knauf and co-workers (26, 49) have demonstrated that when conditions are such that the transport sites would be in the inward-facing conformation (by imposing ion gradients), the outside-facing transport and modifier sites are less accessible to interaction with DIDS and NAP-taurine. Anion gradients also modulate the binding affinity for reversible inhibitors (50). These observations, demonstrating transport-altered accessibility of specific binding ligands of the peptide, strongly support the conclusion that transport involves a conformational change that changes the relative location of specific peptide entities. The kinetic behavior of the transport is consistent with the same conclusion (26, 34, 35, 37, 49). Because the rate of turnover is so high and because anion exchange is a spontaneous reaction driven only by anion gradients, the conformational change presumably involves a small perturbation in peptide arrangement, involving no major energy barriers. It is presumably a localized region of peptide that is rearranged. Because the kinetic evidence indicates that translocation of the unloaded site does not occur, it is reasonable to postulate that the conformational change can only occur after an anion binds to the transport site. It can therefore be assumed that neutralization of a positive charge by anion binding to the site reduces an energy barrier and allows equilibration of conformational states.

Interaction of DIDS or other disulfonic stilbenes with band 3 results in substantial conformational changes not necessarily related to transport (for discussion see refs. 10, 39, 45, 56, 65). Some of these reflect themselves at a distance or even across the bilayer at the cytoplasmic side of band 3 or in substantial changes in the responses of band 3 to temperature changes. The rate of conformational change is influenced by lipid-soluble substances (22). As previously discussed, DIDS is a large molecule that interacts with multiple ligands in the peptide, at least one of which appears to be an essential part of the transport site—the Cl^- binding site. The architecture of band 3 is complex, involving multiple folding of the peptide to form an assembly of bilayer-crossing strands and loops exposed to the outside and to the cytoplasm. The cross-linking experiments discussed previously (38, 42) indicated that the DIDS-binding site involves at least two of the exposed loops. It is not

[225]

surprising therefore that DIDS produces extensive changes that are measurable in distant parts of the peptide.

The general picture of band 3 that emerges is a peptide with two major domains, a cytoplasmic N-terminal segment of about 40,000 daltons that is not essential for transport and a membrane-bound segment of about 55,000 daltons that is essential. The latter segment winds back and forth across the bilayer up to twelve times for each monomer and up to twenty-four times for the dimer that is the predominant form in the bilayer. The crossing strands, largely α-helical, are presumed to form an assembly that provides the pathway for transport, with an aqueous core and with a hydrophobic exterior associated with the lipid core of the bilayer. It is clear that the peptide does not provide a continuous open-diffusion channel for anions but that passage requires binding to a transport site followed by a local conformational change. Various gate and pore models have been evolved to visualize the proceedings. Each monomer behaves independently in transport so the role of the dimer is not clear although, as noted, recent studies indicate that the stilbene-binding sites are only evident in the dimer form. One point is, however, quite clear. Band 3 functional structure is exceedingly complex and will require much more research before we understand the detailed molecular nature of anion transport.

DIRECTION FOR FUTURE RESEARCH

Research related to band 3 expanded rapidly over a ten-year period. My own interest has tapered off and I have become more involved in other areas of research (cell volume and pH regulation), but others have maintained a steady output of publications. A new exciting phase of band 3 research appears to be imminent, encompassing methods derived from molecular-biology approaches including sequence determinations, examination of cell-specific gene expression, and examination of the possible role of proteins similar to band 3 in anion-exchange function in other cell types.

It has been clear for a long time that sequence data would be important in order to understand the functional architecture of band 3. Early attempts at sequencing defined fragments were relatively unsuccessful and unreported because the sequenators and technology available at the time did not perform well with hydrophobic peptides. Partial sequences have been reported, but these constituted less than 10% of the peptide. However, breakthroughs are upon us. Mouse band 3 sequence has been reported (53) and cDNA clones have been isolated (15, 53). The sequence of human band 3 should be available soon. From the sequence data, it should be possible to locate all of the reported specific binding sites and proteolytic cleavage sites in the primary structure and to construct detailed maps of the geographic arrangement of band 3 in the bilayer. We can anticipate that the functional architecture of band 3 will soon be elucidated in much greater detail. We can have new hope that the molecular mechanism of transport will become apparent.

The availability of probes to identify the messenger and the gene and of antibodies to identify the peptide should allow examination of the genetic control of the peptide. It is tantalizing to read that the protein is expressed in the kidney and that its expression is cell specific (14). Cells, in general, possess anion exchangers that are inhibitable by DIDS. Will these transporters turn out to be band 3 homologues? The anion-exchange system of renal cells have been implicated in salt reabsorption and acid secretion. In many cells, anion exchange has been implicated in pH and cell volume regulation. In the latter function, control of volume during cell growth remains to be studied. It is clear that the anion-exchange function that is specialized for CO_2 transport in the red blood cell has important general functions in all other cells. Thus I forsee a serious and expanding effort to extend and amplify the findings related to band 3–anion exchange in red blood cells to anion exchange in all cells.

The question of physiological regulation of anion exchange is also open. It is interesting to note that anion-transport activity is increased in old red blood cells (104) and is decreased in ATP-depleted cells (7), suggesting that some form of metabolic regulation exists.

In Retrospect

The discovery that band 3 is the anion-transport protein attracted considerable research effort. It is an abundant peptide, easily available for biochemical studies. The red blood cell allows easy study of transport, uncomplicated by internal structures, and both sides of its membrane are accessible for the application of impermeant inhibitors and proteolytic enzymes. Anion movement across the membrane is uncomplicated by coupling to enzyme reactions or to potentials. The exchange process contrasts with ionic flows through channels (the domain of nerve physiologists and patch-clampers). It represents a prototype for the simplest type of carrier-mediated process and an opportunity to understand the functional structure of a transport protein.

For a time, progress was very rapid because many new investigators were attracted to studies of band 3 and anion transport. The cross fertilization between biochemists interested in membrane protein structure and physiologists and biophysicists interested in the nature of transport was particularly stimulating. At first there was little common ground because the technologies and research objectives were so different, but the two groups could hardly avoid each other and soon were appearing together in symposia. The connecting links were primarily covalent probes and proteolytic enzymes that could perturb the transport and also interact at specific locations in band 3.

After several hundred papers, many symposia, and numerous reviews, what has been accomplished? A skeptic might suggest that we have kept a certain number of scientists occupied, provided entertainment to those who attended band 3 symposia, and kept the journals full of band 3 papers so that a high rejection rate for other subjects could be maintained. From my perspective we have made considerable progress but not as much as I thought we would in the excitement of the early discoveries. My expectations were perhaps too high. For years we accepted the concept of carriers as hypothetical entities and maintained the attitude that if only we could get our hands on one, we would soon understand how it worked. When Cabanchik showed me the first gels demonstrating that band 3 was the predominant site of DIDS binding, I figuratively yelled, "*Eureka, we have found it.*" The mysteries will disappear. Alas that turned out to be a naive view. Identification of band 3, although an accomplishment, was only the first step that opened up a whole new set of questions related to membrane-protein biochemistry. It is now thirteen years later and despite hundreds of additional papers related to band 3 and anion transport, we still don't really know how the transport works!

We used to hide our ignorance by using the word *carrier*. Now we use the words *conformational change*. We have certainly come a lot closer to understanding but not the whole way. We have taken a large black box, the whole bilayer, and reduced it to a small black box, a local region in the band 3 structure. Thus in research we seem to be in the business of reducing the size of black boxes. At other times, I feel that we are peeling an endless onion, layer after layer. When will we reach the last layer?

I greatly appreciate the efforts of Dr. P. Knauf and Dr. S. Grinstein in reading and criticizing the manuscript. My contributions to band 3 research were supported by the Medical Research Council of Canada, Grant MT4665. This support is gratefully acknowledged.

BIBLIOGRAPHY

1. BENDER, W. W., H. GARAN, and H. C. BERG. Proteins of the human erthrocyte membrane as modified by pronase. *J. Mol. Biol.* 58: 783–797, 1971.
2. BENNETT, V., and P. J. STENBUCK. The membrane attachment protein for spectrin is associated with band 3 in human erthrocyte membranes. *Nature Lond.* 280: 468–473, 1979.
3. BERG, H. C., J. M. DIAMOND, and P. S. MARFEY. Erythrocyte membrane: chemical modification. *Science Wash. DC* 150: 64–66, 1965.
4. BODEMANN, H., and H. PASSOW. Factors controlling the resealing of the membrane of human erythrocyte ghosts after hypotonic hemolysis. *J. Membr. Biol.* 8: 1–26, 1972.

5. BOODHOO, A., and R. A. REITHMEIER. Characterization of matrix-bound band 3, the anion transport protein from human erythrocyte membranes. *J. Biol. Chem.* 259: 785–790, 1984.

6. BRETSCHER, M. S. Human erythrocyte membranes: specific labelling of surface proteins. *J. Mol. Biol.* 58: 775–781, 1971.

7. BURSAUX, E., M. HILLY, A. BLUZE, and C. POYART. Organic phosphates modulate anion self-exchange across the human erythrocyte membrane. *Biochim. Biophys. Acta* 777: 253–260, 1984.

8. CABANTCHIK, Z. I., M. BALSHIN, W. BREUER, and A. ROTHSTEIN. Pyridoxal phosphate. An anionic probe for protein amino groups exposed on the outer and inner surfaces of intact human red blood cells. *J. Biol. Chem.* 250: 5130–5136, 1975.

9. CABANTCHIK, Z. I., P. A. KNAUF, T. OSTWALD, H. MARKUS, L. DAVIDSON, W. BREUER, and A. ROTHSTEIN. The interaction of an anionic photoreactive probe with the anion transport system of the human red blood cell. *Biochim. Biophys. Acta* 455: 526–537, 1976.

10. CABANTCHIK, Z. I., P. A. KNAUF, and A. ROTHSTEIN. The anion transport system of the red blood cell. The role of membrane protein evaluated by the use of "probes." *Biochim. Biophys. Acta* 515: 239–302, 1978.

11. CABANTCHIK, Z. I., and A. ROTHSTEIN. The nature of the membrane sites controlling anion permeability of human red blood cells as determined by studies with disulfonic stilbene derivatives. *J. Membr. Biol.* 10: 311–330, 1972.

12. CABANTCHIK, Z. I., and A. ROTHSTEIN. Membrane proteins related to anion permeability of human red blood cells. I. Localization of disulfonic stilbene binding sites in proteins involved in permeation. *J. Membr. Biol.* 15: 207–226, 1974.

13. CABANTCHIK, Z. I., and A. ROTHSTEIN. Membrane proteins related to anion permeability of human red blood cells. II. Effects of proteolytic enzymes on disulfonic stilbene sites of surface proteins. *J. Membr. Biol.* 15: 227–248, 1974.

14. CABANTCHIK, Z. I., D. J. VOLSKY, H. GINSBURG, and A. LOYTER. Reconstitution of the erythrocyte anion transport system: in vitro and in vivo approaches. *Ann. NY Acad. Sci.* 341: 444–454, 1980.

15. COX, J. V., R. T. MOON, and E. LAZARIDES. Anion transporter: highly cell-type-specific expression of distinct polypeptides and transcripts in erythroid and non-erythroid cells. *J. Cell Biol.* 100: 1548–1577, 1985.

16. DALMARK, M. Chloride in the human erythrocyte: distribution and transport between cellular and extracellular fluids and structural features of the cell membrane. *Prog. Biophys. Mol. Biol.* 31: 145–164, 1976.

17. DEUTICKE, B. Properties and structural basis of simple diffusion pathways in the erythrocyte membrane. *Rev. Physiol. Biochem. Pharmacol.* 78: 1–97, 1977.

18. DRICKAMER, K. Arrangement of the red cell anion transport protein in the red cell membrane: investigation by chemical labelling methods. *Ann. NY Acad. Sci.* 341: 419–432, 1980.

19. FAIRBANKS, G., T. L. STECK, and D. F. H. WALLACH. Electrophoretic analysis of the major polypeptides of the human erythrocyte membrane. *Biochemistry* 10: 2606–2617, 1971.

20. FALKE, J. J., and S. I. CHAN. Evidence that anion transport by band 3 proceeds via a ping-pong mechanism involving a single transport site. A ^{35}Cl NMR study. *J. Biol. Chem.* 260: 9537–9544, 1985.

21. FALKE, J. J., K. J. KANES, and S. I. CHAN. The minimal structure containing the band 3 anion transport site. A ^{35}Cl NMR study. *J. Biol. Chem.* 260: 13294–13303, 1985.

22. FORMAN, S. A., A. S. VERKMAN, J. A. DIX, and A. K. SOLOMON. N-alkanols and halothane inhibit red cell anion transport and increase band 3 conformational rate change. *Biochemistry* 24: 4859–4866, 1985.

23. FORTES, P. A. Anion movements in red cells. In: *Membrane Transport in Red Cells*, edited by J. C. Ellory and V. L. Lew. New York: Academic, 1977, p. 175–195.

24. FRÖHLICH, O. The external anion binding site of the human erythrocyte anion transporter: DNDS binding and competition with chloride. *J. Membr. Biol.* 45: 111–123, 1982.

25. FRÖHLICH, O. Relative contributions of the slippage and tunneling mechanisms to anion net efflux from human erythrocytes. *J. Gen. Physiol.* 84: 877–893, 1984.

26. FURUYA, W., T. TARSHIS, F. Y. LAW, and P. A. KNAUF. Transmembrane effects of intracellular chloride on the inhibitory potency of extracellular H$_2$DIDS. Evidence for two conformations of the transport site of the human erythrocyte anion exchange protein. *J. Gen. Physiol.* 83: 657–681, 1984.

27. GARFIELD, E. ABCs of cluster mapping. II. Most-active fields in the physical sciences in 1978. *Curr. Contents* 41: 5–12, 1980.

28. GRINSTEIN, S., L. McCULLOCH, and A. ROTHSTEIN. Transmembrane effects of irreversible inhibitors of anion transport in red blood cells. Evidence for mobile transport sites. *J. Gen. Physiol.* 73: 493–514, 1979.

29. GRINSTEIN, S., S. SHIP, and A. ROTHSTEIN. Anion transport in relation to proteolytic dissection of band 3 protein. *Biochim. Biophys. Acta* 507: 294–304, 1978.

30. GUIDOTTI, G. The structure of intrinsic membrane proteins. *J. Supramol. Struct.* 7: 489–497, 1977.

31. GUNN, R. B. Considerations of the titratable carrier model for sulfate transport in human red blood cells. In: *Membrane Transport Processes*, edited by D. C. Tosteson, A. Yu, and R. L. Ovchinnikov. New York: Raven, 1978, p. 61–77.

32. GUNN, R. B. Transport of anions across red cell membranes. In: *Membrane Transport in Biology*, edited by G. Giebisch, D. C. Tosteson, and H. H. Ussing. Berlin: Springer-Verlag, 1979, vol. 2, p. 59–79.

33. GUNN, R. B., and O. FRÖHLICH. Asymmetry in the mechanism for anion exchange in human red cell membranes. Evidence for reciprocating sites that react with one transported ion at a time. *J. Gen. Physiol.* 74: 351–374, 1979.

34. GUNN, R. B., and O. FRÖHLICH. The kinetics of the titratable carrier for anion exchange in erythrocytes. *Ann. NY Acad. Sci.* 341: 384–393, 1980.

35. GUNN, R. B., and O. FRÖHLICH. Arguments in support of a single transport site on each anion transporter in human red cells. In: *Chloride Transport in Biological Membranes*, edited by J. A. Zaidunaisky. New York: Academic, 1982, p. 33–59.

36. Ho, M., and G. GUIDOTTI. A membrane protein from human erythrocytes involved in anion exchange. *J. Biol. Chem.* 250; 675–683, 1975.

37. JENNINGS, M. L. Stoichiometry of a half-turnover of band 3, the chloride transport protein of human erythrocytes. *J. Gen. Physiol.* 79: 169–185, 1982.

38. JENNINGS, M. L. Oligomeric structure and the anion transport function of human erythrocyte band 3 protein. *J. Membr. Biol.* 80: 105–117, 1984.

39. JENNINGS, M. L. Kinetics and mechanism of anion transport in red blood cells. *Annu. Rev. Physiol.* 47: 519–533, 1985.

40. JENNINGS, M. L., M. ADAMS-LACKEY, and G. H. DENNY. Peptides of human erythrocyte band 3 protein produced by extracellular papain cleavage. *J. Biol. Chem.* 259: 4652–4660, 1984.

41. JENNINGS, M. L., R. MONAGHAN, M. S. DOUGLAS, and J. S. MICKNISH. Functions of extracellular lysine residues in the human erythrocyte anion transport protein. *J. Gen. Physiol.* 86: 653–669, 1985.

42. JENNINGS, M. L., and H. PASSOW. Anion transport across the erythrocyte membrane, in situ proteolysis of band 3 protein, and cross-linking of proteolytic fragments by 4,4'-diisothiocyano dihydrostilbene-2,2'-disulfonate. *Biochim. Biophys. Acta* 554: 498–519, 1979.

43. JULIANO, R. L. The proteins of the erythrocyte membrane. *Biochim. Biophys. Acta* 300: 341–378, 1973.

44. KAPLAN, J. H., M. PRING, and H. PASSOW. Band-3 protein-mediated anion conductance of the red cell membrane. Slippage vs. ionic diffusion. *FEBS Lett.* 156: 175–179, 1983.

45. KNAUF, P. A. Erythrocyte anion exchange and the band 3 protein: transport kinetics and molecular structure. *Curr. Top. Membr. Transp.* 12: 251–363, 1979.

46. KNAUF, P. A., G. F. FUHRMANN, S. ROTHSTEIN, and A. ROTHSTEIN. The relationship between anion exchange and net anion flow across the human red blood cell membrane. *J. Gen. Physiol.* 69: 363–386, 1977.

47. KNAUF, P. A., and F. Y. LAW. Relationship of net anion flow to the anion exchange system. In: *Membrane Transport in Erythrocytes*, edited by V. V. Lassen, H. H. Ussing, and J. O. Wieth. Copenhagen: Munksgaard, 1980, p. 488–493. (Alfred Benzon Symp., no. 14.)

48. KNAUF, P. A., F. Y. LAW, and P. J. MARCHANT. Relationship of net chloride flow across the human erythrocyte membrane to the anion exchange mechanism. *J. Gen. Physiol.* 81: 95–126, 1983.

49. KNAUF, P. A., F. Y. LAW, T. TARSHIS, and W. FURUYA. Effects of the transport site conformation on the binding of external NAP-taurine to the human erythrocyte anion exchange system. Evidence for intrinsic asymmetry. *J. Gen. Physiol.* 83: 683–701, 1984.

50. KNAUF, P. A., and N. A. MANN. Use of niflumic acid to determine the nature of the asymmetry of human erythrocyte anion exchange system. *J. Gen. Physiol.* 83: 703–725, 1984.

51. KNAUF, P. A., and N. A. MANN. Location of the chloride self-inhibitory site of the human erythrocyte anion exchange system. *Am. J. Physiol.* 251 (*Cell Physiol.* 20): C1–C9, 1986.

52. KNAUF, P. A., and ROTHSTEIN, A. Chemical modification of membranes. I. Effects of sulfhydryl and amino reactive reagents on anion and cation permeability of the human red blood cell. *J. Gen. Physiol.* 58: 190–210, 1971.

53. Kopito, R. R., and H. F. Lodisch. Primary structure and transmembrane orientation of the murine anion exchange protein. *Nature Lond.* 316: 234–238, 1985.

54. Lassen, U. V., L. Pape, and B. Vestergarrd-Bogind. Chloride conductance of the *Amphiuma* red cell membrane. *J. Membr. Biol.* 39: 27–48, 1978.

55. Lepke, S., H. Fasold, M. Pring, and H. Passow. A study of the relationship between inhibition of anion exchange and binding to the red blood cell membrane of 4,4'-diisothiocyanostilbene-2,2'-disulfonic acid (DIDS) and of its dihydro derivative (H_2DIDS). *J. Membr. Biol.* 29: 147–177, 1976.

56. Macara, I. G., and L. C. Cantley. The structure and function of band 3. In: *Cell Membranes: Methods & Review*, edited by E. Elson, W. Frazier, and L. Glaser. New York: Plenum, 1983, vol 1., p. 41–87.

57. Maddy, H. A fluorescent label for the outer components of the erythrocyte membrane. *Biochim. Biophys. Acta* 88: 390–399, 1964.

58. Matsuyama, H., Y. Kawano, and N. Hamasaki. Anion transport activity in the human erythrocyte membrane modulated by proteolytic digestion of the 38,000 dalton fragment in band 3. *J. Biol. Chem.* 258: 15376–15381, 1983.

59. Mond, R. Umkehr der Anionenpermeabilitat der roten Blutkörperchen in eine elektive Durchlässigkeit für Kationen. *Pfluegers Arch. Gesamte Physiol. Menschen Tiere* 217: 618–630, 1927.

60. Motais, R., and J. L. Cousin. A structure activity study of some drugs acting as reversible inhibitors of chloride permeability in red cell membranes: influence of ring substituents. In: *Cell Membrane Receptors for Drugs and Hormones: A Multidisciplinary Approach*, edited by R. W. Straub and L. Bolis. New York: Raven, 1978, p. 219–225.

61. Oikawa, K., D. M. Lieberman, and R. A. F. Reithmeier. Conformation and stability of the anion transport protein of human erythrocyte membranes. *Biochemistry*, 24: 2843–2848, 1985.

62. Passow, H. Passive ion permeability of the erythrocyte membrane: an assessment of the scope and limitations of the fixed charge hypothesis. *Prog. Biophys. Mol. Biol.* 19: 423–467, 1969.

63. Passow, H. Effects of pronase on passive ion permeability of the human red blood cell. *J. Membr. Biol.* 6: 233–258, 1971.

64. Passow, H. The binding of 1-fluoro-2,4-dinitrobenzene and of certain stilbene-2,2'-disulfonic acids to anion permeability-controlling sites on the protein in band 3 of the red blood cell membrane. In: *Cell Membrane Receptors for Drugs and Hormones: A Multidisciplinary Approach*, edited by R. W. Straub and L. Bolis. New York: Raven, 1978, p. 203–218.

65. Passow, H. Anion-transport-related conformational changes of the band 3 protein in the red blood cell membrane. In: *Membranes and Transport*, edited by A. N. Martonosi. New York: Plenum, 1982, vol. 2, p. 451–460.

66. Passow, H. Molecular aspects of band 3 protein mediated anion transport across the red blood cell membrane. *Rev. Physiol. Biochem. Pharmacol.* 103: 61–203, 1986.

67. Passow, H., H. Fasold, S. Lepke, M. Pring, and B. Schuhmann. Chemical and enzymic modification of membrane proteins and anion trans-

port in human red blood cells. In: *Membrane Toxicity*, edited by M. W. Miller and A. E. Shamoo. New York: Plenum, 1977, p. 353–377.

68. PASSOW, H., H. FASOLD, L. ZAKI, B. SCHUHMANN, and S. LEPKE. Membrane proteins and anion exchange in human erythrocytes. In: *Biomembranes: Structure and Function*, edited by G. Gardos and I. Szasz. Amsterdam: North-Holland, 1975, p. 197–214.

69. PASSOW, H., L. KAMPMANN, H. FASOLD, M. JENNINGS, and S. LEPKE. Relations between function and molecular structure. In: *Membrane Transport in Erythrocytes*, edited by V. V. Lassen, H. H. Ussing, and J. O. Wieth. Copenhagen: Munksgaard, 1980, p. 354–367. (Alfred Benzon Symp., no. 14.)

70. PASSOW, H., A. ROTHSTEIN, and T. W. CLARKSON. The general pharmacology of the heavy metals. *Pharmacol. Rev.* 13: 185–224, 1961.

71. PASSOW, H., and P. G. WOOD. Current concepts of the mechanism of anion permeability. In: *Drugs and Transport Processes*, edited by B. A. Callingham. London: Macmillan, 1974, p. 149–171.

72. PASSOW, H., and L. ZAKI. Studies on the molecular mechanism of anion transport across the red blood cell membrane. In: *Molecular Specialization and Symmetry in Membrane Function*, edited by A. K. Solomon and M. Karnovsky. Cambridge, MA: Harvard Univ. Press, 1978, p. 229–250.

73. RAMJEESINGH, M., A. GAARN, and A. ROTHSTEIN. The amino acid conjugate formed by the interaction of the anion transport inhibitor 4,4′-diisothiocyano-2,2′-stilbenedisulfonic acid (DIDS) with band 3 protein from human red blood cell membranes. *Biochim. Biophys. Acta* 641: 173–182, 1981.

74. RAMJEESINGH, M., A. GAARN, and A. ROTHSTEIN. Pepsin cleavage of band 3 produces its membrane-crossing domains. *Biochim. Biophys. Acta* 769: 381–389, 1984.

75. REITHMEIER, R. A. F. Fragmentation of the band 3 polypeptide from human erythrocyte membranes. Size and detergent binding of the membrane-associated domain. *J. Biol. Chem.* 254: 3054–3060, 1979.

76. ROTHSTEIN, A. Sulfhydryl groups in cell membrane structure and function. *Curr. Top. Membr. Transp.* 1: 1–76, 1970.

77. ROTHSTEIN, A. The functional roles of band 3 protein of the red blood cell. In: *Molecular Specialization and Symmetry in Membrane Function*, edited by A. K. Solomon and M. Karnovsky. Cambridge, MA: Harvard Univ. Press, 1978, p. 128–159.

78. ROTHSTEIN, A. Functional structure of band 3, the anion transport protein of the red blood cells, as determined by proteolytic and chemical cleavages. In: *Membranes and Transport*, edited by A. N. Martonosi. New York: Plenum, vol. 2, 1982, p. 435–440.

79. ROTHSTEIN, A. Membrane mythology: technical versus conceptual developments in the progress of research. *Can. J. Biochem. Cell Biol.* 62: 1111–1120, 1984.

80. ROTHSTEIN, A., and Z. I. CABANTCHIK. Protein structures involved in the anion permeability of the red blood cell membrane. In: *Comparative Biochemistry and Physiology of Transport*, edited by L. Bolis, K. Bloch, S. E. Luria, and F. Lynen. Amsterdam: North-Holland, 1974, p. 354–362.

81. ROTHSTEIN, A., Z. I. CABANTCHIK, M. BALSHIN, and R. JULIANO. Enhance-

BAND 3—
ANION
EXCHANGE

ment of anion permeability in lecithin vesicles by hydrophobic proteins extracted from red blood cells. *Biochem. Biophys. Res. Commun.* 64: 144–150, 1975.

82. ROTHSTEIN, A., Z. I. CABANTCHIK, and P. KNAUF. Mechanisms of anion transport in red blood cells: role of membrane proteins. *Federation Proc.* 35: 3–10, 1976.

83. ROTHSTEIN, A., S. GRINSTEIN, S. SHIP, and P. A. KNAUF. Asymmetry of functional sites of the erythrocyte anion transport protein. *Trends Biochem. Sci.* 3: 126–128, 1978.

84. ROTHSTEIN, A., and M. RAMJEESINGH. The functional arrangement of the anion channel of red blood cells. *Ann. NY Acad. Sci.* 358: 1–12, 1980.

85. ROTHSTEIN, A., and M. RAMJEESINGH. The red cell band 3 protein: its role in anion transport. *Philos. Trans. R. Soc. Lond. B Biol. Sci.* 299: 497–507, 1982.

86. SCHNELL, K. F., W. ELBE, J. KÄSBAUER, and E. KAUFMANN. Electron spin resonance studies of the inorganic anion-transport system of the human red blood cell: binding of a disulfonatostilbene spin label (NDS-TEMPO) and inhibition of anion transport. *Biochim. Biophys. Acta* 732: 266–275, 1983.

87. SCHNELL, K. F., S. GERHARDT, and A. SCHÖPPE-FREDENBURG. Kinetic characteristics of the sulfate self-exchange in human red blood cells and red blood cell ghosts. *J. Membr. Biol.* 301: 319–350, 1977.

88. SHAMI, Y., J. A. CARVER, S. SHIP, and A. ROTHSTEIN. Inhibition of Cl^- binding to anion transport protein of the red blood cell by DIDS (4,4′-diisothiocyano-2,2′-stilbene disulfonic acid) measured by (^{35}Cl)NMR. *Biochem. Biophys. Res. Commun.* 76: 429–436, 1977.

89. SHAMI, Y., A. ROTHSTEIN, and P. A. KNAUF. Identification of the Cl^- transport site of human red blood cells by a kinetic analysis of the inhibitory effects of a chemical probe. *Biochim. Biophys. Acta* 508: 357–363, 1978.

90. SHIP, S., Y. SHAMI, W. BREUER, and A. ROTHSTEIN. Synthesis of tritiated 4,4′-diisothiocyano-2,2′-stilbene disulfonic acid ($[^3H]$DIDS) and its covalent reaction with sites related to anion transport in human red blood cells. *J. Membr. Biol.* 33: 311–324, 1977.

91. SOLOMON, A. K., B. CHASSON, J. A. DIX, M. F. LUKAVIC, M. R. TOON, and A. S. VERKMAN. The aqueous pore in the red cell membrane: band 3 as a channel for anions, cations, nonelectrolytes, and water. *Ann. NY Acad. Sci.* 414: 97–124, 1983.

92. STAROS, J. V., and F. M. RICHARDS. Photochemical labelling of the surface proteins of human erythrocytes. *Biochemistry* 13: 2720–2726, 1974.

93. STECK, T. L. The organization of proteins in the human red blood cell membrane. *J. Cell Biol.* 62: 1–19, 1974.

94. STECK, T. L. Preparation of impermeable inside-out vesicles from erythrocyte membranes. In: *Methods in Biology*, edited by E. D. Korn. New York: Plenum, 1974, vol. 2, p. 245–281.

95. STECK, T. L. The band 3 protein of the human red cell membrane: a review. *J. Supramol. Struct.* 8: 311–324, 1978.

96. TANNER, M. J. A. Isolation of integral membrane proteins and criteria for identifying carrier proteins. *Curr. Top. Membr. Transp.* 12: 279–325, 1979.

97. TANNER, M. J. A., D. G. WILLIAMS, and R. E. JENKINS. Structure of the

erythrocyte anion transport protein. *Ann. NY Acad. Sci.* 341: 455–464, 1980.

98. TOSTESON, D. C. Halide transport in red blood cells. *Acta Physiol. Scand.* 46: 19–41, 1959.

99. WIETH, J. O. The selective ion permeability of the red cell membrane. In: *Oxygen Affinity of Hemoglobin and Red Cell Acid-Base Status,* edited by M. Rorth and P. Astrup. Copenhagen: Munksgaard, 1972, p. 265–278.

100. WIETH, J. O., O. S. ANDERSON, J. BRAHM, P. J. BJERRUIN, and C. C. BORDERS, JR. Chloride-bicarbonate exchange in red blood cells: physiology of transport and chemical modification binding sites. *Philos. Trans. R. Soc. Lond. B Biol. Sci.* 299: 383–399, 1982.

101. WIETH, J. O., J. BRAHM, and J. FUNDER. Transport and interactions of anions and protons in the red blood cell membrane. *Ann. NY Acad. Sci.* 341: 394–418, 1980.

102. WIETH, J. O., and M. T. TOSTESON. Organotin-mediated exchange diffusion of anions in human red cells. *J. Gen. Physiol.* 73: 765–788, 1979.

103. ZAKI, L., and T. JULIEN. Anion transport in red blood cells and arginine-specific reagents. Interaction between the substrate binding site and the binding site of arginine-specific reagents. *Biochim. Biophys. Acta* 818: 325–332, 1985.

104. ZANNER, M. A., and W. R. GABY. Aged human erythrocytes exhibit increased anion exchange. *Biochim. Biophys. Acta* 818: 310–315, 1985.

BAND 3–
ANION
EXCHANGE

VIII

The Unfinished Story of Secondary Active Transport

ERICH HEINZ

I recall those five years in the fifties when I, feeling like a "provincial" as far as membrane transport was concerned, had the opportunity to do research in Boston, and I gratefully remember the eminent scientists with whom I collaborated and to whom I owe so much concerning my further career. In particular I think of H. N. Christensen, then of Tufts University Medical School, of A. K. Solomon and the late R. P. Durbin of Harvard Medical School, and others. This, however, is not the place to enlarge on personal thanksgivings. Rather, in this chapter I describe and, where possible, characterize the state and the development of the research in membrane transport I was witnessing during that time.

AMINO ACID TRANSPORT

Periods of scientific research are usually characterized by their discoveries and elucidations. Just as characteristic, probably, are their errors and controversies, especially if viewed in the light of present knowledge.

The decade of the 1950s was particularly important for the research on membrane transport. Not only was this field promoted by important findings and concepts, it can also be said that the field "came of age" in the sense that it gained more general recognition among the traditional physiological sciences, especially by biochemists. In this chapter I try to illustrate this development in the field of biological membrane transport through one particular topic: "secondary active transport," that is, the (active) cotransport of organic substrates such as amino acids with special inorganic cations such as Na^+. This kind of active transport, which is supposed to be energized by an electrochemical ion gradient (gradient hypothesis), was a true newcomer in membrane transport. I am not aware that anything like this had ever been suggested before, not even by the most prophetic of pioneers in the field. The roots of this concept of cotransport, particularly of sugars and neutral amino acids, can be traced back to the 1950s, although the full development took place over subsequent decades. Before turning to details, let me first try to "set the stage," that is, to briefly outline the general state of our knowledge about membrane transport in the early 1950s.

At that time biochemists were not very interested in membrane transport. I was once asked by a distinguished biochemist, "What is interesting in a process in which a molecule does not undergo any chemical change?" In view of the impressive picture of cellular metabolism that emerged at that time, of the harmonious complexity of intertwined biochemical reactions coordinated by the concerted action of a multitude of regulated enzymes, this attitude was understandable. It seems that biochemists then looked at a cell as something comparable to a concert hall in which an excellent orchestra is performing. Of course, a concert hall has walls and doors and needs doormen to let in the performers and ticketholders and to keep out disturbances. Who would consider the doormen as important or even as interesting and exciting as the performers? This might change if the doormen gained the power and determination to decide on their own which of the performers would be admitted to the hall. Then the quality of the concerts would be different from what it is now. Such a situation seems utterly absurd but isn't it similar to what the membrane physiologists want to make us believe, namely that the entry of the cell by a substrate is just as important and interesting as, or even more so than, the metabolic alterations it may subsequently undergo inside the cell?

Yet there is some truth in this idea. By now quite a few metabolic pathways are known or suspected to be controlled by the rate at which their substrates penetrate the cellular membranes. One of the first-known and most typical examples is the facilitated diffusion of glucose into the muscle cell. The catabolic rate of this substance appears to be primarily controlled by the rate at which the "doorman" insulin admits it to the cell. This discovery, made in the early 1950s by Levine and associates as reviewed by Levine (28), appears to mark a turning point in the recognition of membrane transport as a highly important function in cellular metabolism. After all, the metaphor of a concert hall, however well it may fit the way classic biochemists tended to look at the cell, is inadequate from the present vantage point. Unlike a concert hall, a cell is a self-containing organism, and its membrane does not merely serve as a boundary with entrance gates but controls most interactions of the cell with its environment. It not only protects the cell from its environment but it also enables the cell to exploit or even dominate that environment. We are no longer surprised that the plasma membrane is a very important and potent part of the cell, equal if not superior to other organelles. Underestimating its importance appears to have led to underestimating its complexity.

MEMBRANES AND TRANSPORT THIRTY-FIVE YEARS AGO

In the early 1950s it was not even generally accepted that a cellular membrane existed at all and, if it did, that it had any

[238]

significance as a barrier for solutes entering the cell. For instance some physical chemists seemed to compare the cell to a nonaqueous phase, like a drop of oil, so that permeation could be treated as a phase transition. Some biologists, on the other hand, regarded the cell as a little gel, formed by proteins or other peptide chains, with or without a membrane to hold it together. Solutes were taken up either by binding to specific sites or, especially in the case of ions, owing to a reduction of their activity coefficient in "structured" water within the gelatinous network. On the opposite side were biologists who considered the cell to be like a bag, formed by a tight membrane and containing a free solution that, except for a higher protein concentration, physicochemically resembled the extracellular fluid. This latter view, though surely too simplistic, may come closest to the cellular model on which many transport studies were based in the subsequent years.

The picture of the cell membrane, to the extent that it was accepted as a significant barrier, was still that of a porous lipid bilayer similar to the kind devised by J. F. Danielli and H. Davson in 1935 (6), who based their model on the findings of E. Gorter and F. Grendel in 1926 (13). Whereas lipophilic solutes apparently pass the membrane by dissolving in the lipid bilayer, hydrophilic solutes pass supposedly by only two different functional devices: "pores" and "carriers."

The pores were considered static rigid holes, like those of a Millipore filter, with finite calibers to be estimated from the permeability to molecules of distinct sizes (equivalent pore size) (38). The apparent specificity of these pores with respect to passing solutes was attributed to mechanical and electrical interactions between the passing solute and the pore wall (9, 31).

The carriers, by contrast, were assumed to be highly mobile parts of the membrane that could freely shuttle substrates back and forth across the lipid barrier, like a ferryboat across a river. Some low molecular weight molecules, such as acetylcholine, pyridoxal, and cholic acid had temporarily been considered candidates for this function.

It is interesting to compare these past views with our present ones. Many pores have become "channels" that have at least some mobile parts so that they can open and close. The carriers, on the other hand, have become "alternating gates" that differ from the channels only in that they are open only to one side of the membrane at a time. Obviously pores and carriers have meanwhile come much closer to each other: the pores have become less static and rigid, assuming carrier-like properties, and the carriers had to give up much of their mobility, assuming porelike properties. Possibly the two may come still closer to each other in the future, eventually fusing into the same thing, which, depending on the conditions, will display either carrier-like or channel-like properties.

I add in parentheses that the concept of alternating gates, as different from a freely mobile carrier, was already offered by C. S. Patlak as early as 1957 (33), but obviously its time had not come yet. At its first presentation, it was completely rejected by the transport world. It was aggravating that Patlak had chosen the fatal term *demon*, which staunch physical chemists mistook to indicate a covert attempt to overthrow the second law of thermodynamics.

Active Transport of Organic Substrates Emerges

The concept of active transport as far as electrolyte ions were concerned had already been developed and strongly advanced in the 1940s, mainly by the classic work of H. H. Ussing (42) and T. Teorell (39), based on the discovery of active Na^+ transport in the frog skin in 1935 by E. Huf (24). By contrast rather little was known at that time about active transport of organic substrates. Apparently the intracellular metabolism of these substances, widely considered the domain of biochemists, had attracted most of the attention.

One of the pioneers who introduced the concept of active transport into the biochemistry of metabolism, undiscouraged by the general disregard among biochemists, was H. N. Christensen. He found that Ehrlich cells suspended in a medium containing glycine or some other amino acid for one or two hours, presumably then in the steady state, apparently accumulated the amino acid as the cellular content of glycine relative to the cellular content of water greatly exceeded its concentration in the medium (concentrative uptake). The uptake was found to be saturable, subject to competitive inhibition by analogues, and sensitive to metabolic inhibition as well as to changes in the ionic environment (2). This was perhaps one of the most decisive findings in the field because it was to become the basis of numerous fruitful studies on the active transport of nutrient substrates. Christensen's early interpretation of concentrative uptake as active transport, though still far from being established at that time, is now generally considered correct.

Shortly thereafter E. Heinz found also that the initial rate of uptake displayed saturability and the various kinds of inhibitability and sensitivity to the ionic environment with parameters similar to those of the corresponding phenomena found by Christensen for the steady state (14). Hence the latter could be taken to reflect the uptake process.

In order to prove active transport, one would have to demonstrate that the amino acid taken up is not simply adsorbed to the cellular surface but absorbed by the cell and furthermore, that it is in the cell in free aqueous solution, at least to the extent that its activity in the cytoplasm exceeds its activity in the medium. An outside adsorption could be dismissed as extremely unlikely on the basis of several observations such as the sensitivity to metabolic inhibition, the swelling of the cells in the presence of excessive uptake of amino acids, and the failure to obtain concentrative uptake with broken cells. The definite proof that the intracellular amino acids were indeed free, however, remained a tough problem for many years to come and may still not be complete even today, according to some skeptics. For instance metabolic inhibitability does not argue as strongly against intracellular binding as it does against extracellular binding because the former might be endergonic. Meanwhile evidence against a major extent of intracellular binding has been supplied. There is the aforementioned swelling, which, however, could also have had a different cause, for instance a disturbance of cellular volume regulation. Furthermore, there is the preloading effect.

To exclude binding more conclusively, Heinz studied the influx of labeled glycine into preloaded cells that had been presaturated with unlabeled glycine. It was expected that the preloading solute, to the extent that it occupied the intracellular binding sites, would inhibit the subsequent binding of the labeled solute and hence slow down its uptake (transinhibition). The result came as a surprise: preloading not only failed to inhibit but rather markedly stimulated the uptake of the labeled solute (transstimulation). The failure of the preloading to slow down the subsequent uptake of label has been taken as strong evidence that the amino acid used for preloading had really entered the cell without being significantly bound inside (14). Beyond this, in interpreting the phenomenon of transstimulation, one may still make the somewhat farfetched objection that the binding of a labeled substrate to an enzyme also has been reported to be stimulated after presaturation of the enzyme with the same but unlabeled substrate (23), presumably owing to the reduction of the activation energy of this binding. Notwithstanding, the transstimulatory effect of preloading has given us some information on the mechanism of this type of membrane transport. It was already interpreted then as a kind of exchange caused by the differing mobilities of loaded and unloaded carrier, respectively (22). When transstimulation was subsequently tested with different amino acids, evidence was found that the mechanism of transstimulation was the same whether the cells had been preloaded by the same (isoexchange) or by a different amino acid (heteroexchange), provided that both were served by the same transport mechanism (22). These observations thus became part of the basis of what is now called *tracer coupling* or *coupling by antiport*, respectively.

Further evidence against binding was later supplied by S. Udenfriend (40) who showed that the fluorescence of an artificial derivative of tryptophan was identical with that in free, extracellular solution and distinctly different from that of tryptophan when attached to protein.

Despite some weaknesses in the argument, there is now little doubt that the amino acids concentrated by the Ehrlich cell are predominantly free. This is especially true because the proof to the contrary, namely that the amino acids are bound to a major extent, is inconclusive.

Active transport, whenever postulated in those days, was more or less explicitly assumed to be what we now call *primary*, that is, directly coupled to an endergonic chemical reaction such as ATP hydrolysis. The same principle applied to the active transport of amino acids into Ehrlich cells and more so because it was so readily inhibited by dinitrophenol and other uncouplers of oxidative phosphorylation. Whereas a direct transduction of energy from a chemical reaction to a solute flow was considered, though somewhat uneasily, there was less readiness to accept a direct coupling between two solute flows. It is interesting that coupling by antiport was recognized before coupling by symport; the first time for neutral solutes such as glycine in Erlich cells in the form of transstimulation, under conditions that electric effects could safely be excluded (14). This transstimulation is related to what was previously called *exchange diffusion*, which was discovered in 1947 by Ussing (41) for the exchange between labeled and unlabeled Na^+ in the membrane of muscle cells and applied by P. Mitchell (29) in 1954 to the exchange between phosphate and arsenate in *Staphylococcus fecalis* and by Heinz and Durbin (18) in 1957 to the self-exchange of Cl^- in the frog gastric mucosa.

In antiport the situation is simpler than in symport, because only one binding site need be involved for which the antiporting solutes compete on the *trans* side. In symport, by contrast, the interacting solutes must be bound to the mediator at two distinctly different sites (unless they are able to associate directly without a mediator, which is very unlikely at the low concentrations involved). It may be for this reason that symport was conceived later than antiport.

Gradient Hypothesis

As early as 1952, Christensen and co-workers (3) had already observed that the uptake of amino acids by Ehrlich cells depends on the presence of Na^+ in the medium and is associated with the loss of K^+ and a gain of Na^+. These observations, which were followed by analogous ones with sugars in the intestinal epithelial cells (4, 36), were at first difficult to interpret but would become the basis of the concept of Na^+-linked secondary active transport. In view of transstimulation, T. R. Riggs et al. (35) later hypothesized that the active transport of (neutral) amino acids was driven by an outward K^+ gradient. This hypothesis of what actually would amount to a K^+-amino acid antiport was widely rejected. Indeed, little evidence of such an antiport could be provided for the Ehrlich cell.

Soon afterward, based on the fundamental observations and interpretations by R. K. Crane (5), evidence was supplied by the laboratories of A. A. Eddy (8), S. G. Schultz (37), and others that at least part of the driving force for amino acids in Ehrlich cells and many other cells and tissues is an inward electrochemical potential gradient of Na$^+$ (gradient hypothesis) that functions in essentially the same way as in sugar transport. From then on, the active transport of either substrate and of other compounds studied later, based on the same principle, was studied as a unit.

Many Na$^+$-linked cotransport (symport) systems are now known, mostly for various neutral and anionic organic solutes (32) but rarely for electrolyte ions only. The only certain case is the Na$^+$-K$^+$-2Cl$^-$ cotransport system (12). The known countertransport (antiport) systems appear to promote the extrusion of certain electrolyte ions from the cell, such as H$^+$ (Na$^+$-H$^+$ antiport) and Ca^{2+} (Na$^+$-Ca^{2+} antiport). For the Na$^+$-linked transport of glutamate and analogues, a combination of Na$^+$-linked symport with K$^+$-linked antiport has been shown (1).

The gradient hypothesis became appealing to some, but others viewed this new concept with everything from skepticism to outright disbelief. Many laboratories, including our own, set out to disprove the gradient hypothesis in any possible way but without permanent success, while the gradient hypothesis, little by little, became widely accepted. Notwithstanding, there are still some important problems to be solved, among them problems concerning energetic adequacy and presumed molecular mechanism.

ENERGETIC ADEQUACY

One of the most serious arguments against the gradient hypothesis from the beginning was the seeming energetic inadequacy of the available driving force of the Na$^+$-gradient. To identify the available driving force we have to know two things: *1*) the actual driving force and *2*) the efficiency of coupling. The actual driving force or, more precisely, the electrochemical potential difference of Na$^+$, as derived from the overall Na$^+$ concentration in the cell and from the most creditable measurements of the electrical membrane potential, clearly falls short of the requirement for at least the highest accumulation ratios of glycine observed, even assuming 100% efficiency of coupling. However, it has been suspected that the actual driving force, derived as mentioned above, might be markedly underestimated and certain corrections of the underlying face values of Na$^+$-activity and electrical potential have been suggested.

Some proponents of the gradient hypothesis tried to solve this problem by implementing the outward K$^+$ gradient, thus postulating a combination of Na$^+$ symport and K$^+$ antiport. We now know that such a combination exists, for instance in glutamate transport in renal brush-border membranes (1) but probably does not in the systems for which it was originally postulated. It is still doubtful whether an outward K$^+$ gradient adds to the driving force in Ehrlich cells other than through a diffusion potential, which should be contained within the measured electrical membrane potential.

[243]

Expecting to deal the gradient hypothesis the coup de grâce, at least from the energetic point of view, we embarked on testing experimentally the efficiency of coupling between amino acid transport with the nonmetabolizable glycine analogue diaminobutyrate (DAB) and Na^+ flow (10). This efficiency has been defined as a function of the degree of coupling, in terms of the thermodynamics of irreversible processes (25). The result yielded an efficiency of only 8%–10%, which was grossly inadequate, even after the most favorable correction of the gradient. This should have finished the gradient hypothesis, at least its claim that the Na^+ gradient be the sole driving force. However, we soon became aware (fortunately before the criticism came from elsewhere) that our procedure, although formally correct, was inappropriate; it had not taken into account that a major part of Na^+ entry is not coupled to amino acid transport (19).

What was needed here was the "intrinsic efficiency," which relates the amino acid transport to the coupled Na^+ entry only. The revised evaluation of our experimental data gave, assuming 1:1 stoichiometry, an intrinsic efficiency of ~60%, which means that the electrochemical potential difference (PD) of Na^+ must be ~170% of the corresponding (inverse) difference of the amino acid, in order to be approximately adequate. In view of this rather favorable efficiency, the energetic adequacy of the Na^+ gradient no longer looks to be out of the question, provided that the previously mentioned estimate of the actual driving force is too low. One may therefore look at the corrections of the Na^+ gradient that have been suggested to provide the true electrochemical PD of Na^+ during active amino acid accumulation. The proposition that the cytoplasmic Na^+ activity is much lower than previously estimated, owing to sequestration or compartmentalization of Na^+ within the cellular nucleus or elsewhere, had been postulated by C. Pietrzyk and Heinz (34) on the basis of subcellular fractionation and, more recently, by W. D. Dawson and T. C. Smith (7) on the basis of the Na^+-microelectrode measurements. Both studies arrive at a cytoplasmic Na^+ activity that is sufficiently low to render the driving force adequate in connection with the previously mentioned efficiency. The sequestration has, however, not been confirmed by the electronic probe, according to which the Na^+ is rather equally distributed throughout the intracellular space. If the latter method were correctly monitoring the freely dissolved Na^+—the final answer does not seem to be available yet—the electrochemical potential gradient of Na^+ could be adequate only if the electrical potential were considerably higher than the widely accepted value of 25 mV for Ehrlich cells (27). There are perhaps some reasons to doubt this low value, at least to the extent that it is derived from the distribution of Cl^-. Chloride for a major part passes the membrane through the (electroneutral) Na^+-K^+-$2Cl^-$ cotransport system (12), so that the equilibrium Cl^- distribution ratio is bound to be significantly lower than would correspond to the electrical PD according to the Nernst equation. Also the PD directly measured by intracellular microelectrodes (27), apart from confirming the presumably underestimated value of 25 mV derived from the Cl^- distribution, may be too low. Immediately after the impalement the PD has been found to drop continuously, presumably because of damage to the membrane, the true PD value had to be estimated by linear extrapolation to time zero. It may be argued that the linearity of the decay might be preceeded by a short but steep exponential phase. This might result from the fact that the electrical PD of the undamaged cell membrane is mostly electrogenic in nature, that is, directly generated by the Na^+-K^+ pump (20). It is likely that at least this part drops much more rapidly than the rest of the PD and thus escapes analysis (17), provided that the electrogenic pump stopped after the impalement. Electrical PD values determined with lipophilic cations, for example tetraphenylphosphonium, are much higher than values obtained by microelectrode; but there is good reason to believe that the values are overestimated. It thus appears that the true PD is situated somewhere between the two extremes. There still appears to be considerable uncertainty as to its precise magnitude, especially during the time of maximal amino acid accumulation.

Another attempt to solve the problem is by postulating that besides the Na^+ gradient, ATP hydrolysis may be involved. What would amount to a combination of primary and secondary active transport in this system does not appear promising either, because by the same technique used for efficiency of coupling, a direct coupling between α-aminoisobutyric acid influx and ATP hydrolysis could be shown to be absent (11).

The final verification that the Na^+ gradient alone is energetically adequate for this transport thus hinges on the validity of at least one of two as yet uncertain assumptions: 1) either that the Na^+ activity in the cytoplasm is as low as is claimed, owing to intracellular sequestration or 2) that the electrical PD is, at least transiently, as high as postulated because of the electrogenic pump. Only one of these two assumptions has to be verified in order to affirm the energetic adequacy of the Na^+ gradient. To negate this adequacy, however, both have to be disproved.

MOLECULAR MECHANISM: AFFINITY VERSUS VELOCITY

Besides the uncertainty concerning energetic adequacy, there still is uncertainty concerning the molecular mechanism of secondary active transport. It has been shown that there are two fundamentally different types of mechanisms, each of which alone could fully account for coupling. They have been called "affinity-type" and "velocity-type" mechanisms, respectively (21). However, which of these two types is actually present, exclusively or predominantly, in a given system of cotransport? The pure affinity-type model implies

[245]

that binding of the one solute to the translocator more or less strongly increases the affinity of the other and vice versa, presumably by positive cooperativity, whereas the mobility of the translocator remains unaffected. The pure velocity-type model, by contrast, implies distinct differences between the mobilities of the various translocator species but no change in affinity. If the product of the mobility of the fully loaded translocator times that of the unloaded translocator exceeds the corresponding product of the two partially loaded translocator species energetic coupling would be warranted without any cooperativity effect. In the beginning, researchers in the field postulated merely one of the two types of coupling without considering the other. Meanwhile the two types have been clearly characterized and mathematically analyzed (21, 30) but few experimental tests have been attempted so far in any system of secondary active transport.

One may ask whether there are reliable kinetic criteria to distinguish between the two types experimentally. It is helpful to visualize the characteristic differences between the two types for the simplest (minimal) model in which the flow of one solute, a, is (energetically) coupled with that of a second solute, b, by cotransport (Fig. 1). It can be seen that active accumulation of a, for instance, can occur only if

$$r \cdot P_{ab} \cdot P_o > P_a \cdot P_b$$

where r is the cooperativity coefficient between a and b for binding to the respective translocator sites, P_{ab}, P_o, P_a, and P_b the mobilities of the four translocator species abx, x, ax, and bx, respectively. This condition holds for both affinity-type and velocity-type mechanisms. Clearly r has to exceed unity only for the former, whereas the latter merely requires that $P_{ab} \cdot P_o$ exceed $P_a \cdot P_b$ (16). By contrast, for active accumulation of a solute by countertransport (antiport), the reciprocal relation has to hold.

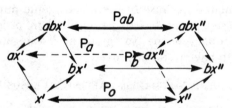

Figure 1. Simplest model of 1:1 cotransport (symport) between solutes a and b. Symmetry, quasi equilibrium (translocation rate-limiting), and electroneutrality are assumed. x, Translocator; ax and bx, translocator's (binary) complexes with a and b, respectively; abx, translocator's (ternary) complex with both a and b; P_o, P_a, P_b, and P_{ab}, corresponding mobilities. *Heavy arrows*, transport-effective pathways (P_{ab} and P_o); *thin arrows* and *broken arrows*, leakage pathways (P_a and P_b); ' and " refer to the two sides of the barrier.

Before looking for experimental tests, one may first ask whether there are any a priori arguments that favor one type of coupling over the other. For instance the two types may be compared with respect to presumable energetic efficacy and mechanistic plausibility. As to the energetic aspect, the efficacy of accumulation (ϵ_α) may be a suitable parameter to serve this purpose (15). It has been defined as

$$\epsilon_\alpha = \frac{\log \text{HR}}{\log(b'/b'')}$$

where HR is the Haldane ratio, b' and b'' are the activities of the driving solute b on the *cis* and *trans* side of the barrier, respectively. It is, of course, difficult to compare the two models under equivalent conditions, but intuitively one would hardly predict a major difference in ϵ_α, except if b' and b'' were far below or far above the respective Michaelis constants. In the first case, the velocity-type model may provide the higher efficacy, whereas in the second case the affinity-type model may do so. This apparent discrepancy may be the basis for an experimental test.

Because cooperativity effects are very familiar from enzymology, the affinity type seems the more plausible of the two models. The velocity type, by contrast, requires some seemingly unplausible assumptions, namely that at least one of the two different ligands, if bound alone, slows down the translocator but both ligands bound simultaneously reaccelerate it. The following working hypothesis has been invoked to explain the observed inhibition of Na^+-linked glucose transport across renal brush-border membranes by Ca^{2+} by velocity effects only: All translocator species move intrinsically fast, except for one of them, possibly the binary complex with b, xb, which is very slow or even immobile. For such a situation one would merely have to postulate a specific blocking device for the xb complex. Because the inhibitory effect of Ca^{2+} is associated with an increased uncoupled entry (inner leakage) of Na^+, it may occur through a removal of such a blocking device, here one for the Na^+-linked translocator species. This would be consistent with the observation that Ca^{2+} inhibition can be prevented by neomycin, which is known to inhibit phospholipase C, an enzyme normally activated by Ca^{2+}. It therefore looks as if Ca^{2+} exerts its inhibitory effect by activating this enzyme, which, we may speculate further, removes the (possibly phospholipase C–sensitive) blocking device. As a consequence Na^+ could freely enter through this pathway, in line with the experimental findings (J. T. Lin, E. Windhager, and E. Heinz, unpublished observations). This hypothesis is based solely on a velocity-type coupling but it does not exclude affinity effects. Characteristic kinetic differences between affinity and velocity effects become apparent in the parameters of the initial rate of transport.

In summary, there seems very little known about the contribution of affinity effects and velocity effects, respectively, in any system of secondary active transport. For some symport systems, there is evidence of cooperativity between symported solute species, suggesting that affinity effects are involved, but in these cases it cannot be said yet whether and to what extent velocity effects contribute as well (26). On the other hand, the inhibition of Na^+-linked glucose symport by Ca^{2+} in the renal brush-border membrane can apparently be explained solely and sufficiently on the basis of a velocity-type coupling. This strongly suggests that velocity effects are significantly, if not predominantly involved but does not exclude additional affinity effects.

Regarding antiport systems, the mere observation that the presence of the antiporting solute on the *trans* side affects the initial rate of the transport strongly suggests a velocity-type coupling. This appears to be true for the glutamate-K^+ antiport found in the renal brush-border membrane, whereas little is known in this respect for the other antiport systems for electrolyte ions (1).

BIBLIOGRAPHY

1. BURCKHARDT, G., R. KINNE, G. STANGE, and H. MURER. The effects of potassium and membrane potential on sodium-dependent glutamic acid uptake. *Biochim. Biophys. Acta* 599: 191–201, 1980.
2. CHRISTENSEN, H. N., and T. R. RIGGS. Concentrative uptake of amino acids by the Ehrlich mouse ascites carcinoma cell. *J. Biol. Chem.* 194: 57–68, 1952.
3. CHRISTENSEN, H. N., T. R. RIGGS, H. FISCHER, and I. M. PALATRUE. Amino acid concentration by a free cell neoplasm: relations among amino acids. *J. Biol. Chem.* 198: 1–16, 1952.
4. CRANE, R. K. Hypothesis for mechanism of intestinal active transport of sugars. *Federation Proc.* 21: 891–895, 1962.
5. CRANE, R. K. Na-dependent transport in the intestine and other animal tissues. *Federation Proc.* 24: 1000–1006, 1965.
6. DAVSON, H., and J. F. DANIELLI. *The Permeability of Natural Membranes* (2nd ed.). London: Cambridge Univ. Press, 1952.
7. DAWSON, W. D., and T. C. SMITH. Intracellular Na^+, K^+ and Cl^- in Ehrlich ascites tumor cells. *Biochim. Biophys. Acta* 866: 293–300, 1986.
8. EDDY, A. A. The effects of varying the cellular and extracellular concentrations of sodium and potassium ions on the uptake of glycine by mouse ascites-tumour cells in the presence and absence of sodium cyanide. *Biochem. J.* 108: 489–498, 1968.
9. EISENMANN, G. On the elementary atomic origin of equilibrium ionic specificity. In: *Membrane Transport and Metabolism*, edited by A. Kleinzeller and A. Kotzyk. New York: Academic, 1961, p. 608.
10. GECK, P., E. HEINZ, and B. PFEIFFER. The degree and the efficiency of coupling between the influxes of Na^+ and α-aminoisobutyrate in Ehrlich cells. *Biochim. Biophys. Acta* 288: 486–491, 1972.
11. GECK, P., E. HEINZ, and B. PFEIFFER. Evidence against direct coupling between amino acid transport and ATP hydrolysis. *Biochim. Biophys. Acta* 339: 419–425, 1974.
12. GECK, P., C. PIETRZYK, B. C. BURCKHARDT, B. PFEIFFER, and E. HEINZ.

Electrically silent cotransport of Na^+, K^+ and Cl^- in Ehrlich cells. *Biochim. Biophys. Acta* 600: 432–447, 1980.

13. GORTER, E., and F. GRENDEL. On bimolecular layers of lipids on the chromocytes of the blood. *J. Exp. Med.* 41: 439–443, 1925.

14. HEINZ, E. Kinetic studies on the influx of glycine-C^{14} into the Ehrlich mouse ascites carcinoma cell. *J. Biol. Chem.* 211: 781–790, 1954.

15. HEINZ, E. Coupling and energy transfer in active amino acid transport. In: *Current Topics in Membranes and Transport*, edited by F. Bronner and A. Kleinzeller. New York: Academic, 1974, vol. 5, p. 137–159. (Proc. Int. Conf. Hydrogen Ion Transport in Epithelia, Frankfurt am Main, FRG.)

16. HEINZ, E. *Mechanics* and ENERGETICS OF BIOLOGICAL TRANSPORT. New York: Springer, 1978.

17. HEINZ, E. Electrogenic and electrically silent proton pumps. In: *Hydrogen Ion Transport in Epithelia*, edited by I. Schulz. Amsterdam: Elsevier/North-Holland, 1980, p. 41–46.

18. HEINZ, E., and R. P. DURBIN. Studies of the chloride transport in the gastric mucosa of the frog. *J. Gen. Physiol.* 41: 101–117, 1957.

19. HEINZ, E., and P. GECK. The efficiency of energetic coupling between Na^+ flow and amino acid transport in Ehrlich cells—a revised assessment. *Biochim. Biophys. Acta* 339: 426–431, 1974.

20. HEINZ, E., P. GECK, and C. PIETRZYK. Driving forces of amino acid transport in animal cells. *Ann. NY Acad. Sci.* 264: 428–441, 1975.

21. HEINZ, E., P. GECK, and W. WILBRANDT. Coupling in secondary active transport. Activation of transport by co-transport and/or counter-transport with the fluxes of other solutes. *Biochim. Biophys. Acta* 255: 442–461, 1972.

22. HEINZ, E., and P. M. WALSH. Exchange diffusion, transport and intracellular level of glycine and related compounds. *J. Biol. Chem.* 233: 1488–1493, 1958.

23. HOLZER, H., and K. BEAUCAMP. Nachweis und Charakterisierung von Zwischenprodukten der Decarboxylierung und Oxydation von Pyruvat: "aktives Pyruvat" und "aktives Acetaldehyd." *Angew. Chem. Int. Ed. Eng.* 71: 776, 1959.

24. HUF, E. Versuche über den Zusammenhang zwischen Stoffwechsel, Potentialwirkung und Funktion der Froschhaut. *Pfluegers Arch. Gesamte Physiol. Menschen Tiere* 235: 655–673, 1935.

25. KEDEM, O., and S. R. CAPLAN. Degree of coupling and its relation to efficiency of energy conversion. *Faraday Soc. Trans.* 61: 1897–1911, 1965.

26. KINNE, R., and E. HEINZ. Role of K^+ in cotransport systems. *Curr. Top. Membr. Transp.* 28: 73–85, 1987.

27. LASSEN, U. V., M. T. NIELSEN, L. PAPE, and L. O. SIMONSEN. The membrane potential of Ehrlich ascites tumor cells, microelectrode measurements and their critical evaluation. *J. Membr. Biol.* 6: 269–288, 1971.

28. LEVINE, R. Some mechanisms for hormone effects on substrate transport. In: *Metabolic Transport*, edited by L. E. Hokin. New York: Academic, 1972, p. 627–639.

29. MITCHELL, P. Transport of phosphate through an osmotic barrier. *Soc. Exp. Biol. Semin. Ser.* 8: 254–261, 1954.

30. MITCHELL, P. Translocations through natural membranes. *Adv. Enzymol. Relat. Areas Mol. Biol.* 29: 33–87, 1967.

AMINO ACID
TRANSPORT

31. MULLINS, L. J. The penetration of some cations into muscle. *J. Gen. Physiol.* 42: 817–819, 1959.

32. MURER, H., and R. KINNE. The use of isolated membrane vesicles to study epithelial transport processes. *J. Membr. Biol.* 55: 81–95, 1980.

33. PATLAK, C. S. Contributions to the theory of active transport. II. The gate-type non-carrier mechanism and generalizations concerning tracer flow efficiency, and measurement of energy expenditure. *Bull. Math. Biophys.* 19: 209–235, 1957.

34. PIETRZYK, C., and E. HEINZ. The sequestration of Na^+, K^+ and Cl^- in the cellular nucleus and its energetic consequences for the gradient hypothesis of amino acid transport in Ehrlich cells. *Biochim. Biophys. Acta* 352: 397–411, 1974.

35. RIGGS, T. R., L. M. WALKER, and H. N. CHRISTENSEN. Potassium migration and amino acid transport. *J. Biol. Chem.* 233: 1479–1484, 1958.

36. RIKLIS, E., and J. H. QUASTEL. Effect of cations on sugar absorption by isolated surviving guinea pig intestine. *Can. J. Biochem. Physiol.* 36: 347–362, 1958.

37. SCHULTZ, S. G., and P. F. CURRAN. Coupled transport of sodium and organic solutes. *Physiol. Rev.* 50: 637–718, 1970.

38. SOLOMON, A. K. Pores in the cell membrane. *Sci. Am.* 203: 146–156, 1960.

39. TEORELL, T. Zur quantitativen Behandlung der Membranpermeabilität. *Z. Elektrochem.* 55: 460–469, 1951.

40. UDENFRIEND, S., P. ZALTZMAN-NIRENBURG, and G. GUROFF. A study of cellular transport with the fluorescent amino acid aminonaphthylalanine. *Arch. Biochem. Biophys.* 116: 261–270, 1966.

41. USSING, H. H. Interpretation of the exchange of radio-sodium in isolated muscle. *Nature Lond.* 160: 262–263, 1947.

42. USSING, H. H. Transport of ions across cellular membranes. *Physiol. Rev.* 29: 127–155, 1949.

IX

Electrodiffusion in Membranes

DAVID E. GOLDMAN

OF the many membrane transport processes now known, those of diffusion and electric current flow were the earliest to be studied, are the simplest, and together are especially suited to the analysis of problems of ionic movement. Electrodiffusion is a passive process by which charged particles are transported through a medium under the influence of two simultaneously acting forces, namely, the gradients of particle density and electric potential. The particle flows are given as the sum of these two forces, each multiplied by a coefficient that serves as a facilitating factor or conductance. When the particles are relatively small and the media are very dilute aqueous solutions, the diffusional force is the particle concentration; otherwise the thermodynamic activity, as defined by G. N. Lewis, takes its place. The complexity of the conductance factor depends on the nature of the ions and the structure of the medium. These conductances are not well understood in any but the simplest cases. In addition, in real systems additional forces are usually present, for example, thermal, mechanical, and magnetic; fortunately, in most systems of practical interest to the physical chemist and the biologist, these can be neglected or treated as minor perturbations. The equation expressing the electrodiffusion process is known as the Nernst-Planck equation. There is one such equation for each participating ion species, but because the electric field strength is also a variable, another equation is needed to complete the system. This is the Poisson equation, which relates the electric field strength to the net charge density at any point. The charge density in turn depends on all the ions present, whether mobile or fixed. This equation is the differential form of Gauss's law of electrostatics. In addition, there are both physical and geometrical boundary conditions that must be met.

By the early twentieth century these equations had been formulated and their meanings had become reasonably clear when applied to simple electrolytes. The Nernst-Planck equation may be written in its simplest form as

$$I_j = z_j u_j \left(RT \frac{dC_j}{dx} + z_j FC_j \frac{dE}{dx} \right)$$

where for the jth ion species, I_j is the current density, z_j is the valence, u_j is the mobility, and C_j is the concentration; R and T are the gas constant and absolute temperature, respectively, E is the potential, and F is the Faraday constant. The first term expresses Fick's law of diffusion with the Einstein relation $\alpha_j = (RT/F)u_j$ between the diffusion constant and the ionic mobility. Now we would use the activity instead of the concentration. The second term is Ohm's law incorporating the Arrhenius-Kohlrausch relation between conductance and concentration. The Poisson equation has the form

$$\frac{d^2E}{dx^2} = -\left(\frac{4\pi}{\epsilon} \Sigma_j z_j^2 C_j \right)$$

where ϵ is the permittivity (dielectric constant) of the medium. To these equations must be added the Nernst equilibrium equation

$$\Delta E_j = z_j \left(\frac{F}{RT} \ln \frac{a_{j2}}{a_{j1}} \right)$$

which is the potential difference between two concentrations of an electrolyte, a special case of the Boltzmann distribution law. This set of equations formed the basis for studies of practical problems involving ion flow in electrolytic solutions. However, access to the electrical properties of a solution required the presence of solution-metal interfaces, that is, electrodes whose properties turned out to be unpleasantly complicated. After considerable study the concept of electrode reversibility was developed so that interpretable electrochemical measurements could be made. This in turn led to problems involving liquid junctions between solutions of different composition.

The application of electrodiffusion theory to liquid junctions was made by Max Planck in 1890, who, with the aid of simplifying assumptions, was able to integrate the equations for a one-dimensional system and obtain a liquid-junction potential when the ions were all univalent. The basic assumptions were 1) the ions behaved normally, showed no interspecies interaction (this has become known as the independence principle), and had fixed mobilities and diffusion coefficients; 2) no significant ion accumulation occurred, and therefore the solutions were electrically neutral everywhere; and 3) the liquid junction was constrained by having the concentrations at its ends maintained at the levels of the adjacent solutions. Some years later, Henderson gave a simpler method of integrating the equations for a liquid junction with concentrations that varied linearly between its end values. In this case the ions could be polyvalent. The other assumptions were the same as those used by Planck. These results and some further modifications by other workers proved to be quite satisfactory for the systems for which they were developed. However, extensions to more complex systems such as those of importance to the biologist were exceedingly difficult. This was true not only because so little was known of biological structure but because of the problems of dealing with complex geometries. Furthermore the equations are nonlinear because of the terms of Ohm's law, which contain the product of the concentration and the electric field strength (both of which are dependent variables). Still more complexities awaited those who, not satisfied with an attempt to obtain steady-state solutions, wished to deal with transients, as indeed was sometimes desirable. In fact, because the equations themselves are evidently approximations, the entire process of formulation and solution depends on the skill and good fortune with which the approximations, both physicochemical and mathematical, can be made. It is easy to understand why physical chemists have a strong tendency to shy away from the complications of biological systems. Unfortunately this combination of common sense and cowardice is a luxury that biologists can ill afford.

When these developments were first attracting wide attention, the electrical properties of cells and tissues were, if not well understood, at least perceived as being of great importance in biology and as being based somehow on the same phenomena as those observed and dealt with in physical chemistry. It was being accepted that cells had surface regions or membranes that controlled the passage of materials in and out of cells. Thus it followed that if one knew the structure of membranes and the composition of the surrounding media, the desired understanding of many bioelectrical phenomena could be reached. Of course, the magnitude of the "if" has turned out to be immense, and many of the critical problems remain unsolved. The conditions existing in biological systems are vastly different from those of simple, dilute electrolytes, and membranes have an exceedingly complex structure. To obtain useful results from the application of the Nernst-Planck equations to a living membrane, three sets of conditions must be met. 1) Given that the membrane is reasonably homogeneous, the structure of the system has to be such that the electrodiffusion process is not overwhelmed by the presence of discrete energy barriers, by active transport, or by chemical reactions affecting the moving particles or the structure of the membrane. 2) Simplifying approximations must be made for ion mobilities, activity coefficients, and dielectric constants. 3) The boundary conditions must be manageable by the mathematical techniques available. It does not take very much complexity to consume the resources of a medium-size computer.

A basic problem for the biologist is understanding the electrochemical behavior of a membrane between two dissimilar solutions. Such a system contains three phases and two interfaces (once the electrode regions have been removed from consideration). Each interface, however, has two boundary layers, one on either side, that are not the same even if the membrane is uniform because of surface barriers, distribution coefficients, and solution differences. They are electric double layers and were originally visualized by H. Helmholtz as analogous to thin, parallel plate condensers and later by A. Gouy as diffuse layers through which the ion concentrations varied from some set value at the boundary to the value in the far solution in an approximately exponential way. The space constant (κ) of the layer, which is the reciprocal of what is called the thickness of the layer, can be shown to depend on the temperature and on the ratio of the ionic strength to the dielectric constant of the medium

$$\kappa^2 = \frac{4\pi F^2}{\epsilon RT} \Sigma_j z_j^2 C_j$$

The same concept was later used by P. Debye for the calculation of activity coefficients in dilute solutions. The presence of dilute solutions in biological systems is rare but it is hoped that the effective concentrations within membranes are very small. In any case, what the biologist wants is a value for the membrane potential and an expression for the current-voltage relation.

As to membrane structures, it had become accepted that all membranes were very thin. They were believed by some to have pores that could be traversed by molecules up to a certain size and by others to be lipid in nature, permitting molecules to pass in order of their lipid solubility. A combination of these two viewpoints was considered an acceptable compromise. Several studies in plant cells showed a fairly good correlation of plasmolysis and penetration rate with oil-water distribution coefficients. It was known that small molecules generally could penetrate more rapidly than large ones, that polar molecules were slower than neutral molecules of the same size, and that ions were slower still. Early studies on the electrical properties of membranes carried out by H. Fricke and his collaborators showed that membranes generally had large electrical capacitances and small electrical conductances. With a capacitance of almost 1 $\mu F/cm^2$ and a lipid dielectric constant of ~3, membrane thicknesses could be estimated to be ~3.5 nm, which corresponded roughly to a double layer of lipid or phospholipid molecules. E. Gorter and F. Grendel spread erythrocyte ghost lipid extracts as a monolayer and found that there was enough material to form a bilayer in the membrane. J. F. Danielli and others later postulated that many cell membranes had an oriented layer at each interface, although there could be some unspecified lipid material inside. A protein coating was also considered. The development of the electron microscope ultimately provided direct evidence of the existence of the bilayer structure and revealed membrane thicknesses estimated at ~7.5 nm. Recently a number of protein elements have been identified within the membrane that are associated either with the control of ion flows or with mechanisms of active transport.

The two important sets of coefficients in the Nernst-Planck equations are the mobilities and the activity coefficients of the participating ions. Debye had formulated an elegant physicochemical theory for activity coefficients based on considerations of interionic electrostatic attraction, which had been thoroughly analyzed and experimentally tested for small ions of low valence. This theory has proved very reliable for dilute solutions but attempts to extend it to concentrated solutions have had only limited success. A ray of hope is found in the observation that although activity coefficients at high dilution decrease steadily as the concentration increases, they seem in many cases to reach a minimum and then increase again—sometimes exceeding unity. L. Onsager applied the interionic-attraction theory to the calculation of ion mobilities; again the results have been quite satisfactory for dilute solutions but much less so for concentrations of interest to the biologist. Of course, all this refers to the environmental solutions in which the membranes are bathed. Because most interest is in the membrane itself, one can only rely on what measurements may be possible, try to fit them into the general theoretical framework, and then hope for the best. The relatively low values of the overall dielectric constant of the membrane (5–10) suggests that in the absence of interfering structures the ion concentrations may be rather low. The fact that the results based on such guesses are often well within an order of magnitude of reality is often a cause of amazement, at least to me.

Generally the concentrations of the membrane's environmental solutions are rather high. Where permeability or conductance measurements have been made, it has been found that permeability to ions is quite small and may vary considerably from membrane to membrane and from ion species to ion species. These often wide differences in permeability for a single membrane provide a semipermeability whose nature is necessarily closely related to the function of the cell involved. For example, erythrocytes are quite permeable to anions but excitable cells are not. Sometimes these differences are due to such things as ion hydration; sometimes they are due to the presence of special channels with high specificity. This semipermeability can give rise to a Donnan situation characterized by a marked asymmetry of ion concentrations, an osmotic pressure difference, and a pseudoequilibrium potential whose existence actually depends on the presence of metabolically driven ion pumps, as does the electrodiffusion potential. These potentials provide a reference for the cell membrane about which much electrochemical activity can take place. An early use of the concept of semipermeability was made by J. Bernstein, who proposed that in the resting state the membrane of the nerve axon was permeable primarily to potassium but that during activity this semipermeability was lost, thus reducing its membrane potential to zero. Although it has been greatly modified, this idea remains basic to nerve excitation. It also contains the notion that membrane permeability can be markedly changed by electrical influences.

This was the state of the art when I became a graduate student in Kacy Cole's laboratory at Columbia University in the late 1930s. Cole, with several collaborators, had been working on the electrical impedance measurements of a variety of cells and tissues and was beginning his studies on the squid giant axon, which had recently been publicized by J. Z. Young. My original interest in membranes stemmed from my early training in physics, modified by a few years as a technician in a medical research laboratory. I brought with me some ideas about measuring the water permeability of surface layers. I had worked out a method for doing so, but my research was deferred while I took required courses. By the time these were completed, the experiment had already been done in a much better manner in the laboratory of I. Langmuir. Cole told me that he had spent a summer as an assistant in Langmuir's laboratory and passed on to me many words of laboratory wisdom, which he claimed to have learned. I turned my attention to some simple DC electrical measurements on some natural and artificial membranes. This rapidly taught me the pitfalls of electrode polarization. I also worked at trying to understand as much theoretical physical chemistry as I could under Cole's tutelage and discovered that he had spent a postdoctoral year in Leipzig with Debye. Again I was treated to many wise and penetrating observations. Thus I imbibed the best (to me at least) of relevant science from Cole and also became in a sense a descendent of both Langmuir and Debye, to say nothing of those whose works had provided me with much background.

The DC measurements were not going well, so I switched to an AC bridge technique and tried out some membranes using the technique of R. B. Dean, which consisted of pushing together two oil-water interfaces dotted with protein. These were of limited stability and reproducibility, so I settled for membranes made of collodion laced with phospholipid, which were easy to handle, had resistances within the range of the bridge, and (still better) showed small but definite changes in conductance with applied DC, that is, rectification. This was what I had hoped for, so I was naturally very pleased. In the meantime Cole had gone to Princeton for a sabbatical at the Institute for Advanced Study and then to the Metallurgical Laboratory at the University of Chicago to work on the biological effects of ionizing radiation. Before he left he had introduced me to various ideas about the possible origin of electrical rectification in membranes (especially that of the squid giant axon) and had suggested that I look into them. This led me to the Mott theory of the copper–copper oxide rectifier, in which electrodiffusion occurs in a very thin layer with electrons as the charge carriers. The seeming resemblance to a liquid junction could not be overlooked. However, the Planck approach was clearly not suitable for biological cell membranes. A liquid junction is essentially an electrical resistance, whereas a membrane is more nearly a leaky condenser. In other words, the charge accumulation near the membrane boundaries made it very unlikely that microscopic electroneutrality could occur, especially when the membrane parameters were such that the double layers occupied most of the membrane thickness. Under these circumstances the nonlinearity of the equations threatened to become a serious barrier to further progress. However, the overlap of the two double layers could tend to straighten out the internal potential profile, and thus the electric field could be nearly constant. The Poisson equation states that the curvature of the potential profile at any point is proportional to the net charge density there with a very large constant of proportionality. The use of the constant field had been mentioned by U. Behn and by Sitte but had not been pursued because of the character of the liquid junction in electrolytes. It had been applied to the copper–copper oxide rectifier and was likely to be useful for a membrane system both as a physical approximation and as a means of linearizing the equations. A possible alternative was to introduce a fixed-charge distribution into the membrane. This had been proposed by T. Teorell and by K. H. Meyer and J. F. Sievers and appeared to be very useful for certain artificial membranes. I solved the relevant equations by using microscopic electroneutrality, but this also introduced an unknown distribution of charges into a membrane already badly bruised by rough approximations, so I did not follow it through. As it was, the constant-field approximation provided a very simple, workable formula. As part of the integration process, the steady-state current-voltage curves had to be calculated, and these showed rectification of the right kind, although with numerical values that seemed too small. On the other hand, the potential-concentration curves were quite satisfactory. The calculations were carried out with the findings of Cole and H. Curtis for the squid giant axon, which at that time was the only preparation for which usable data existed.

By this time it was 1943, so after several months as a member of a group working on the physics of aerosols in the laboratory of V. K. LaMer, I joined the Navy as an aviation physiologist and for the next several years almost forgot about electrophysiology. Then came Cole's development of the voltage clamp and the work of Hodgkin, Huxley, and Katz. At Cole's suggestion, they had taken up the constant-field formulation, and after making modifications that introduced permeabilities for the diffusion coefficients and the ion partition coefficients, they applied it successfully to their squid axon data. As their views of excitability spread, so did the use of the constant-field equation by many investigators, who found it very helpful. As the equation became popular, a number of cases were found in which it did not seem to fit. As I have already mentioned, among the conditions for applicability is the absence of other transport mechanisms in significant quantity. Nearly all cells were known to have active-transport systems to one degree or another, so when the constant-field equation did not apply, the discrepancy could usually be explained by the presence of the active transport. A striking example was found in 1970 by A. L. F. Gorman and M. F. Marmor, who used temperature changes as a means of controlling ion-pump activity in a molluscan neuron. These and other similar findings raised the question of how active transport and electrodiffusion interacted. Using the quantitative data of L. J. Mullins and F. J. Brinley on the active transport of sodium and potassium in squid axons, I was able to calculate the effect of active transport on membrane potential. In the squid axon the pump activity was controlled by internal perfusion both with and without the presence of ATP. These calculations required the modest use of a computer and gave results that, although not numerically accurate, were within a reasonable factor of the observed data. Other transport mechanisms, especially if passive like some forms of exchange diffusion, could be included as modifications. However, the presence of other metabolically generated transport would, if possible, require more elaborate treatment. In 1949 H. H. Ussing pointed out that the expression for the flux ratio of two penetrating ions was much less dependent on specific assumptions than that for the membrane potential. This has been extremely helpful in ion flux studies of a number of unexcitable cells. Teorell added Donnan potentials at membrane-solution interfaces, thus calling attention to important discontinuities in the system. However, numerical data were lacking.

An alternative, very general attack on transport problems arose in the form of the discipline of nonequilibrium thermodynamics, which had been developed by Lord Rayleigh, Onsager, Prigogine, and others. It expressed ion fluxes as the sum of a series of terms, each involving one of the possible driving forces and did not require the use of the independence principle. To the extent to which the forces were known, study of the equations brought to light an entire set of flux elements; these were expressed as the sum of terms, each of which represented one of the forces acting within the system. The matrix of coefficients had been shown by Onsager to be antisymmetrical; that is, if the jth flux was the sum of terms, each of which represented the ith force element, then the coefficients L_{ij} and L_{ji} were equal. This was an important simplification, particularly because many of the coefficients are extremely difficult to measure and are often themselves variable, thus making the system nonlinear. The method is generally a powerful means of obtaining an overview of transport processes, but to paraphrase Einstein, although God may not be mean, the biologist gets the impression that God is very

[258]

tricky indeed. Thus far, progress in biological applications of this approach has been limited.

There have been several attempts to modify and generalize the constant-field formulation. These have usually been encouraging in that they have helped to remove some of the original restrictions. Unfortunately some of these improvements have required additional assumptions and parameters and have produced vexing mathematical complications. The Nernst-Planck system can also be reduced to a nonlinear differential equation in the field strength without the use of any linearizing assumptions. For a single univalent salt this has the form of the Painlevé type II transcendent, which (though complicated) at least does not have critical points whose positions vary with boundary values. A computer solution has been obtained, but its use in membrane studies does not appear feasible at present. The analogous equation for higher valences or for a number of univalent ions seems to be beyond the reach of current mathematical techniques. Several mathematicians have carried out some analytical studies on the Nernst-Planck system and have been quite helpful in bringing to light difficulties of the simpler approximations. However, the equations are themselves approximations of physical reality, and the extent to which an exact solution of approximate equations is an improvement over a simplified treatment must be evaluated carefully. As pointed out earlier, membrane structures are not known well enough to permit the use of reliable mobility and activity coefficient values or of precise boundary conditions. Recent developments in the analysis of specialized membrane channels and transfer proteins have introduced new ways of looking at the molecular aspects of membrane transport. It would be nice to be able to lay out the various membrane-transport mechanisms in a comprehensible quantitative fashion, such as a staunch biophysicist would like, but the day when membrane transport is that well understood is far off, although I hope not impossibly so.

BIBLIOGRAPHY

1. COLE, K. S. *Membranes, Ions, and Impulses.* Berkeley: Univ. of California Press, 1968.
2. DAVSON, H. *A Textbook of General Physiology.* London: Churchill, 1970.
3. HÖBER, R. *Physical Chemistry of Cells and Tissues.* Philadelphia: Blakiston, 1945.
4. KATCHALSKY, A., and P. F. CURRAN. *Non-Equilibrium Thermodynamics in Biophysics.* Cambridge, MA: Harvard Univ. Press, 1965.
5. MACINNES, D. A. *The Principles of Electrochemistry.* New York: Reinhold, 1939.
6. LAKSHMINARAYANAIAH, N. *Transport Phenomena in Membranes.* New York: Academic, 1969.

X

Reflections on Selectivity

CLAY M. ARMSTRONG

I entered the field of biophysics well after the conclusion of what
B. Hille (13) has aptly called the heroic era, when *"the membrane-
ionic theory of excitation was transformed from untested hypothesis
to experimental fact."* By that time it was clear that the basic outlines
of the Hodgkin and Huxley formulation were there to stay and that
anyone wishing to do useful work in the area must accept that. In
fact K. S. Cole told me that he thought it would be twenty-five years
before enough new information was available to require a new
formulation. Given this framework, the research questions worth
pursuing dealt with the physical mechanisms underlying the
phenomena that A. L. Hodgkin and A. F. Huxley had described with
such power. There were three major questions. 1) How do ions pass
through membranes? 2) How does the membrane select among
monovalent cations? 3) How does the membrane change its dominant
permeability in a fraction of a millisecond? All of the questions are
closely related or, as in the case of the first two, virtually inseparable.
Here I deal only with the question of selectivity. The scope of this
chapter is mainly limited to a discussion of the ideas of Lorin Mullins
and George Eisenman. Their theories are quite different in character,
with Mullins emphasizing the importance of steric considerations,
whereas Eisenman virtually eliminates steric factors in his theory
and deals only with electrostatics and hydration energies. Even with
this limitation of scope, I found that as the chapter progressed I had
much to learn.

In distinguishing between alkali metal cations, the membrane's
only clues are size and properties such as hydration that are directly
related to size. The cations of this series have a charge of +1, and
because only their *s* orbitals are filled, they are spherically
symmetrical. Differences in size determine the strength of their
interactions with water: the intense electric field at the surface of

the small cations causes them to hydrate more strongly than do the larger cations of the series. This is reflected in the free energy of hydration, which is given by Robinson and Stokes (15) as 114.6 kcal/mol for Li^+ and 60.8 kcal/mol for Cs^+. The hydrated radius of the ions is somewhat ill defined, but it seems clear that Li^+, for example, carries with it a larger number of firmly attached water molecules than does Cs^+.

Given these facts, how can one imagine that ion selectivity arises? When the hypotheses to be discussed were brought forward, the discussion was complicated by uncertainty about the means by which ions cross the membrane—through pores or by means of carriers. A further complication was uncertainty about the state of hydration of ions passing through the membrane. An obvious first guess is that a sieving effect prevents the larger ion from going through the pathway selective for the smaller ion; but what factors exclude the smaller ion from the pathway of the larger ion?

Close-Fit Hypothesis

The first good hypothesis that I am aware of that can answer this question was proposed by Mullins (14). He imagined that Na^+ and K^+ crossed the membrane by a single set of pores that changed from Na^+ selective to K^+ selective. In Na^+-selective form, the radius was 3.67 Å, which is the radius of an Na^+ with one complete hydration shell. When K^+ selective, the radius was 4.05 Å, the radius of a K^+ with one hydration shell. The preferred conformation of the pores was K^+ selective, but they could contract to a radius near 3.67 Å through binding Ca^{2+} in the pore mouth. On depolarization the pore opened and the occupying Ca^{2+} was dislodged. The pore then remained Na^+ selective for a short time before reverting to its preferred K^+-selective state. There is much in this proposal that merits thinking about, but the question at hand is what prevents the smaller hydrated Na^+ from going through a pore large enough for a hydrated K^+? Mullins pointed out what seems obvious in retrospect, that in passing through a pore there is a substantial interaction energy between the ion and the pore wall and that the ion must be a good fit to the pore for this interaction to be favorable. Sodium is too small to be a good fit in the K^+-selective pore, and its potential energy is high because it does not bind favorably to the pore walls.

The close-fit hypothesis emphasizes what I think must be an important element in selectivity, although subsequent evidence has made it necessary to modify the details of the proposal. It now seems highly probable that ions lose much of their hydration in passing through a pore. To many this seemed unlikely because the hydration energy is so large, but the experimental evidence is convincing. For one thing, gramicidin A pores are extremely efficient conductors of ions, and the gramicidin pore is only about 4 Å in diameter, much too small to accommodate a hydrated Na^+ or K^+. Unless D. W. Urry's model of the pore (16) is wrong, which seems unlikely, one must conclude that a K^+, for which the pores show a modest preference, can lose hydrating water molecules very rapidly. Further, Hille (10) has provided strong evidence that the Na^+ channel of excitable membranes is about 3×5 Å. Similar evidence indicates that the diameter of a K^+ channel is about equal to the diameter of a dehydrated K^+ (5, 11). In light of these findings, the close-fit hypothesis was redrawn as in Figure 1 to explain the exclusion of Na^+ by K^+ pores (2). The upper part of the figure shows a K^+ and an Na^+ in water, each surrounded by a single hydration shell. The dipolar water molecules are attracted to the cations and pack closely around them. The lower part of the figure shows a hypothetical pore with dipolar groups surrounding the ion. For the K^+, the dipolar groups replace the waters of hydration effectively and the energy of the ion in the pore is about the same as in water. Conformational restraints, however, prevent the dipolar groups from closely approaching the Na^+ (lower right), and its energy in the pore is much higher than in water. An estimate of how much higher the energy in the pore is can be obtained by a simple electrostatic calculation. Suppose that, in a vacuum, a water molecule were pulled away from an Na^+, from close fit to a separation of 1 Å. What energy is required? The number is startlingly high, about 8.6 kcal/mol, corresponding to an energy barrier that only one in two million ions would have the energy to

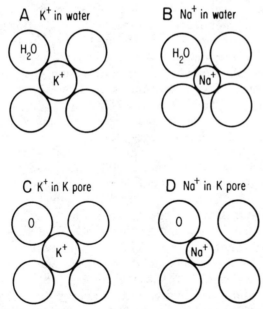

Figure 1. Schematic representation of K^+ and Na^+ in water and in hypothetical K^+ pore. Pore walls are too rigid to close about Na^+, and the ion bonds effectively to only two of the four oxygen atoms, with the result that its potential energy in the pore is higher than in water.

surmount. Thus even a minor perturbation of a cation's hydration or its interaction with a pore wall could give rise to a very large selectivity ratio.

Although not essential to the story, it is of interest to relate current thinking about some of the other ideas in Mullins' paper. Mullins now agrees that the membrane contains two separate sets of pores. The crucial evidence to his mind was the demonstration that destruction of Na^+ inactivation did not prevent the development of a full-size potassium current, which would be impossible if the Na^+ pores, by "inactivating," were actually becoming K^+ pores as he had quite reasonably hypothesized. The idea that pores are deformed by the ions passing through them remains an attractive one, but such an effect has not, to my knowledge, been demonstrated. Finally, the idea that closed pores are occupied by Ca^{2+}, although not original to Mullins (see ref. 8), is receiving new attention and I believe the idea to be true at least for squid K^+ channels (3).

FIELD-STRENGTH THEORY

The next idea to explain selectivity was presented by Eisenman (6) and was based on his studies of the glasses used to make ion-selective electrodes. Eisenman pointed out that a useful construct for thinking about selectivity is to consider the interactions of a naked alkali-metal cation with an anionic site in a vacuum. This model has an appealing abstractness, rather like the thermodynamic cycles of the nineteenth century and is useful in averting the manifold complications encountered when trying to calculate the energies of ions in water. The anionic site was simplified to the essence: a sphere with a charge of $-1e$. The energy of a cation near such a site is plotted in Figure 2 as a function of the center-to-center distance from cation to site. The left curve is for a small site with a radius of 0.8 Å, and the energy at the point of closest approach is indicated for Na^+ and K^+. The binding energy for Na^+ is greater, and it is clear that on an absolute basis the site always prefers Na^+ to K^+ regardless of the radius of the site. Eisenman had the important insight that the base line or reference level is not, however, the energy of the cation at infinite distance from the site (the quantity given by the horizontal axis) but instead is the energy of the ion in water. This hydration energy is plotted for Na^+ and K^+ by the dots. The relevant quantity for selectivity is not the absolute binding energy but the difference between the binding energy and the hydration energy, which is 91.2 kcal/mol for Na^+ and 73.6 kcal/mol for K^+. [These numbers were obtained by adding the hydration-free energy for Cs^+, as given by Robinson and Stokes (15), to the hydration-energy differences between Na^+ and Cs^+ and K^+ and Cs^+, as quoted in Eisenman (6). The precise hydration energies turn out to be important, as discussed next.] In moving from water to site, Na^+ thus falls into a deeper

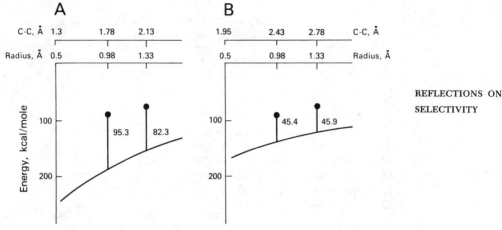

Figure 2. Site-cation interactions predicted by field-strength theory. Interaction energies between cation of specified radius and site of 0.95 Å (*A*) or
1.45 Å (*B*) are given by curves. C-C, center-to-center distance from cation
to site; dots, hydration energy for Na$^+$ (0.98 Å radius) and K$^+$ (1.33 Å radius);
vertical lines, net depth of energy well experienced by cation when completely dehydrated and then allowed to approach a negative site of given
radius in a vacuum.

energy well (95.3 kcal/mol) than does K$^+$ (82.3 kcal/mol), and the
site is Na$^+$ selective.

The binding energy curve for a larger site, 1.45 Å in radius, is
given in Figure 2B. At point of closest approach the binding energy
for both ions is less than for the smaller site, but on an absolute basis
the site still prefers Na$^+$. When the hydration energy is taken into
account, however, it is seen that K$^+$ falls into a deeper well (45.9
kcal/mol) on going from water to site than does Na$^+$ (45.4 kcal/mol),
so considering the ions to be initially solvated, the site is K$^+$ selective.

When he calculated the selectivity ratios for five alkali metal
cations interacting with sites of various diameters, Eisenman discovered that only thirteen sequences were predicted. These thirteen
are listed in Table 1. The small number of predicted sequences
seems remarkable in view of the fact that there are 120 permutations
of five ions, and thus potentially 120 selectivity sequences. Even
more remarkable, and a point that provided strong support for the
theory, the predicted sequences correspond in most cases to the
limited number of sequences that have been observed experimentally with a wide variety of membranes.

The fact that nature pruned the number of sequences to be used
in membrane channels from 120 to roughly thirteen seems less
remarkable after a bit of thought. Consider a site that is selective for
Li$^+$ over the other four cations. There are twenty-four (4!) possible
sequences beginning with Li$^+$ at the head of the list. One of these
would be the sequence Li$^+$ > K$^+$ > Cs$^+$ > Rb$^+$ > Na$^+$, as diagrammed

TABLE 1
Affinity Sequences Predicted by Eisenman's
Field-Strength Theory

I	$Cs^+ > Rb^+ > K^+ > Na^+ > Li^+$		
IIa	$Cs^+ > K^+ > Rb^+ > Na^+ > Li^+$	II	$Rb^+ > Cs^+ > K^+ > Na^+ > Li^+$
IIIa	$K^+ > Cs^+ > Rb^+ > Na^+ > Li^+$	III	$Rb^+ > K^+ > Cs^+ > Na^+ > Li^+$
IV	$K^+ > Rb^+ > Cs^+ > Na^+ > Li^+$		
V	$K^+ > Rb^+ > Na^+ > Cs^+ > Li^+$		
VI	$K^+ > Na^+ > Rb^+ > Cs^+ > Li^+$		
VII	$Na^+ > K^+ > Rb^+ > Cs^+ > Li^+$		
VIII	$Na^+ > K^+ > Rb^+ > Li^+ > Cs^+$		
IX	$Na^+ > K^+ > Li^+ > Rb^+ > Cs^+$		
X	$Na^+ > Li^+ > K^+ > Rb^+ > Cs^+$		
XI	$Li^+ > Na^+ > K^+ > Rb^+ > Cs^+$		

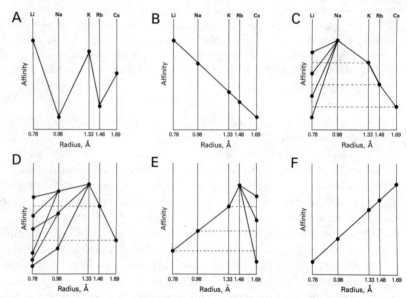

Figure 3.   Affinity sequences and monotonicity. *A*: one of the twenty-four
(4!) possible sequences of the five alkali metal cations that begin with Li^+.
Sequence seems unreasonable because there is no clear relation between
affinity and cation radius. *B*: postulating a monotonic relation between
affinity and radius reduces the twenty-four sequences to one. *C–F*: monotonicity applied to sequences beginning with Na^+, K^+, Li^+, and Cs^+ reduces
number of sequences from 120 (5!) to sixteen.

in Figure 3A. The sequence seems inherently unreasonable, as there
is no simple relation between affinity and radius, which is the only
distinguishing characteristic of alkali-metal cations. If one postulates
that affinity is related to size and that cation-site interaction is less
favorable as a cation's radius increasingly diverges from Li^+ (i.e.,

[266]

there is a monotonic relationship between affinity and radius), there is only one possible sequence: $Li^+ > Na^+ > K^+ > Rb^+ > Cs^+$ (Fig. 3B). This simple argument reduces the number of possible sequences from twenty-four to one. Only one sequence beginning with Li^+ is predicted by Eisenman's model and it is precisely the sequence given in the figure.

Similar arguments can be made for a sequence beginning with Na^+ as the preferred ion. In theory there are again twenty-four possible sequences, but this number is again greatly reduced by postulating a monotonic relation between selectivity and radius on either side of the preferred ion. The possibilities for an Na^+ site are given in Figure 3C. To the right of Na^+ in the plot there is a single curve indicating that selectivity decreases as the cations get progressively larger relative to Na^+. The curve is monotonic. The slope of its three segments were picked arbitrarily and make no difference to the argument. To the left, there is only Li^+. There is no a priori way of determining how rapidly the selectivity-radius curve should drop off from Na^+ to Li^+. One can readily imagine situations in which the penalty for being smaller than Na^+ might be either mild or severe. All in all there are four possibilities as diagrammed in the figure. These correspond to the sequences $Na^+ > Li^+ > K^+ > Rb^+ > Cs^+$; $Na^+ > K^+ > Li^+ > Rb^+ > Cs^+$; $Na^+ > K^+ > Rb^+ > Li^+ > Cs^+$; and $Na^+ > K^+ > Rb^+ > Cs^+ > Li^+$. These are exactly the four sequences beginning with Na^+ that are predicted by the Eisenman theory.

Diagrams for sequences beginning with the three other cations of the series are given in Figure 3D–F. There is only one predicted sequence that begins with Cs^+, six that begin with K^+, and four that begin with Rb^+. Figure 3D, which is the most complicated, requires explanation. On both sides of K^+ the relation is monotonic, but there is again no way of knowing in advance whether affinity will decrease steeply or slowly with radius for the smaller cations or whether the slope will be steep or shallow in going from Na^+ to Li^+. All of the possibilities that are consistent with monotonicity are given in the figure, and the total number of possible sequences beginning with Na^+ is six. Another way of stating the problem is to ask for all the sequences beginning with K^+ in which Na^+ precedes Li^+ and Cs^+ precedes Rb^+. Again the answer is six.

In total there are sixteen sequences that are consistent with monotonicity. The field-strength hypothesis is more restrictive and reduces the number to thirteen.

What evidence is there that nature used field-strength considerations rather than monotonicity as a criterion for selecting sequences? One might ask how well the field-strength hypothesis performs quantitatively in predicting the selectivity ratios and binding energies for ions in the Na^+ and K^+ channels of nerve. A first problem is that the affinity or "equilibrium selectivity" predicted by the field-strength hypothesis has no clear relation to selective permeation through a membrane channel (see refs. 1, 2, 5, 7, 9, 12, 13). I simply set this question aside and ask how the predicted affinities compare with the selectivity of membrane channels because it was originally such comparisons that gave support to the field-strength theory. Quantitatively there are undoubtedly problems. Beginning with the Na^+ channel, permeability to Li^+ and Na^+ are about equal and exceed the K^+ permeability of the channel by a factor of about ten (see the last two rows of Table 3, taken from ref. 13). For widely spaced sites (6) Li^+ and Na^+ are about equal in affinity for a site with a radius of 0.8 Å. The affinity for K^+ is lower by a factor of about a billion, rather than the experimentally observed factor of about ten. When, on the other hand, the affinities for Na^+ and K^+ have a ratio of 10:1 ($Na^+:K^+$), the affinity for Li^+ is less by a factor of about 10 million, rather than the observed factor of about one. Thus the affinity ratios predicted by the theory are very unlike the experimental observations on Na^+ channels.

Another problem is that the binding energies are enormous. The absolute enthalpy of the Na^+-site bond for a site of 0.8 Å is 186 kcal/mol, making the bond much stronger than most covalent bonds. When the hydration energy is subtracted out, the energy well is still more than 90 kcal/mol. Thus binding of an Na^+ to the site would be essentially irreversible. Clearly this mechanism does not seem appropriate for even a very slow transporter of ions, let alone for a pore that we know carries many ions per microsecond.

Other problems arise on applying the theory to a K^+-selective channel. Experimentally, the permeability of a K^+ channel has the ratios $< 0.01:1: < 0.77$ (11). For widely spaced sites (6) the most appropriate site radius is 1.55 Å, yielding affinity ratios of 0.06:1:1; for larger sites, Cs^+ affinity is too high, and for smaller sites the $K^+:Na^+$ affinity, which is too low for the 1.55 Å site, is even lower. Thus at the optimum radius, the $K^+:Na^+$ affinity is substantially lower than required, by a factor of about six.

Perhaps more serious, the predictions of the theory with regard to K^+-selective sites are very sensitive to the choice of crystal radius and hydration energy, as illustrated by the following two examples. 1) If Pauling radii (Na^+, 0.95 Å; K^+, 1.33 Å; Cs^+, 1.69 Å) are used together with the hydration-free energies given in Robinson and Stokes (15), the theory predicts no sequences in which affinity is highest for K^+, as illustrated by the data in Table 2. The table contains the relative affinities of Na^+, K^+, and Cs^+ for a site of the specified radius at which the affinity of the preferred ion is taken as 1.0. To determine affinities, the interaction energy (in kcal/mol) of cation and site was calculated from the formula (see ref. 6)

$$-332/(r_{cation} + r_{site})$$

where r is the radius using the Pauling crystal radii. Using the hydration energies given in Robinson and Stokes (15), the affinities were then calculated by the formula

affinity = exp[−(interaction energy − hydration energy)/RT]

where exp is exponent and RT is gas constant × absolute temperature. The sequences predicted are $Na^+ > K^+ > Cs^+$ for the smaller radii in the table, with a transition to $Na^+ > Cs^+ > K^+$ at 1.56 Å. For a site radius of 1.62 Å or larger, Cs^+ is the preferred ion. Sequences beginning with K^+ are not predicted. 2) If Goldschmidt radii (Na^+, 0.98 Å; K^+, 1.33 Å; Cs^+, 1.65 Å) are used together with the hydration data of Rossini (6), there are also no sequences that begin with K^+. (Eisenman used radii of 0.98, 1.33, and 1.69 Å together with the hydration data of Rossini.) To recapitulate, the predictions regarding dominantly K^+-selective sites are extremely sensitive to the choice of radii and hydration energies. In one case a decrease of only 0.04 Å in the radius assigned to Cs^+ is enough to eliminate dominantly K^+-selective sites altogether. One must note that estimates for the radius of Cs^+ vary rather widely, from 1.63 Å to 1.86 Å (see ref. 13, p. 164).

Can the field-strength theory be altered to reduce the binding energy for an Na^+ site and to provide more satisfactory predictions for K^+-selective sites? One avenue for modification is mentioned by Hille (13), who points out that only some rather than all water molecules surrounding an Na^+ need to be removed when it binds to a site. In line with this suggestion I have performed calculations based on the model shown in Figure 4. The figure shows a cation surrounded by six water molecules. In the process of binding to the site, one or two water molecules are removed a distance of 3 Å and the cation then moves 3 Å until it contacts a negative site of radius to be specified. Of course there are many possible modifications of this model, and the selection of six water molecules is a somewhat arbitrary simplification, but it seems a plausible starting point for bringing the Eisenman construct closer to life.

TABLE 2
Relative Affinities Predicted by Field-Strength Theory
Using Pauling Crystal Radii and Hydration Energies

Site Radius, Å	Relative Affinities		
	Na^+	K^+	Cs^+
1.51	1.00	0.05	0.02
1.55	1.00	0.011	0.09
1.56	1.00	0.14	0.13
1.57	1.00	0.17	0.19
1.58	1.00	0.21	0.26
1.61	1.00	0.39	0.74
1.62	0.64	0.46	1.00

* Pauling crystal radii and hydration energies from Robinson and Stokes (15).

[269]

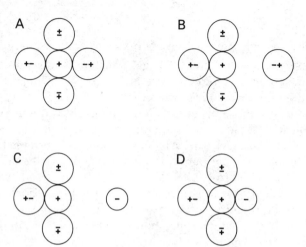

Figure 4. Simplified model for partial dehydration of an ion and its interaction with a negative site. *A*: cation surrounded by six water molecules (one
in and one above page). Each water molecule is treated as a dipole, with
negative charge separated from positive by 0.389 Å. *B*: one (or two) water
molecule is withdrawn distance of 3 Å. *C, D*: cation approaches negative
site from distance of 3 Å. Model approximates energy changes seen upon
exchanging a water molecule for a negative site.

In performing the calculations, a water molecule was treated as a
dipole consisting of a negative charge precisely at the center of the
molecule and a positive charge 0.389 Å away, yielding a dipole
moment of 1.87 Debye. This constellation of charges, one on the
site, one on the cation, and two on each water molecule, was then
handled in accordance with the usual laws of electrostatics, with
energy of interaction between two charges determined by the reciprocal of the distance between them. Entropy was not considered in
the calculations, nor were interactions between the removed water
molecule(s) and the site. Affinity ratios were calculated from the
interaction energies as previously described.

The results, assuming the removal of one or two water molecules,
are shown in Table 3. Each cation is assigned a number that is its
affinity relative to the preferred ion, which has affinity of 1.0. The
right column in the table is the energy of the preferred cation when
bound to the site relative to its energy with all six water molecules
in place. This number thus indicates the depth of the energy well
that an ion enters when it binds to the site. The two bottom rows in
the table give the selectivities of Na^+ and K^+ channels (13).

The predictions for removal of either one or two water molecules
fail to provide a good match to the experimental selectivity data. *1)*
The model generates Na^+:K^+ affinity ratios similar to an Na^+ channel,
that is, about ten to one, for one removed water and a 2.0 Å site, or
for two removed waters and a 1.55 Å site. In both cases, however,
the Na^+:Li^+ ratio is too large, three or four to one rather than about

TABLE 3
Affinitites From Field-Strength Theory Assuming Removal
of One or Two Water Molecules

Site Radius	Affinity Relative to Preferred Ion					
	Li$^+$	Na$^+$	K$^+$	Rb$^+$	Cs$^+$	ΔH$^+$
Removal of one water molecule						
2.0	1.000	0.296	0.029	0.011	0.003	−20.67
2.6	0.936	1.000	0.753	0.597	0.393	−11.83
2.8	0.597	0.846	1.000	0.920	0.766	−9.93
3.5	0.051	0.140	0.480	0.693	1.000	−5.63
Removal of two water molecules						
1.55	1.000	0.212	0.016	0.005	0.001	−17.74
1.80	0.819	1.000	0.779	0.617	0.400	−10.06
1.85	0.540	0.861	1.000	0.890	0.682	−9.12
2.00	0.064	0.209	0.670	0.846	1.000	−6.83
Na$^+$ channel*	0.93	1.0	0.086	<0.012	<0.013	
K$^+$ channel	<0.018	<0.010	1.0	0.91	<0.77	

* From Hille (13).

one to one, and the K$^+$:Rb$^+$ ratio is too small, about three rather than more than seven. 2) When the Na$^+$:Li$^+$ ratio is near one (one water, 2.60 Å site; two waters, 1.80 Å site), the Na$^+$:K$^+$ ratio is only slightly more than one, and the same is true for the K$^+$:Rb$^+$ ratio—both too small to agree with the Na$^+$ channel data. Thus the site is either Li$^+$ selective or almost unselective. 3) When affinity for K$^+$ is highest (one water, 2.8 Å site; two waters, 1.85 Å site), the site is almost nonselective, with K$^+$:Na$^+$ ratio near one. Making the site larger (one water, 3.5 Å; two waters, 2.0 Å) predicts a Cs$^+$-selective site that is no closer to the data.

There are of course many permutations of the model that have not been considered, for example, if the site were moved 4 Å away rather than 3 Å (this has in fact been tried, and it does not help). On one point, however, the failures here parallel those of the original Eisenman calculations and seem to point to a weakness inherent in this type of model. Specifically, neither here nor in the original work do the predictions for dominantly K$^+$-selective sites show adequate selectivity. This suggests a general and perhaps irremediable weakness of the field-strength hypothesis with respect to K$^+$ channels.

Regarding Na$^+$ channels, removing only one or two water molecules is in some respects an improvement. Interaction energies are more nearly acceptable, although still quite large (−17.74 and −20.67 kcal/mol). It is conceivable that other considerations, for example, image forces, could reduce this to a few kcal/mol, in better agreement with experimental data. But for removal of one, two, or all the surrounding water molecules, the theory seems to have serious quantitative deficiencies in predicting affinity ratios similar to the

[271]

experimental selectivity ratios. For removing one or two water molecules, the site is almost nonselective when the $Na^+:Li^+$ ratio is acceptably near one. On removing all waters (6) the $Na^+:K^+$ ratio is approximately 2.5×10^9 when $Na^+:Li^+$ is near one, as noted above, and the interaction energy of Na^+ with the site is unacceptably large.

Can these problems be fixed while retaining the general framework of the field-strength theory? For the K^+ channel, I should say almost certainly not: the ability of the field-strength theory to predict dominantly K^+-selective sites seems tenuous indeed. In general I believe it is more profitable to think about K^+ channels along the lines of the close-fit hypothesis. For Na^+-channel modifications, suitable modifications of the field-strength hypothesis remain to be demonstrated. Looking at it pessimistically, the approximately ten to one selectivity of the Na^+ channel may be difficult to explain even when one knows the disposition of every atom in the protein.

This work was supported by USPHS Grant No. NS08951.

BIBLIOGRAPHY

1. ALMERS, W., E. W. McCLESKEY, and P. T. PALADE. A non-selective cation conductance in frog muscle membrane blocked by micromolar external calcium ions. *J. Physiol. Lond.* 353: 565–583, 1984.
2. ARMSTRONG, C. M. Ionic pores, gates, and gating currents. *Q. Rev. Biophys.* 7: 179–210, 1975.
3. ARMSTRONG, C. M., and J. LOPEZ BARNEO. External calcium ions are required for potassium channel gating in squid neurons. *Science Wash. DC* 236: 712–714, 1987.
4. ARMSTRONG, C. M., and R. MATTESON. The role of calcium ions in the closing of K channels. *J. Gen. Physiol.* 87: 817–832, 1986.
5. BEZANILLA, F., and C. M. ARMSTRONG. Negative conductance caused by entry of sodium and cesium ions into the potassium channels of squid axons. *J. Gen. Physiol.* 60: 588–608, 1972.
6. EISENMAN, G. Cation selective glass electrodes and mode of operation. *Biophys. J.* 2, Suppl. 2: 259–323, 1962.
7. EISENMAN, G., and R. HORN. Ionic selectivity revisited: The role of kinetic and equilibrium processes in ion permeation through channels. *J. Membr. Biol.* 76: 197–225, 1983.
8. FRANKENHAEUSER, B., and A. L. HODGKIN. The action of calcium on the electrical properties of squid axons. *J. Physiol. Lond.* 137: 218–244, 1957.
9. HESS, P., and R. W. TSIEN. Mechanism of ion permeation through calcium channels. *Nature Lond.* 309: 453–456, 1984.
10. HILLE, B. The permeability of the sodium channel to organic cations in myelinated nerve. *J. Gen. Physiol.* 58: 599–619, 1971.
11. HILLE, B. Potassium channels in myelinated nerve. Selective permeability to small cations. *J. Gen. Physiol.* 61: 669–686, 1973.
12. HILLE, B. Ionic selectivity of Na and K channels of nerve membranes.

In: *Membranes: A Series of Advances. Lipid Bilayers and Biological Membranes: Dynamic Properties*, edited by G. Eisenman. New York: Dekker, 1975, vol. 3, 255–323.

13. HILLE, B. *Ionic Channels of Excitable Membranes*. Sunderland, MA: Sinauer, 1984.
14. MULLINS, L. J. An analysis of conductance changes in squid axon. *J. Gen. Physiol.* 42: 817–829, 1959.
15. ROBINSON, R. A., and R. H. STOKES. *Electrolyte Solutions* (2nd ed.). London: Butterworths, 1959.
16. URRY, D. W. The gramicidin A transmembrane channel: A proposed helix. *Proc. Natl. Acad. Sci. USA* 68: 1907–1911, 1971.

XI

Propagation of Electrical Impulses

LORIN J. MULLINS

W HEN he asked me to write this chapter, our editor, Daniel Tosteson, also asked that I try to convey some impression of the scientists with whom I interacted. This necessarily involves discussing topics that have interested me for many years and that have in large measure circumscribed the nature of this review.

I received my PhD from the University of California at Berkeley. I studied ion transport in *Nitella* under S. C. Brooks, who (although a professor of zoology) was primarily interested in how ions and molecules penetrated cells. He had an immense knowledge of the literature and was a pioneer in utilizing radioisotopes, which were just being produced in the Berkeley cyclotron, for permeability studies.

ION SELECTIVITY

My first job was as an instructor in medical physiology at the University of Rochester. The department was presided over by Wallace Fenn, a man of tremendous intellectual talents and of wide-ranging interests. Although trained in myothermal measurements in muscle in A. V. Hill's laboratory, he came to Rochester and began at once to investigate a wide variety of permeability problems. He was interested in electrolyte balance in muscle and in red blood cells. He set me to making measurements of Na^+ uptake by red cells—measurements that clearly showed an Na^+ permeability of these cells. From Fenn I learned the need for careful advance analysis of the nature of a problem and the need for planned experiments to test a given hypothesis.

Although World War II cut short my stay in Rochester, I got a job at Purdue University immediately afterward. My review starts with the work I did there.

ION SELECTIVITY OF EXCITABLE MEMBRANE

Classic electrophysiology of axons provided scientists with a coherent body of facts that they could use to explain excitation and conduction by combining the idea that depolarization led to both an enhanced excitability and (more slowly) to a diminished excitability. This two-factor theory of excitation was quite successful in predicting the effects of long prepulses on the subsequent response of an axon to test excitations but it did not explain how a small depolarization of the membrane was amplified into a large one that led to an action potential that was propagated along the fiber.

Hodgkin and Huxley's studies in 1952 (22) that led to a complete understanding of how the action potential was generated depended greatly on two developments: the use of radioisotopes to confirm the gain in Na^+ that was theoretically expected to underlie excitation, and the use of an electronic feedback system and an internal current wire, as originally developed by K. S. Cole and by G. Marmont, that effectively prevented impulse conduction. With these techniques it was possible to compute a reasonable action potential having only tabular values of the conductance parameters m, h, and n and to obtain a knowledge of their kinetics. Thus the problems of under-standing excitation moved from understanding electrical parameters to understanding *1*) how the selectivity of the membrane changes from one favoring K^+ over Na^+ by ~25:1 to the reverse, *2*) how membrane potential controls the kinetics of the selectivity processes, and *3*) how the dissipative process that leads to an Na^+ entry and an equivalent K^+ loss is reversed.

The initiation of the action potential involves dissipative processes, questions of ion selectivity, and questions about the way that the electric field of the membrane modifies the conformation of mem-brane proteins. In any event, my interest was directed at the nature of ion selectivity.

One of the startling facts that was emerging as an explanation of how nerve excitation took place was that the excitable membrane, long considered a K^+-selective structure, underwent a dramatic change to become Na^+ selective when the membrane was depolar-ized. The usual explanation of why cell membranes were K^+ selective was that K^+ was smaller than Na^+. Measurements in aqueous solutions showed that K^+ had a higher limiting conductance and the application of Stoke's law to a structureless solvent made it possible to visualize apparent ionic radii as shown in Figure 1. According to a strict application of such notions, ions like Rb^+ and Cs^+ ought to be substantially more permeable than K^+ because their apparent hy-drated radii were less than that of K^+. I decided to do some experi-ments. It was relatively simple to measure the influx of ions such as Na^+ and Cs^+ into frog skeletal muscle fibers and to show that both Cs^+ and Na^+ influx were small when compared with that of K^+ (29). Clearly there was something wrong.

It seemed necessary to think about the nature of ions in solution and to develop a physically reasonable view of how ions might escape from their aqueous environment into some new region (the mem-

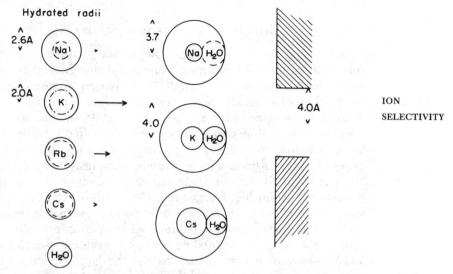

Hydrated radii

Figure 1. Calculated "hydrated" radii of alkali metal cations (left) compared to calculated size of these ions with a single shell of hydration (right). Note that for Cs^+ there is virtually no contribution from water to its hydrated radius (cf. water molecule immediately below the ion) and indeed only Na^+ has appreciable hydration. Length of *arrows*, proportional to permeability of membrane to ion. Note that the largest hydrated ion (Na^+) and the smallest (Cs^+) have similar values for permeability. On the extreme right is a hypothetical pore of radius 4.0 Å into which K^+ will fit. The Na^+ will be excluded because it does not fit closely and Cs^+ because it is too large. [From Mullins (31).]

brane) while still requiring ions to obey physical laws. The idea that ions could dissolve in a low dielectric constant liquid such as the bilayer could be rejected out of hand given the hydration energies of ions in aqueous solution (of the order of 100 kcal/mol or the strength of many chemical bonds). This left the idea that a cation might either combine with a fixed anionic charge in the membrane and so become an electroneutral entity that might move across the membrane or enter some region in the membrane that is sufficiently polarizable to serve as a substitute for hydration and thus allow an ion to escape from water and enter the membrane phase. Two arrangements seemed possible to me: *1*) a relatively small polypeptide that had a polarizable core and space for the ion and *2*) a protein that formed a channel across the membrane. I emphasize that these notions were developed long before valinomycin or gramicidin were known. Any model also had to address the now-known fact that an excitable membrane could be highly K^+ permselective at the resting potential, yet become highly Na^+ selective upon depolarization. Moreover, this Na^+ selectivity had to be a transient, given the reality of Na^+ inactivation.

[277]

How Ions Get Through Membrane

MEMBRANE
TRANSPORT

One of the things that seemed important to study was how the selectivity of the excitable membrane could change from the resting state where a permeability for K^+ was favored to an excited state where the permeability of Na^+ was paramount. To understand this problem, it was essential to look at the state of ions in aqueous solution and to understand the physical forces that were involved. I knew that hydration energies for substances like KCl were of the order of 100 kcal/mol, a force stronger than many chemical bonds. With the aid of Robert Platzman, my colleague in the physics department at Purdue, I began to understand that it is impossible to move ions such as K^+ in aqueous solution to a low dielectric medium such as the phospholipid membrane without an intervening energy source. Calculating the binding energy of water molecules to Na^+ and K^+ showed that more than half of the large hydration energy was in layers of water beyond the first shell so that it was possible to visualize an alkali cation with one complete water shell as an entity that could penetrate through a pore in the plasma membrane (28). A totally dehydrated ion could penetrate but only if the pore replaced the hydration that the ion enjoyed in water with a solvation of a similar magnitude.

This exercise in theory promptly suggested experiments. If Na^+ and K^+ penetrated either as the dehydrated ion or as the ion with one water shell then, because the crystal radius of Na^+ was smaller than that of K^+, Na^+ would not fit a K^+-size pore as well as K^+ itself. Therefore the rate of penetration of an ion would be proportional to how well it fit a pore of a particular size. Thus, large ions would have a low permeability because they were large, and small ions would have a low permeability because they could not escape from aqueous hydration. Measurement on frog muscle fibers showed that (29) both Na^+ and Cs^+ had a low permeability (0.043 and 0.11), while K^+ was maximal and Rb^+ somewhat lower in its permeability (1.0 and 0.54).

A second study was necessary to look for an ion that was quite different from K^+ in its chemical properties but very similar in its crystal radius. This ion proved to be Tl^+. A study of the transfer rate into frog muscle showed that Tl^+ was slightly faster than K^+ (33). It appeared, therefore, that one could predict the rate of penetration of monovalent cations if one assumed that there was a mean size for the channel or pore through which they must pass and that ions larger or smaller than this mean pore radius would be poorly permeable. This is the case for all known monovalent cations that do not interact chemically either with the membrane or with water.

These studies then suggested that if one could change the mean pore size in the membrane from a large size that was suited to the passage of K^+ across the membrane to a smaller one that would pass

Na$^+$, it would be possible to explain nerve excitation on the basis of this change (30). It has turned out that instead of one population of pores with a variable size, there are actually two populations, each with a different size, but this finding in no way invalidates the idea behind partial ionic hydration and ion permeability.

These suggestions were received by my associates with something less than enthusiasm. George Eisenman had come out with a detailed and quantitative explanation of how glasses discriminate between ions (13) and his theory was widely applied to biological problems even though the connection between glass membranes and plasma membranes remains obscure at best. The following quote from a review (5) provides a sample of how partial ion hydration was viewed in 1969.

Mullins (1956) has argued that a pore might select a partially hydrated ion, with a certain number of hydration shells, of an optimal size and might discriminate against both larger and smaller ions: "If an ion at a specified level of hydration closely approximates the size and shape of a pore, it may exchange all its hydration, beyond the specified level for the solvation afforded by the walls of the membrane pore. . . . If K$^+$ approaches a pore that is precisely the same size as this ion with its first hydration shell (denoted (K$^+$)$_1$), it may, as indicated previously, exchange hydration, for water shells from 2 to infinity, for a similar attraction with the structure lining the pore. If the pore is somewhat smaller than (K$^+$)$_1$, penetration cannot occur for steric reasons, while if the pore is somewhat too large, penetration likewise cannot occur because the attraction of the ion for water shells of 2 and greater is not compensated by a solvation of similar magnitude in the pore."

The erroneous assumption implicit in the . . . expressions "similar attraction" and "similar magnitude" is that any nearest neighbor at a given distance, whether water or membrane substance, will exert similar attractive forces upon the ion. In fact, the attractive force exerted by the membrane may be much more or less than that exerted by water, depending upon whether its field strength is much more or much less than that of water, and it is this variation in electrostatic forces which determines selectivity. In addition, one of the earliest reasons why physical chemists discarded Gregor's selectivity theory and Ling's 1952 theory, which were founded on the concept of hydrated size, was the ambiguity of this concept, and Mullins' discussion of the precise fit of a pore to specific hydration shells of an ion is even less defensible physically.

A more accurate and realistic assessment of the idea that partial ionic hydration is important and that Na$^+$ is a smaller ion than K$^+$ is provided by Hille (21).

This idea that K$^+$ and Cl$^-$ are smaller ions than Na$^+$ and that their *permeability is consistent with a small-pore theory is repeated in Krogh's encyclopedic Croonian lecture (1946) and in an identical manner in Hodgkin and Katz's pivotal paper (1949) on the sodium hypothesis for action potentials in squid axons. Both papers, however,*

conclude mistakenly that cases of selective Na^+ permeability could not be explained by pores "which would require a definitely higher diffusion rate for K than for Na" (Krogh, 1946). We understand now that errors in this thinking lie in the long-held misconception that the hydrated particle is a defined and rigid ion-water complex and in the failure to include interaction energies in addition to mechanical size in predicting what ions would be permeant. Indeed, for this reason pore theory itself may have been an intellectual barrier retarding both the first postulation and the later acceptance of the sodium hypothesis for axons.

A modification of the traditional pore theory was offered by Mullins (1959a,b; 1960). He recognized that the barrier to movement of a heavily hydrated ion into a narrow pore is the energy required to dehydrate the ion. Mullins (1959a) argued that the energy barrier would be eliminated if, as waters are shed from the ion, they are replaced by "solvation of similar magnitude obtained from the pore wall." This idea prevails today.

Hille is responsible for much of the modern work on the passage of ions through the excitable membrane. In his 1971 paper (19) he showed that the organic cation hydroxylamine was about as permeable through the Na^+ channel as Na^+ itself and that hydrazine and ammonium ions were measurably permeant while methylhydrazine and methylamine were not measurably permeant. Hille developed a detailed model for Na^+ pore-selectivity filter as an oxygen-lined pore of dimensions 3×5 Å. These dimensions require that Na^+ be accompanied by at least 1 molecule of water in order to enter the pore.

Parallel developments with model membranes, by Mueller and Rudin (26), showed that it was possible to form phospholipid bilayers and that a variety of substances could stimulate significant ion conductance in such structures. Some of these substances, such as gramicidin, formed channels in the membrane while others formed carriers. As the structure of these compounds became known, it was clear that they had in common a core of polarizable oxygen atoms to furnish substitute hydration for an ion and an outer layer of hydrophobic material to allow the insertion of the molecule into a phospholipid bilayer. Further developments in the study of ionic channels made it possible to characterize the ion selectivity of both Na^+ channels and K^+ channels. Although a great variety of organic cations are permeant via the Na^+ channel, the K^+ channel is substantially more selective because it will pass only Tl^+, K^+, Rb^+, and NH_4 (20). Interestingly, only two ions, Tl^+ and NH_4, are reasonably permeant through both channels in the frog node possibly because the ions assume different hydration patterns as they pass through the two types of channels.

ANESTHETIC ACTION OR OCCUPANCY OF MEMBRANE BY FOREIGN MOLECULES

The usual membrane proteins, such as Na^+-K^+-ATPase, or Na^+ channels are so large that it is not usually necessary to worry about either the entry or loss of such molecules from a phospholipid bilayer. On the other hand, paraffinic molecules such as decane may enter or leave a phospholipid bilayer at a rapid rate. My interest in such matters was generated by a paper by Brink and Posternak (3), who showed rather convincingly that it was possible to block conduction in sympathetic ganglia for synapsing fibers, whereas axons passing through the ganglion were not affected by the blocking agents, which in this case were a homologous series of alcohols up to eight carbons in length.

Ferguson (14) had suggested that organic chemical molecules affected biological functions if they were present as a constant fraction of saturation of the compound in aqueous solution, and he related this to a constant thermodynamic activity of the compound acting in whatever biological phase was being affected. This suggestion was a monumental departure from the usual Meyer-Overton hypothesis that compounds acted in proportion to their partition coefficient (although it is possible to show that this property is directly related to the thermodynamic indices proposed by Ferguson).

As the size (or number of carbon atoms) in any homologous series of compounds increases, so does thermodynamic activity required to produce an anesthetic effect. It is sometimes argued that this is merely a reflection of the increasing difficulty of dissolving compounds in water, which has to be the initial site where compounds are introduced into the biophase. Indeed, a correlation can be made between water solubility and speed with which compounds produce their effects. All such kinetic arguments ignore that if enough time is allowed, the system ought to come to equilibrium and in the case of whole animal experiments, that days, weeks, and even months are available to produce such an equilibrium. Still, compounds beyond a specific number of carbon atoms do not have physiological effects. A second resort is to the supposition that large paraffinic molecules like decane may be susceptible to enzymatic or bacterial attack, which makes equilibrium impossible because there is a constant consumption of the material being supplied. This seems inherently improbable given the wide range of compounds that would have to be attacked at a rate proportional to their water solubility. Some compounds such as perfluoropentane would appear to be immune to such biodegradation, yet this is a compound that is inert insofar as anesthesia is concerned and has a boiling point virtually identical to pentane itself (a powerful anesthetic). Also, the water insolubility argument does not hold up well in this instance because pentane-oxygen mixtures cause anesthesia in mice in tens of seconds and although the perfluoro compound is about 5 times less soluble in water, there is no effect on mice in times of the order of hours.

Chemical thermodynamics suggests that ideal solutions are those where the heat of mixing is zero and the entropy of mixing is $-R \ln X$, where X is the mole fraction of the compound being added to a mixture. This relationship implies that if substances mix ideally with the phospholipid bilayer of the excitable membrane, there is no real way by which the mixture is perturbed by the addition of a substance composed of foreign lipidlike molecules.

It seemed to me that this part of the application of thermodynamics to the phospholipid bilayer was wrong in the following ways. First, there was a large difference in the molecular volume between anesthetic molecules and those of the phospholipid bilayer. Thus one could expect that a nonideality of mixing would arise simply because of this disparity. A second difficulty might be anticipated along the following lines: The phospholipid bilayer is itself a stable structure that can exist between two aqueous phases—it is hardly a bulk phase, however, and it is therefore unreasonable to expect that macroscopic considerations would apply to this structure. In fact, it seemed simpler to consider the phospholipid bilayer as a crystalline structure and the molecules added to it from the aqueous phases as substances that could be incorporated into the interstices of the structure but that such substances could not increase the total volume of the phospholipid bilayer structure. These considerations lead to an analysis that suggested that anesthetic molecules would work if 1) they were not too large to fit into the interstices of the phospholipid structure, 2) the molecule had a reasonable attractive force between itself and the paraffinic structures that it encountered in the bilayer, and 3) anesthetic action could be predicted by using the solubility parameters of the Hildebrand theory of solutions (27).

These suggestions were received with more enthusiasm than I dared hope. It seemed that they fixed up both the Meyer-Overton hypothesis and the subsequent ideas of J. Ferguson in a way that was acceptable to many people. Still, there were difficulties. One of the things that a comprehensive theory of anesthesia would be able to predict was how to design the ideal anesthetic molecule. Practical anesthetics were few and far between and could only be developed by extensive testing on animals. I had hoped that the relatively simple considerations of the Hildebrand theory (which said that molecules mix the most ideally with substances of the same sort) could be used to understand that, for example, heavily chlorinated hydrocarbons would not be practical anesthetics because their high

[282]

solubility parameter would lead them to selectively accumulate in structures such as heart muscle, which seemed to have an affinity for such substances. This idea, however, appeared to fail when details about structures homologous with halothane became available. It was clear that compounds with virtually the same boiling point (and hence with identical internal attractive forces) had quite different thresholds for anesthesia even though molecular volumes were the same. This consideration led me to believe that as molecular size increased, there were steric considerations that now governed the way such molecules could be introduced into the bilayer structure. There is no dispute that substances such as xenon or nitrous oxide are ideal anesthetics—they are small molecules and would be as practical as nitrogen for an anesthetic if not for the inconvenience of having to apply such substances under pressures higher than atmospheric.

For the last fifteen years the study of how anesthetic molecules influence membranes has multiplied and diversified. Both the gramicidin channel in artificial bilayers and the Na^+ channel in squid axons are affected by physiological concentrations of anesthetic molecules.

From this tremendous volume of work, I must pick only a few developments that have in my view offered new insights into how foreign molecules interact with the membrane at the molecular level. The studies comparing the capacitance of the squid axon membrane (hence, its thickness) with the addition of n-alkanes to the membrane show that the membrane thickens with such additions and that Na^+ currents are diminished (17). Such a finding strongly suggests that the thickening of the membrane is brought about because hydrocarbons enter the core of the phospholipid structure, thus expanding membrane thickness.

Other studies using alcohols (18), however, show just as clearly that the alcohols do not thicken the bilayer at concentrations where they inhibit Na^+ currents nor would one expect such a thickening since the polar groups of such substances lie at the membrane-water interface. These findings taken together surely suggest that the inhibition of Na^+ channels may come about by at least two different ways; in all likelihood, there may be even more ways that anesthetic substances act to effectively reduce Na^+ currents.

Swenson et al. (36) described a different sort of action of anesthetics where Na^+ inactivation was greatly enhanced by octanol whereas the maximum Na^+ conductance was unaffected; hence the ability of anesthetics to affect Na^+ channel inactivation is just another way that Na^+ currents can be reduced without blocking the hydrophilic sites of the channel.

Finally, one of the more dramatic effects in anesthesia is to treat tadpoles with urethane at an appropriate dose and note that they become immobile. When hydrostatic pressure is applied, they recover with the drug still present. This pressure reversal has resulted in studies in isolated cervical sympathetic ganglia, and surprisingly there is no reversal of anesthetic block (25). Such a finding makes it clear that although pressure reversal is a dramatic effect it is not a general one in the sense that all excitable cells can regain excitability with the application of pressure.

ACTION POTENTIALS IN PLANT CELLS

One of the things that was difficult to understand about how electrical excitation in single algal cells such as those of *Chara* took place was that there did not appear to be a requirement for a specific ion in the external medium that would *1)* allow excitation in its presence and *2)* lead to a loss of excitability in its absence. Hodgkin and Huxley's definitive studies had appeared in 1952 and these clearly showed the crucial role of external Na^+ in bringing about electrical excitability in nerve. Thus it was reasonable to think that in *Chara* (or *Nitella*) there ought to be some cation analogous to Na^+ that would carry the inward current of depolarization during the action potential.

I had an energetic graduate student (Cornelius Gaffey) who was interested in identifying a cation that would produce excitability in *Chara*. He set to work to show that he could, in fact, work with an external medium—distilled water—with even divalent cations at the level of less than 10 nM and still get an action potential in the cell. He could not rule out that some cations leaking from the cell were available just outside the excitable membrane but still it seemed possible to me that the release of an internal anion during depolarization was a likely explanation of further cell depolarization.

Experimental verification of this suggestion was intrinsically difficult because, aside from a very short half-life isotope of Cl^- that was impractical to work with, the long-life isotope had only a very low specific activity since commercial production of such an isotope only began in 1952. Fortunately, friends in Oak Ridge, Tennessee, were able to supply some ^{36}Cl that had gone into the pile in 1946—hence a large gain in specific activity. With a thirty-day loading of *Chara* cells and a fairly refined counting technique, it proved possible to show that there was a massive outpouring of Cl^- during a single action potential (16). Parallel experiments with ^{42}K showed that there was an equivalent efflux of these ions; hence the view emerged that electrical excitation in the plant cell involved first the efflux of Cl^- that produced the depolarization, followed by an efflux of K^+ that repolarized the cell.

These rather simple and (to me) satisfying findings were not received by the scientific establishment with any great enthusiasm. I was distressed to learn that our paper describing this work was rejected by the *Journal of General Physiology* because at the time

W. J. V. Osterhout was one of its editors. He had spent his life studying *Nitella* and *Chara* in addition to other plant cells, and he had developed a theory that I could never understand of how electrical excitation in these cells takes place. Clearly the findings of our study were not within the bounds of Osterhout's understanding, and an apologetic editor told me on the phone that there was nothing he could do with the manuscript.

A happy ending to this story is that the manuscript was sent to the *Journal of Physiology* and Bernard Katz as a reviewer told me he considered it a very clear demonstration of how electric excitation in plants takes place. Richard Keynes, at that time an editor, said it was undoubtedly the first botanical paper that the *Journal of Physiology* had ever published.

Although our findings that isotope studies clearly showed equivalent quantities of Cl^- and K^+ were released per action potential, and hence the matter might be considered to be quite settled, a few years later there were reports from Australia (15, 24) that the way electrical excitation took place was via the inward movement of Ca^{2+}. These reports were based on the effects that changes in external Ca^{2+} concentration had on membrane currents, and although it is now known that there is a definite Ca^{2+} entry with depolarization, such an entry is far too small to influence the large currents generated by Cl^- efflux.

RECOVERY PROCESSES IN NERVE

With the universal acceptance of the Hodgkin-Huxley idea that each nerve impulse results in the dissipative movement of Na^+ inward and K^+ outward, attention in the 1960s began to focus on how these ionic movements were reversed. The Na^+ pump was a well-accepted entity but details about its actual mode of operation were somewhat sketchy. As flux studies became more developed, it became clear that more Na^+ was extruded than K^+ taken up, so that Na^+ pumping necessarily became electrogenic (i.e., charge-producing). This fact has had important consequences for an understanding of how very small nerve fibers might function, since in such structures, transient changes in Na^+ pumping in response to nerve impulses could affect membrane potential and hence excitability.

Really quantitative studies of Na^+-K^+ transport in squid giant axons were not possible until the development of the technique of internal dialysis (4), although there was the promise that perhaps internal perfusion (2) would be a technique suitable both for a controlled internal environment for both voltage clamp and for transport studies. While internal perfusion has proven indispensable for many voltage-clamp studies, it is not a technique that produces stable isotopic fluxes in the absence of fluoride ion, a substance that effectively poisons most metabolically dependent ion transport. In

[285]

the absence of this toxic substance, Na^+ fluxes with internal perfusion become highly unstable, of large magnitude, and in great measure insensitive to external ouabain or internal ATP. In contrast, with dialyzed squid axons a normal Na^+ efflux can be measured in the presence of ATP supplied by dialysis and such an efflux can be reduced from ~35 to 1 $pmol \cdot cm^{-2} \cdot s^{-1}$ as ATP is washed from the fiber by dialyzing with ATP-free solutions.

Internal dialysis has become a technique of considerable use in looking at a wide variety of transport processes. In addition to looking at the Na^+ pump where it has provided the strongest support for the notion that no substrate other than ATP or deoxy ATP can serve as substrate, it has also been used to measure K^+ transport (32) and to study the movement of Mg^{2+}, Cl^-, H^+, and Ca^{2+} as well.

Ionic Calcium Movement in Nerve

One of the last ions to be studied in nerve was Ca^{2+}. The reasons for this are not hard to discern since Ca^{2+} forms many insoluble complexes and was known to be sequestered in cells. Hodgkin and Keynes (23) showed that an axon soaked in ^{45}Ca required an impossibly long time to clear the isotope from extracellular compartments and they introduced the substantial improvement of microinjecting the isotope into the axon. Even this method, though, was limited as the amount injected raised Ca^{2+} levels materially and ^{45}Ca underwent a continuous change in specific activity as the experiment progressed.

A major advance was made by DiPolo (6), who initiated the dialysis of ^{45}Ca into axons. Here, the internal Ca^{2+} concentration was under control and very reasonable flux measurements were made. Early experiments showed that if $[Ca^{2+}]_i$ were less than 100 nM, Ca^{2+} efflux was relatively unaffected by the composition of the external solution, but at a $[Ca^{2+}]$ of 1 μM most of the efflux was cut off in Na^+-free solutions. Two interpretations were possible: first, that Ca^{2+} efflux in Na^+-free solutions represented an ATP-driven Ca^{2+} pump of the sort known in red cells; and second, that at low $[Ca^{2+}]_i$ many cations could serve as agents for the countertransport of Ca^{2+}.

A separation of these two mechanisms was not helped by the finding that both were activated by ATP although in the absence of external Na^+ the efflux was much more sensitive to ATP than in the presence of external Na^+ (7).

Subsequent studies with the measurement of isotopic fluxes have shown that these are dependent on K^+_i (this activates with a $k_{1/2}$ of 90 mM (11), Na^+_i, Na^+_o, Ca^{2+}_i, Ca^{2+}_o, ATP_i, H^+_i (10), H^+_o, and on membrane potential. Furthermore, many of these variables are interactive in the sense, for example, that Ca^{2+} efflux is increased by hyperpolarization but only if Na^+_i (12) is present. This interesting finding suggests that possibly the coupling ratio of the Na^+/Ca^{2+}

exchange might be variable and go from an electroneutral 2 $Na^+/$ Ca^{2+} when the energy in the Na^+ gradient is high to electrogenic modes of coupling when there is a demand for more energy. If we add the possibility of variable coupling to all the other variables known to affect Na^+/Ca^{2+}, it becomes less plausible to say that because uncoupled Ca^{2+} efflux has a K_m of 0.2 μM while Na^+/Ca^{2+} has values in the range of from 2–20 μM that it is possible to make a clear separation between these systems. Vanadate is known to be an inhibitor of many phosphorylations and it inhibits uncoupled Ca^{2+} efflux with a K_i^+ of 7 μM (9), which is many times the value necessary for inhibition of the Na^+/K^+ pump. It also must be expected to inhibit Ca^{2+} buffering by ATP-driven mechanisms such as the endoplasmic reticulum and such a change may affect the concentration of Ca^{2+} delivered to the surface membrane at the inside.

A most interesting finding made originally by Baker and Mc-Naughton (1) was that in an injected axon the Ca_o^{2+}-dependent Na^+ efflux was abolished if EGTA were injected into the axon. Measurements of Ca^{2+} influx by dialysis (8) showed that with saturating Na_i^+, Ca^{2+} influx was much larger if Ca_i^{2+} were finite than if it were made nominally zero. This suggested that Ca_i^{2+} had a positive feedback on Ca^{2+} entry as did the results with injected axons. Another factor that needs to be considered is whether a low Ca_i^{2+} (and hence a very large Ca^{2+} electrochemical gradient directed inward) might have some influence of itself on the Na^+/Ca^{2+} exchange. We have cited above the finding that internal K^+ promotes Ca^{2+} influx by Na^+/Ca^{2+} exchange and by inference suggests that internal K^+ as well as internal Na^+ can exchange for entering Ca^{2+}. If this is so, then the presence of Ca_i^{2+} might merely make the Na^+/Ca^{2+} exchange more selective for internal Na^+ and in its absence internal K^+ would supplant Na^+.

The reason for doubting that internal Ca^{2+} needs to be of the order of 0.5 μM in order to promote Ca^{2+} entry is that in intact axons with a normal Ca_i^{2+} there is a large Ca^{2+} entry that is Na^+_i dependent and this occurs in Na^+-free conditions or with depolarization (34). Entry of Ca^{2+} as measured optically with arsenazo III as an indicator is half maximal at 28 mM Na_i and the relationship between Na_i and Ca^{2+} entry is very steep with a Hill coefficient of 7 (35). By contrast, Ca^{2+} influx measured in dialyzed axons with Ca_i 60 nM is half maximal at 60 mM and the relationship is much shallower (8). The reason the net Ca^{2+} entry (arsenazo) and the isotopically measured Ca^{2+} influx differ may be that the internal dialysis system does not capture all the entering Ca^{2+} (this has never been rigidly established), or that the use of Ca^{2+} buffers and artificial internal conditions somehow inactivate a good deal of the carrier.

In conclusion, although ATP-driven Ca^{2+} pumps are clearly demonstrated in the red cell and in the SR, as well as in other preparations, the squid axon has a Ca^{2+} efflux system that is sensitive to ATP but one cannot be really certain that this is an ATP-driven Ca^{2+} pump rather than an aspect of the Na^+/Ca^{2+} exchange system when the Na^+ gradient is made very small.

BIBLIOGRAPHY

1. BAKER, P. F., and P. A. McNAUGHTON. Calcium-dependent calcium efflux from intact squid axons: Ca-Ca exchange or net extrusion? *J. Physiol. Lond.* 258: 97P–98P, 1976.
2. BAKER, P. F., A. L. HODGKIN, and T. I. SHAW. Replacement of the axoplasm of giant nerve fibres with artificial solutions. *J. Physiol. Lond.* 164: 330–354, 1962.
3. BRINK, F. J., and J. M. POSTERNAK. Thermodynamic analysis of the relative effectiveness of narcotics. *J. Cell. Comp. Physiol.* 32: 387–396, 1948.
4. BRINLEY, F. J., JR., and L. J. MULLINS. Sodium extrusion by internally dialyzed squid axons. *J. Gen. Physiol.* 50: 2303, 1967.
5. DIAMOND, J. M., and E. M. WRIGHT. Biological membranes: the physical basis of ion and nonelectrolyte selectivity. *Annu. Rev. Physiol.* 31: 581–646, 1969.
6. DIPOLO, R. Calcium efflux from internally dialyzed squid giant axons. *J. Gen. Physiol.* 62: 575–589, 1973.
7. DIPOLO, R. Ca pump driven by ATP in squid axons. *Nature Lond.* 274: 390–392, 1978.
8. DIPOLO, R. Ca influx in internally dialyzed squid axons. *J. Gen. Physiol.* 73: 91–113, 1979.
9. DIPOLO, R., and L. BEAUGÉ. Physiological role of ATP-driven calcium pump in squid axon. *Nature Lond.* 278: 271–273, 1979.
10. DIPOLO, R., and L. BEAUGÉ. The effect of pH on Ca extrusion mechanisms in dialyzed squid axons. *Biochim. Biophys. Acta* 688: 237–245, 1982.
11. DIPOLO, R., and H. ROJAS. Effect of internal and external K^+ on Na^+-Ca^{2+} exchange in dialyzed squid axons under voltage clamp. *Biochim. Biophys. Acta* 776: 313–316, 1984.
12. DIPOLO, R., F. BEZANILLA, C. CAPUTO, and H. ROJAS. Voltage dependence of the Na/Ca exchange in voltage clamped dialyzed squid axons. *J. Gen. Physiol.* 86: 457–478, 1985.
13. EISENMAN, G. Cation selective glass electrodes and their mode of operation. *Biophys. J.* 2: 259–323, 1962.
14. FERGUSON, J. The use of chemical potentials as indices of toxicity. *Proc. R. Soc. Lond. Ser. B* 127: 387–404, 1939.
15. FINDLAY, G. P. Voltage clamp experiments with *Nitella*. *Nature Lond.* 191: 813–814, 1961.
16. GAFFEY, C. T., and L. J. MULLINS. Ion fluxes during the action potential of *Chara*. *J. Physiol. Lond.* 144: 505, 1958.
17. HAYDON, D. A., J. REQUENA, and B. W. URBAN. Some effects of aliphatic hydrocarbons on electrical capacity and ionic currents of the squid axon membrane. *J. Physiol. Lond.* 309: 229–245, 1980.
18. HAYDON, D. A., and B. W. URBAN. The action of alcohols and other non-ionic surface-active substances on the Na current of the squid axon. *J. Physiol. Lond.* 341: 411–428, 1983.

19. HILLE, B. The permeability of the Na channel to organic cations in myelinated nerve. *J. Gen. Physiol.* 58: 599–619, 1971.
20. HILLE, B. K channels in myelinated nerve. *J. Gen. Physiol.* 61: 669–686, 1973.
21. HILLE, B. *Ionic Channels of Excitable Membranes.* Sunderland, MA: Sinauer, 1984.
22. HODGKIN, A. L., and A. F. HUXLEY. A quantitative description of membrane current and its application to conduction and excitation in nerve. *J. Physiol. Lond.* 117: 500–544, 1952.
23. HODGKIN, A. L., and R. D. KEYNES. Movements of labelled calcium in squid giant axons. *J. Physiol. Lond.* 138: 253–281, 1957.
24. HOPE, A. B. The action potential in cells of *Chara. Nature Lond.* 191: 811–812, 1961.
25. KENDIG, J. J., J. R. TRUDELL, and E. N. COHEN. Effects of pressure and anesthetics on conduction and synaptic transmission. *J. Pharmacol. Exp. Ther.* 195: 216–224, 1975.
26. MUELLER, P., and D. O. RUDIN. Translocators in biomolecular lipid membranes. *Curr. Top. Bioenerg.* 3: 157–249, 1969.
27. MULLINS, L. J. Some physical mechanisms in narcosis. *Chem. Rev.* 54: 289, 1954.
28. MULLINS, L. J. *The Structure of Nerve Cell Membranes, Molecular Structure and Functional Activity of Nerve Cells.* Pub. 1, American Institute of Biological Sciences, 1956, p. 123.
29. MULLINS, L. J. The penetration of cations into muscle. *J. Gen. Physiol.* 42: 817–829, 1959.
30. MULLINS, L. J. An analysis of conductance changes in squid axon. *J. Gen. Physiol.* 42: 1013, 1959.
31. MULLINS, L. J. The macromolecular properties of excitable membranes. *Ann. NY Acad. Sci.* 94: 390, 1961.
32. MULLINS, L. J., and F. J. BRINLEY, JR. Potassium fluxes in dialyzed squid axons. *J. Gen. Physiol.* 53: 704–740, 1969.
33. MULLINS, L. J., and R. D. MOORE. The movement of thallium ions in muscle. *J. Gen. Physiol.* 43: 759, 1960.
34. MULLINS, L. J., and J. REQUENA. The "late" Ca channel in squid axons. *J. Gen. Physiol.* 78: 683–700, 1981.
35. MULLINS, L. J., T. TIFFERT, G. VASSORT, and J. WHITTEMBURY. Effects of internal sodium and hydrogen ions and of external calcium ions and membrane potential on calcium entry in squid axons. *J. Physiol. Lond.* 338: 295–319, 1983.
36. SWENSON, R. P., G. S. OXFORD, and T. NARAHASHI. Enhancement of Na channel inactivation by octanol and decanol. *Biophys. J.* 21: 41a, 1978.

XII

Membrane Transport in Excitation-Contraction Coupling

RICHARD J. PODOLSKY

EXCITATION-contraction coupling is the sequence of largely membrane-dominated processes that follow depolarization of the outer membrane of a muscle cell and leads to activation of the contractile mechanism. The major steps in the sequence are listed in Figure 1. The definition of these events came about through the interaction of many investigators following many different lines of research [see reviews by Ebashi and Endo (6), Weber and Murray (36), Peachey and Franzini-Armstrong (27), Baylor (2), and Martonosi and Beeler (25)]. Several contributions to this story have come out of my laboratory, and this chapter describes how they came about.

EXCITATION-
CONTRACTION
COUPLING

I was introduced to the idea of excitation-contraction coupling at University College, London, where in 1955 I went on a postdoctoral fellowship. This was my second postdoctoral experience. My first was spent at the Naval Medical Research Institute in Bethesda, Maryland, with Manuel Morales, where the emphasis of the laboratory was on the biochemistry of the muscle proteins. My project there was to study the hydrolysis of ATP by myosin (31). At University College, my sponsor was A. V. Hill, who led me to frog sartorius muscles and classical muscle physiology.

My first task was to make solutions for bathing dissected muscles, and I puzzled over the role played by each of the ions in the traditional Ringer solution. It was surprising to me that such a simple solution could keep the muscle functional for many days. Following up these thoughts, I replaced the Cl^- in the Ringer solution with Br^-, NO_3^-, and I^-, repeating experiments done earlier by Hill and MacPherson (15) and Kahn and Sandow (24), and saw the twitches increase in magnitude while the tetanus tension remained the same. D. R. Wilkie had an apparatus in which muscle contractile force could be measured at elevated pressure, and with it I saw that pressure had the same graded effect on twitch tension as the anion series. I wondered whether the effect of Cl^- (possibly in positioning negative charge in some critical site involved in muscle activation) could be augmented by pressure to simulate the effect of the other ions. To see whether this effect of pressure could be demonstrated in other anion-sensitive systems, I searched the literature for another preparation that showed a graded anion effect and found it in the cat erythrocyte Na^+ transport system. Na^+ efflux from cat erythrocytes was reported to be anion dependent (5). It was maximum when Cl^- was the predominant anion in the bathing solution, and decreased progressively in Br^-, NO_3^-, and I^-. This work had been done by Hugh Davson, a pioneer in membrane studies, who fortunately was at that time working at University College. He showed me how to draw blood from a cat and helped me reproduce his erythrocyte experiments. Then I tried the effect of pressure on this system and found that increased pressure decreased the Na^+ efflux. More interestingly I found that the amount of pressure required to make the Na^+ efflux in Cl^- the same as in Br^- was equal to that required in the muscle system to make the Cl^- twitch the same magnitude as the Br^- twitch. In other words, the sensitivity of the cat erythrocyte to pressure, as scaled to the anion effect, was the same as that of the twitch amplitude in muscle. I reported this finding to the Physiological Society (UK) at its University College meeting in 1956 (28) and it was well received. I felt that I had made contact with scientific lines that were current in the Society.

These were lively times for muscle research. The sliding filament model for contraction had just been put forward in a pair of back-to-back papers in *Nature* by A. F. Huxley and R. Niedergerke (17) and by H. E. Huxley and J. Hanson (22). H. E. Huxley had visualized the two sets of interdigitating filaments by electron microscopy (20). A. F. Huxley had worked out his crossbridge model (16) and copies of his manuscript were being circulated for comment. In addition, with R. E. Taylor, another American postdoctoral fellow, he was doing the local activation experiments (19) that led to an understanding of the role of the transverse tubules in conducting the electrical activation signal from the surface membrane of a muscle fiber to the interior of the cell (step 1 in Fig. 1).

Toward the end of 1956, I returned to the Naval Medical Research Institute, where I had arranged for a permanent position. I planned to continue working on the projects started in London, with major effort being given to study of the contraction mechanism in muscle fibers. To do this, of course, required that the fibers be activated in some way, so I remained alert to developments in the excitation-contraction coupling field. I knew about Heilbrunn's microinjection studies, described in his 1943 book on general physiology (11), which showed that Ca^{2+} injected into a muscle fiber with a glass

Steps in excitation-contraction coupling in striated muscle fibers.

Depolarization of the outer membrane
1 ↓
Depolarization of the transverse tubules
2 ↓
Increase of calcium permeability of the sarcoplasmic reticulum
3 ↓
Increase in calcium concentration in the myofibril space
4 ↓
Binding of calcium to troponin
5 ↓
Movement of tropomyosin on actin
6 ↓
Actomyosin crossbridge formation and turnover
7 ↓
Movement of calcium back into the sarcoplasmic reticulum
8 ↓
Relaxation

Figure 1

micropipette produced a strong contraction while other ions had practically no effect. Therefore Heilbrunn thought that entry of Ca^{2+} into the fiber volume, across the surface membrane, was the physiological activator. However, having been with A. V. Hill, I also knew about his experiments on the delay between surface membrane depolarization and contraction (13, 14). He found that the time between the action potential and activation was too short for activation to be caused by diffusion of a substance from the surface to the middle of the fiber, and concluded that a process, not a substance, carried the signal inward from the surface of the fiber.

SKINNED MUSCLE FIBERS AND CALCIUM ACTIVATION

I became directly involved in this question when in 1960 I heard about the skinned fiber preparation of Natori and began to repeat the experiments described in his famous 1954 paper (26). This paper described a very interesting preparation in which a segment of a single muscle fiber was removed from a muscle, covered with oil, and then the surface membrane of the fiber was peeled back, exposing the inner parts of the fiber. Natori, applying test solutions with sharpened wooden sticks, found very little specificity as far as ability to elicit the contractile response was concerned: distilled water seemed to be as effective as Ca^{2+}. This result was clearly at variance with Heilbrunn's microinjection experiments.

The resolution of this apparent conflict turned out to depend on the method used to apply the test solution in the two studies. When I repeated Natori's experiment, I found the lack of specificity he

[293]

reported. However, I found it difficult to score the intensity of the contractile response with his procedure because the fiber generally moved a little when the pointed stick made contact with it. I thought the results might be clearer if test solutions were applied with a more delicate technique. Taking advantage of the fact that the fiber was immersed in oil, I used a glass micropipette to form a small droplet of test solution ($\sim$30 μm diam) in the oil and then moved the droplet toward the fiber until it touched, transferring a known amount of solute to the fiber. With this procedure the specificity of the contractile response for Ca^{2+} could be demonstrated in Natori's preparation (30). Natori's report of no Ca^{2+} specificity could be reconciled with my observation if wood moistened with water leached out Ca^{2+}. I mentioned this possibility to Annemarie Weber, who was studying Ca^{2+} binding to the regulated actomyosin system (35) and had techniques for measuring low levels of Ca^{2+}. Together we measured the Ca^{2+} present in the thin film of water on the surface of a wooden stick that had been dipped into distilled water and found very high Ca^{2+} concentrations. I concluded that the apparent lack of specificity in Natori's experiments was caused by the presence of significant amounts of Ca^{2+} in all of his applied test solutions, and that Heilbrunn was correct in attributing a special role for Ca^{2+} in the activation mechanism.

INTERNAL MEMBRANE SYSTEM

The experiments with skinned fibers introduced me to the fascinating world of microscopy and micromanipulation. I decided to press on with skinned fiber experiments because they were too much fun to put aside. I had noticed that the duration of the contractile response elicited by an applied droplet depended on the amount of Ca^{2+} in the droplet. When Roy Costantin came to my laboratory as a postdoctoral fellow, I suggested that we study this effect quantitatively, hoping to get information about the way the fiber handled added Ca^{2+}.

Initially, we added various amounts of Ca^{2+} to droplets containing 140 mM KCl, and were surprised to find a qualitative difference between the effects of small and large droplets. While small droplets required the presence of Ca^{2+} to elicit a contractile response, large droplets could produce a definite response in the absence of Ca^{2+}. There was also a marked difference in the nature of the responses. The small-droplet–Ca^{2+}-specific response was localized to a region of the fiber close to the applied droplet and was symmetric within this region. The large-droplet response was spotty and often asymmetric relative to the site of droplet application; it appeared to be determined by the inherent responsiveness of the fiber itself (4). Further study showed that the response elicited by a large droplet depended on the monovalent ions in the droplet. We found evidence for the presence of membranes within the fiber (called *internal membranes* to distinguish them from the surface membrane) that responded to depolarization caused by changes in monovalent ion concentration. Large droplets could perturb the concentration of ions within the cell, by either dilution of the intracellular ions or addition of new ions, depending on the chemical composition of the droplet. We concluded that depolarization of some element of the internal membrane system was responsible for the large-droplet responses. This conclusion was buttressed by the demonstration that similar responses could be elicited by direct electrical stimulation of the skinned fiber. We could not decide whether the affected structure was part of the transverse tubules or the sarcoplasmic reticulum (SR), although some of the evidence collected at that time pointed to the tubules. This result agreed with the observation of A. F. Huxley and his collaborators (18, 19) that in intact fibers the transverse tubules carry the effect of surface membrane depolarization to the inner parts of the fiber volume.

Location of Intracellular Calcium

While our experiments with skinned fibers were being carried out, the biochemists were characterizing the membrane-bounded vesicles that were isolated from muscle cells (6, 25). In the presence of ATP, these vesicles accumulated Ca^{2+} and reduced the concentration of Ca^{2+} in the medium to very low levels. In addition, they could be made to release Ca^{2+} by treatment with caffeine (36), a procedure known to activate intact muscle fibers. The introduction of glutaraldehyde as a fixative for electron microscopy enabled the fine structure of the internal membrane system to be visualized. In frog fibers, the transverse tubules penetrated the cell at the Z line (21, 27) and the SR formed cisternae that contacted the transverse tubules at the Z line and were connected within the sarcomere by longitudinal connecting tubules.

These biochemical and morphological results gave rise to the idea that muscle activation and relaxation were due to intracellular movement of Ca^{2+} between the SR space and the myofilament space. The first direct evidence that this was the case came from a nice experiment by F. F. Jobsis and M. J. O'Connor (23), who injected murexide, a Ca^{2+}-sensitive dye, into intact frog fibers and monitored the optical absorption. Following electrical activation the absorption shift showed a transient increase in Ca^{2+} level. Regarding the source of this Ca^{2+}, information came from electron micrographs made by Clara Franzini-Armstrong of relaxed skinned fibers that had been treated by Roy Costantin with oxalate to precipitate the intracellular calcium and form electron-dense precipitates that could be visualized in the electron microscope (3). The micrographs showed precipitate only in the terminal cisternae of the SR. There was no precipitate in either the transverse tubules or in the myofilament space. These findings were consistent with the idea that the activation cycle was controlled by intracellular translocation of Ca^{2+}, and were the first evidence that a specific region of the SR might serve as the source of activator Ca^{2+}.

The impetus for doing the oxalate experiment came from a suggestion of A. F. Huxley to try this reagent as an in situ Ca^{2+} precipitant in a skinned muscle fiber. I have been fortunate to have my laboratory in Bethesda, Maryland, where scientists from all over the world come to visit. I have benefited enormously from the helpful suggestions of a steady stream of visitors. I have found these contacts far more useful than the scientific literature. At any particular time there seems to be a body of unpublished information in the air that one must have access to in order to function effectively. While this information can be picked up at meetings, one-on-one visits have been more useful for me. The suggestion from Huxley was an example of this.

Perfused Skinned Fibers

One of the difficulties in sorting out the processes involved in the activation of muscle fibers has been the extreme sensitivity of troponin, the Ca^{2+} receptor in vertebrate striated muscle, to Ca^{2+}. The concentration required to saturate the troponin is ~ 1 μM, while the contaminant Ca^{2+} in the salts used to make solutions of physiological ionic strength is ~ 10 times greater than this. Before specific chelators for Ca^{2+} became available, all test solutions contained more than enough Ca^{2+} to completely saturate the physiological regulation mechanism, which made it virtually impossible to study the influence of Ca^{2+} on this system with equilibrium measurements.

The situation became manageable when the specific Ca^{2+}-chelating agent ethyleneglycol-bis(2-aminoethylether)-tetraacetic acid (EGTA) became available. This reagent has an apparent stability constant for Ca^{2+} of 6.7 M^{-1} at pH 7.0, which makes it ideal for buffering the concentration of ionized Ca^{2+} in the micromolar range. It was first used as a Ca^{2+} buffer to titrate the ATPase activity of regulated actomyosin by Weber (35) and eventually became widely used in work with fibers where the surface membrane had either been removed by microdissection or made permeable by chemical treatment. A special property of EGTA is that it binds Ca^{2+} 10^5 times more strongly than Mg^{2+}, which makes it possible to control Ca^{2+} independently of Mg^{2+}. This contrasts with ethylenediaminetetraacetic acid (EDTA), which binds Ca^{2+} only 10^2 times more strongly than Mg^{2+} and therefore cannot be used to independently buffer the Ca^{2+} level.

The first set of experiments in my laboratory with the EGTA Ca^{2+} buffer system was made by Duane Hellam to measure the relation between Ca^{2+} concentration and isometric force (12). Until then the contractile response had been measured by applying droplets con-

[296]

taining Ca^{2+} to skinned fibers in oil and monitoring the resulting local movement with a light microscope, either visually or with a motion picture camera (4, 29). These local movements were difficult to measure quantitatively. The new method involved measuring the force of a several-millimeter-long segment of skinned fiber attached to a force transducer at one end and clamped at the other, and perfused with a physiological solution in which the level of Ca^{2+} was controlled with EGTA. The test solutions were contained in an array of wells arranged mechanically so that the bathing solution could be changed in a few seconds. This procedure enabled test solutions at controlled Ca^{2+} concentration levels to be applied uniformly to the fiber surface, helped provide axial homogeneity of experimental conditions, and presumably contributed to the reproducibility of the force measurement.

Delay in Force Development

An unexpected feature of the force records from these experiments was a delay of many seconds between the time the Ca^{2+} concentration in the bathing solution was raised above the contraction threshold and the time force appeared (12). This delay was many times greater than the time required for the main components of the Ca^{2+} buffer system at pH 7.0 ($CaEGTA^{2-}$ and H_2EGTA^{2-}) to diffuse into the fiber. We guessed that the delay was caused by two factors: 1) slow dissociation of Ca^{2+} from the EGTA complex and 2) preferential uptake of Ca^{2+} by the SR rather than troponin. This interpretation was supported by the fact that treatment of the fiber with deoxycholate, an SR inhibitor, decreased the length of the delay. In addition, the rate of release of Ca^{2+} from the EGTA complex, using rate data obtained as described in the next section, appeared to be slower than the rate at which the SR could take up Ca^{2+}.

EGTA Kinetics

Because understanding the details of skinned fiber force measurements required knowledge of the kinetics of Ca^{2+} interaction with EGTA, I undertook to measure these rates directly. I enlisted the cooperation of Robert Berger, a friend who had a stopped-flow apparatus in his laboratory. Our first approach was to measure the rate at which a more tightly binding ion displaced Ca^{2+} from the complex, thinking that the rate-limiting step in the displacement reaction might be dissociation of Ca^{2+} from the complex. This strategy was unsuccessful, probably because the displacement reaction mechanism was more complex than we thought.

While we were puzzling over this, we had another of those fortunate meetings with a visiting scientist, this time George Czerlinski, a very insightful chemical kineticist. He suggested that we measure the rate constant for the Ca^{2+} binding reaction rather than

[297]

the release reaction, and that we make use of the fact that the ratio of these two rate constants is equal to the apparent stability constant. This approach worked (32). We measured the rate of the reaction

$$Ca^{2+} + H_2EGTA^{2-} \rightarrow CaEGTA^{2-} + 2H^+$$

by following proton release with a pH indicator. The rate constant for the reaction was found to be $2 \times 10^6 \cdot M^{-1} \cdot s^{-1}$. The ratio of this number and the apparent stability constant gave $0.4 \, s^{-1}$ as the rate constant for Ca^{2+} release from the complex at pH 7.0. The product of $0.4 \, s^{-1}$ and the concentration of $CaEGTA^{2-}$ inside the fiber in a typical experiment gave release rates that were slow compared with the rate at which the SR was believed to take up Ca^{2+}. This set of measurements confirmed one part of our hypothesis regarding the cause of the delay in force development: the Ca^{2+} supplied by dissociation of $CaEGTA^{2-}$ within the fiber is initially taken up by the SR rather than by the regulatory proteins. However, the capacity of the SR for Ca^{2+} is limited. The delay would be expected to end when this limit is reached because at this stage the Ca^{2+} released by the EGTA system would be shunted to the regulatory proteins and force would develop. Because the SR uptake rate for Ca^{2+} is also limited, a similar shunting effect toward the regulatory proteins presumably takes place when, following the action potential, a large amount of Ca^{2+} is suddenly released by the SR into the myofilament space.

CALCIUM-INDUCED CALCIUM RELEASE

Another interesting phenomenon presented itself when the delay in force development was studied. It came about in the following way. I reasoned that if the rate-limiting step in the loading of the SR in a skinned fiber with Ca^{2+} were the dissociation of $CaEGTA^{2-}$, the length of the delay at a given Ca^{2+} concentration (above the contraction threshold) should be proportional to the concentration of EGTA. When Lincoln Ford came to my laboratory, I suggested that he check this by measuring the delay as a function of EGTA concentration at a given pCa $[-\log_{10} (Ca^{2+})]$. He found that, on average, the delay increased when the EGTA concentration was decreased, but that the delay in a given solution was rather variable (9). More striking, however, was the variability in the time course of the force development when the concentration of EGTA was less than the usual several millimolar. In some cases, the rise in force was monotonic, as expected. However, in most cases there was an overshoot of force above the steady level. The force patterns suggested that the end of the delay was associated with release of Ca^{2+} from the SR as well as a gradual increase in the Ca^{2+} concentration in the myofilament space. Further investigation of the force overshoots showed that they were caused by Ca^{2+} release from the SR, and that this release was triggered by the presence of Ca^{2+} in the myofilament space (8,

[298]

10). These results implied that the Ca^{2+} release mechanism was regenerative.

The full implications of this process are not yet known. Whether or not it can be demonstrated in skinned fibers depends on experimental conditions (degree of SR loading with Ca^{2+}, concentration of Mg^{2+} in the myofilament space, temperature) and one naturally would like to know whether it occurs in intact fibers under physiological conditions. If it did, the next question would be whether Ca^{2+} entering the myofilament space after the action potential acts as a trigger for Ca^{2+} release from the SR. Trigger Ca^{2+} could come from within the transverse tubules [although this seems unlikely because addition of EGTA to the solution bathing intact fibers does not inhibit the activation process (1)] or from a source within the cell (such as the inner surface of the transverse tubules). In any event, the potential of the SR Ca^{2+}-release process to be regenerative is an important fact waiting to be incorporated into the activation mechanism.

In studying Ca^{2+}-induced Ca^{2+} release, an experimental complication is the diffusion delay encountered when the bathing solution is changed, which is probably slow compared with the propagation of a regenerative process from the surface to the inside of a skinned fiber. It would be very worthwhile to reexamine this effect using "caged" Ca^{2+} to change the Ca^{2+} level within the muscle fiber quickly and homogeneously, eliminating diffusion effects.

An interesting postscript to the Ca^{2+}-induced Ca^{2+} release story concerns the publication of the result. While Ford and I were convinced of the effect from the force changes that were seen following various solution changes, we thought it would be more elegant to directly demonstrate that Ca^{2+} in the myofilament space induces Ca^{2+} release from the SR, using radioactive Ca^{2+} (10). However, we were excited about the finding and felt that a short report should be published even before the more definitive study using radioactive Ca^{2+} was completed. This had been a tradition in the lab for most of our discoveries. Our report came out in *Science* in the first issue of 1970 (8). We were pleased when a paper from Endo and his colleagues (7), using different methods but with essentially the same finding, appeared in *Nature* later that year. It was satisfying, and an indication of the way science works, that two labs on opposite sides of the earth had reached the same conclusion at about the same time.

CALCIUM UPTAKE BY SARCOPLASMIC RETICULUM

Relaxation in muscle is caused by the return of calcium that had been released by the SR back into the SR. Careful study of the delays in force development by skinned fibers yielded data that showed that the uptake rate of Ca^{2+} at the low concentrations associated with relaxation is fast enough to account for the physiological relaxation rate (9).

[299]

The experiments that gave this information were set up to see how the uptake rate of Ca^{2+} by the SR depends on the concentration of the EGTA complex in the bathing solution. The method was to use radioactive Ca^{2+} ($^{45}Ca^{2+}$) in the bathing solution and to measure the amount of $^{45}Ca^{2+}$ taken up by the fiber during the delay phase. The fact that $^{45}Ca^{2+}$ was indeed taken up without force development showed directly that the affinity of the SR for Ca^{2+} is greater than that of troponin. The uptake rate at a given pCa was nearly proportional to the $CaEGTA^{2-}$ concentration, which showed that the dissociation of Ca^{2+} from the complex is the rate-limiting step in the uptake process. Because the concentration of the complex could be varied experimentally, the EGTA system was used to present the interior of the fiber with different fluxes of Ca^{2+}, which enabled the rate at which the SR takes up Ca^{2+} to be measured directly. Because no force was developed, the concentration at which the SR worked was known to be below the contraction threshold. These data enabled us to calculate that, even at the low Ca^{2+} concentrations associated with relaxation, the SR can take up the amount of Ca^{2+} bound to the fully activated myofilaments in times close to the physiological relaxation time. This was an important result because the velocity of Ca^{2+} uptake by preparations of fragmented SR in vitro appeared to be too slow to account for the physiological rate of relaxation (6). Apparently the isolation procedure impaired the ability of SR vesicles to accumulate Ca^{2+} rapidly.

CONCLUSION

Our work made clear the following: *1)* The direct contractile response is elicited specifically by Ca^{2+}. *2)* The muscle cell contains electrically polarized internal membranes. Depolarization of these membranes, either ionically or by current flow, leads to Ca^{2+} release from the SR. *3)* In the resting fiber, most of the Ca^{2+} is contained in the terminal cisternae of the SR. *4)* The mechanism by which Ca^{2+} is released from the SR can be regenerative. *5)* Ca^{2+} has a stronger affinity for the membranes of the SR than for the regulatory proteins. *6)* The SR can take up Ca^{2+} quickly enough to account for the relaxation rate.

Some of these results were known before our work, but they were difficult to reconcile with other data. One of our contributions was to resolve these conflicts. The most interesting results came from unexpected effects that occurred in the course of experiments set up with another object in mind. When these unexpected effects occur, the first inclination is to look for an error or a bad solution. It takes energy and imagination to change preconceptions and come up with a new understanding.

In the late 1970s, I began to move away from the excitation-contraction coupling problem. Elizabeth Stephenson, who improved the method of measuring intracellular Ca^{2+} movements and studied the effect of Mg^{2+} on these movements (33, 34), was the last post-doctoral fellow in my laboratory who worked along these lines. I felt that the major steps in the problem had been sorted out. The main unanswered question then was how the excitation signal is transferred from the tubules to the SR (step 2 in Fig. 1); it still is.

BIBLIOGRAPHY

1. ARMSTRONG, C. M., F. BEZANILLA, and P. HOROWICZ. Twitches in the presence of ethylene glycol bis (beta-aminoethylether)-N,N'-tetraacetic acid. *Biochim. Biophys. Acta* 267: 605–608, 1972.

2. BAYLOR, S. M. Optical studies of excitation-contraction coupling using voltage-sensitive and calcium-sensitive probes. In: *Handbook of Physiology. Skeletal Muscle,* edited by L. D. Peachey. Bethesda, MD: Am. Physiol. Soc., 1983, sect. 10, p. 355–379.

3. COSTANTIN, L. L., C. FRANZINI-ARMSTRONG, and R. J. PODOLSKY. Localization of calcium-accumulating structures in striated muscle fibers. *Science Wash. DC* 147: 158–160, 1965.

4. COSTANTIN, L. L., and R. J. PODOLSKY. Depolarization of the internal membrane system in the activation of frog skeletal muscle. *J. Gen. Physiol.* 50: 1101–1124, 1967.

5. DAVSON, H. The influence of the lyotropic series of anions on cation permeability. *Biochem. J.* 34: 917–925, 1940.

6. EBASHI, S., and M. ENDO. Calcium ion and muscle contraction. *Prog. Biophys. Mol. Biol.* 18: 123–183, 1968.

7. ENDO, M., M. TANAKA, and Y. OGAWA.Calcium induced release of calcium from the sarcoplasmic reticulum of skinned skeletal muscle fibres. *Nature Lond.* 228: 34–36, 1970.

8. FORD, L. E., and R. J. PODOLSKY. Regenerative calcium release within muscle cells. *Science Wash. DC* 167: 58–59, 1970.

9. FORD, L. E., and R. J. PODOLSKY. Calcium uptake and force development by skinned muscle fibres in EGTA buffered solutions. *J. Physiol. Lond.* 223: 1–19, 1972.

10. FORD, L. E., and R. J. PODOLSKY. Intracellular calcium movements in skinned muscle fibres. *J. Physiol. Lond.* 223: 21–33, 1972.

11. HEILBRUNN, L. V. *An Outline of General Physiology* (2nd ed.). Philadelphia, PA: Saunders, 1943.

12. HELLAM, D. C., and R. J. PODOLSKY. Force measurements in skinned muscle fibers. *J. Physiol. Lond.* 200: 807–819, 1969.

13. HILL, A. V. On the time required for diffusion and its relation to processes in muscle. *Proc. R. Soc. Lond. B Biol. Sci.* 135: 446–453, 1948.

14. HILL, A. V. The abrupt transition from rest to activity in muscle. *Proc. R. Soc. Lond. B Biol. Sci.* 136: 399–420, 1949.

15. HILL, A. V., and L. MACPHERSON. The effect of nitrate, iodide, and bromide on the duration of the active state in muscle. *Proc. R. Soc. Lond. B Biol. Sci.* 143: 81–102, 1954.

16. HUXLEY, A. F. Muscle structure and theories of contraction. *Prog. Biophys. Biophys. Chem.* 7: 255–318, 1957.

17. HUXLEY, A. F., and R. NIEDERGERKE. Interference microscopy of living muscle fibres. *Nature Lond.* 173: 971–973, 1954.

18. Huxley, A. F., and R. W. Straub. Local activation and interfibrillar structures in striated muscle. *J. Physiol. Lond.* 143: 40–41P, 1958.

19. Huxley, A. F., and R. E. Taylor. Local activation of striated muscle fibres. *J. Physiol. Lond.* 144: 426–441, 1958.

20. Huxley, H. E. The double array of filaments in cross-striated muscle. *J. Biophys. Biochem. Cytol.* 3: 631–648, 1957.

21. Huxley, H. E. Evidence of continuity between the central element of the triads and extracellular spaces in frog sartorius muscle. *Nature Lond.* 202: 1067–1071, 1964.

22. Huxley, H. E., and J. Hanson. Changes in the cross-striation of muscle during contraction and stretch and their structural interpretation. *Nature Lond.* 173: 973–976, 1954.

23. Jobsis, F. F., and M. J. O'Connor. Calcium release and reabsorption in the sartorius muscle of the toad. *Biochem. Biophys. Res. Commun.* 25: 246–252, 1966.

24. Kahn, A. J., and A. Sandow. The potentiation of muscular contraction by the nitrate-ion. *Science Wash. DC* 112: 647–649, 1950.

25. Martonosi, A. N., and T. J. Beeler. Mechanism of Ca^{2+} transport by sarcoplasmic reticulum. In: *Handbook of Physiology. Skeletal Muscle,* edited by L. D. Peachey. Bethesda, MD: Am. Physiol. Soc., 1983, sect. 10, p. 417–485.

26. Natori, R. The property and contraction process of isolated myofibrils. *Jikeikai Med. J.* 1: 119–126, 1954.

27. Peachey, L. D., and C. Franzini-Armstrong. Structure and function of membrane systems of skeletal muscle cells. In: *Handbook of Physiology. Skeletal Muscle,* edited by L. D. Peachey. Bethesda, MD: Am. Physiol. Soc., 1983, sect. 10, p. 23–71.

28. Podolsky, R. J. A mechanism for the effect of hydrostatic pressure on biological systems. *J. Physiol. Lond.* 132: 38–39P, 1956.

29. Podolsky, R. J., and L. L. Costantin. Regulation by calcium of the contraction and relaxation of muscle fibers. *Federation Proc.* 23: 933–939, 1964.

30. Podolsky, R. J., and C. E. Hubert. Activation of the contractile mechanism in isolated myofibrils (Abstract). *Federation Proc.* 20: 301, 1961.

31. Podolsky, R. J., and M. F. Morales. The enthalpy change in adenosine triphosphate hydrolysis II. *J. Biol. Chem.* 218: 945–959, 1956.

32. Smith, P. D., G. W. Liesegang, R. L. Berger, G. Czerlinski, and R. J. Podolsky. A stopped-flow investigation of calcium ion binding by ethylene glycol bis (β-aminoethyl ether)-N,N'-tetraacetic acid. *Anal. Biochem.* 143: 188–195, 1984.

33. Stephenson, E. W., and R. J. Podolsky. Regulation by magnesium of intracellular calcium movement in skinned muscle fibers. *J. Gen. Physiol.* 69: 1–16, 1977.

34. Stephenson, E. W., and R. J. Podolsky. Influence of magnesium on chloride-induced calcium release in skinned muscle fibers. *J. Gen. Physiol.* 69: 17–35, 1977.

35. Weber, A., and R. Herz. The binding of calcium to actomyosin systems in relation to their biological activity. *J. Biol. Chem.* 238: 599–605, 1963.

36. Weber, A., and J. M. Murray. Molecular control mechanisms in muscle contraction. *Physiol. Rev.* 53: 612–673, 1973.

XIII

From Cell Theory to Cell Connectivity: Experiments in Cell-to-Cell Communication

WERNER R. LOEWENSTEIN

"THE *cells are circumscribed autonomous units,*" wrote Mathias Schleiden in 1838, and with this he laid the foundations of cell theory (63). This tenet of cell individuality was to become one of the most influential notions in biology, comparable in importance to Darwin's theory of natural selection and to Mendel's laws of heredity later in that century. Before the century was over, the principle of individuality had to be molded somewhat differently because of the discovery of hormones: the cell unit lost its full autonomy. However, it retained its attribute of circumscription, and in this form the tenet held sway in biological thought to our times. In fact it gained its strongest foothold during the forties and fifties of the present century when permeability studies and electrical measurements on blood, gamete, nerve, and skeletal muscle cells showed that the surface of these cells is a high diffusion barrier all around; soon followed the first electron-microscope observation revealing a seemingly continuous surface membrane in all sorts of cells. Thus the notion of circumscription solidified even more, and recast in this new light, the individuality principle endowed the cell with a complete barrier to the diffusion of hydrophilic molecules; only the smallest molecules, the inorganic ions, could pass through special membrane channels.

CELL CONNECTIVITY

This was the atmosphere in which my generation was brought up, and cell theory was in our blood. Little did I know then that I was to become the spoiler to this precept. This happened in the most unexpected way.

In 1963[1] Yoshinobu Kanno and I were experimenting with an epithelium that, because of its extraordinarily large cells and nuclei, looked promising for our work on nuclei. Yoshinobu had come from Y. Katsuki's laboratory at Tokyo Medical and Dental University to spend a couple of postdoctoral years with me at Columbia. I think he had originally hoped to work on sensory-receptor mechanisms, on which I had worked those past ten years, but I had just started a new line of investigation dealing with the permeability of the nuclear membrane, so we set out to put microelectrodes into the largest possible nuclei we could find. We dabbled with frog and starfish oocytes and then settled on the *Drosophila* salivary gland as our object for study. Theodosius Dobzhansky, the authority on *Drosophila*, was only a few blocks away at our Morningside campus. He gave us *flavorepleta*, the species that held the record for nuclear size, and soon the corridors of the third floor of the College of Physicians and Surgeons were abuzz with these delightful creatures. (Fortunately the offices of Dean Houston Merrit were far enough away on the second floor. We remained friends.)

One afternoon—I remember it as if it were yesterday—we were mapping the electrical field around a *flavorepleta* nucleus when something went amiss with our routine. Usually we had one electrode inside the nucleus, with which we passed current to the bath outside the cells, and another electrode in the cytoplasm of that cell, with which we determined the voltages produced by that current. Unwittingly, however, that electrode had landed in the adjacent cell, and to our astonishment the voltage, instead of disappearing, was almost as high as in the cell containing the current source. We could hardly believe our eyes: it looked as if there was no barrier to the current between the two cells. Had we perforated the membranes between the cells with our electrodes? Was there some cross talk in our voltage-recording circuits? Had the lateral membrane regions lysed (in advance of the eventual dissolution of the gland at the end of the later developmental stages)? Any one of these possibilities seemed to us immediately more likely than such flagrant violation of cell theory.

We worked feverishly the following days. At the end of the third week we no longer had any doubts: there was a path of low resistance between the cells, and this was the normal way of things in the tissue; the cell interiors were interconnected (32).

From then on the sailing was downwind. Within the year we knew that we were not merely dealing with a fluke of an embryonic *Drosophila* salivary gland. Stephen Kuffler and David Potter at Harvard found that neuroglia cells were electrically coupled (35), and in rapid succession Muhamed Nakas, who came to work with us from the University of Sarajevo in Yugoslavia, Soji Higashino from Gunma University in Japan, Sidney Socolar, my long-time colleague, and Richard Penn, then a medical student at Columbia, showed that renal, urinary bladder, sensory epithelial, and liver cells of adult tissues from a wide variety of species were electrically coupled (48, 53). We were dealing with a general form of communication. In the following years, hardly a month went by without reports coming in from other laboratories of cells being coupled in one tissue or another.

In short order we also found out that the small inorganic ions that carried our currents were not the only molecules that move through the membrane junctions. Kanno and I began to inject dyes into cells, and within the same year (1964) we were able to report that an organic molecule of 430 daltons, fluorescein, passes from cell to cell (33). It was thrilling indeed to see this fluorescent substance spreading before our eyes from cell to cell throughout the entire epithelium. I still have the film of the original experiments and show it once a year to our students. It never fails to bring out the sparkle in the eyes or to etch the notion, as it did in us twenty years ago, that cell communities are continuous from within.

In hindsight one may wonder why it took until 1963–1964 to find that out. That it did take so long is itself a bouquet to cell theory. So powerful has been its grip on biological thought that even in the 1940s and 1950s one was only too glad to extend the notion of the continuous pericellular diffusion barrier from a handful of free-floating cell types and the specialized nerve and skeletal muscle cells to all tissues or to read such a barrier into a black line on electron micrographs. There had been inklings of cell connectivity from time to time. There were some in the past century relating to a syncytial nature of tissues; those concerning nerve were snuffed out by Ramón y Cajal's histology, and those regarding heart and smooth muscle (these lingered the longest) were abandoned with the advent of electron microscopy. The first modern foretokens came in 1952 when Silvio Weidmann at Cambridge University showed that Purkinje cells of heart were linked electrically (75) and in 1959 when Edwin Furshpan and David Potter at University College London discovered the electrical synapse where nerve impulses went directly across (20). However, for all their immediate importance to neurobiology, these findings as yet meant no challenge to cell theory. These were cells specializing in the transmission of electrical signals, and the electrical coupling was properly considered a special adaptation for this function. Moreover there was then no evidence that anything more than the small inorganic ions, the carriers of the impulses, passed between the cells. Thus there was no reason to think that the conductive links were anything more than just a close apposition between ordinary membrane regions with a high density of ordinary inorganic ion channels. Indeed, even in the 1960s some workers tried to explain in this way the electrical transmission in heart and nerve synapses.

Our findings left no room for equivocation: the epithelial cells were electrically silent, and fluorescein was too large to fit through the inorganic ion channels. The afterclap kept us busy for the next twenty years.

The next stage led to the postulate of the cell-to-cell channel. By the fall of 1965, Kanno and I had tested enough molecules in *Drosophila* glands to conclude that there is a limit to the sizes of molecules traversing the cell junctions (34). We did not know precisely what the limit was (in fact, we made a mistake, which luckily we detected a few years later), but we knew that a 127,000-M_r fluorescent-labeled polyamino acid did not go through the junctions. At about that time, Seymour Benzer from the California Institute of Technology came by the lab and suggested that we try phage F-2, a 10^6-M_r RNA. Soon our benches were crowded with agar plates to assay the phage. I injected the RNA into *Drosophila* salivary gland cells. Although the molecules were slim, I could detect no trace of them in the cell neighbors. Large or very long molecules evidently were excluded from passing. This provided one guidepost to the postulate of the cell-to-cell channel.

Another guidepost had gone up when Kanno and I mapped the potential field along the gland surface, sniffing, so to speak, for current leaks (from the cell interior) along the intercellular space. The resistance of that route was as high as that of the cell membrane, if not higher. Besides, during the passage of fluorescein from cell to cell, we saw no leakage to the exterior (44). Thus, although highly permeable from cell to cell, the pathway was well insulated from the exterior.

Finally the third pointer came from work with another salivary gland, that of *Chironomus*, which offers an unusually large and simple cell system. Using this system, Nakas, Socolar, and I found that the cell-to-cell pathway was endowed with a closing mechanism that allowed the cell community to seal itself off from an injured member. Our experiment was straightforward: we drilled a hole into a cell in physiological medium. The junctional pathway then closed off, and the cell interiors uncoupled. Searching for the control element, we soon hit upon Ca^{2+}; uncoupling ensued when Ca^{2+} was in the medium, and only then (36).

These three points encouraged me in 1965 to propose the existence of a special channel in the junctions between cells (36). I had a good opportunity to announce our finding and the channel hypothesis in October of that year at a conference at the New York Academy of Sciences, which brought together many of the leading workers in the field of membrane biology, which was then just emerging. I don't recall details of the discussion, but what still stands out in my mind are the exciting comments of George Palade on the structure of epithelial junctional membrane complexes, which he and Marilyn Farquhar had then recently uncovered (17), and those of Aharon Katchalsky on how Ca^{2+} interacts with polymeric molecules. (I was to learn more from him about Ca^{2+} and other things later as we became friends.)

The idea of the cell-to-cell channel was slow in hatching. It took most of 1964. The difficult part for me was to see that Ca^{2+} had two distinct actions: one on the cell-to-cell pathway and another on its

[306]

insulation. We eventually sorted this out experimentally, but a push had come earlier that year through a conversation in a restaurant in Cambridge, Massachusetts.

George Wald had invited me to give a talk at Harvard. Afterward, at the dinner, I sat next to Jerry Lettvin from the Massachusetts Institute of Technology. We talked about Morgenstern, the German poet, and Jerry said that he had translated his poems into English. I had seen many feats of translation, but Morgenstern, I thought, was untranslatable (those who know his delightfully grotesque verse will understand why). Jerry quickly cut me down by reciting "KM 21" bilingually. Morgenstern, of course, led us straight to calcium, and I said that I suspected that Ca^{2+} regulated communication. Besides Morgenstern, Jerry fortunately also knew much about Ca^{2+}, and he told me about its action in various physiological processes. I don't remember exactly what was so reassuring in what he said, but I went home cocksure that I was on the right track. All the same, I would have listened to anyone capable of translating Morgenstern!

By January 1965 the basic pieces fell into place; the conductive pathway between cells, I thought, was satisfactorily representable as a composite of unitary membrane passageways. I could trace out the unit in abstract terms as a special channel made of two symmetrical conductive halves, one contributed by each cell, endowed with a continuous insulation (Fig. 1). The first rendering of the channel

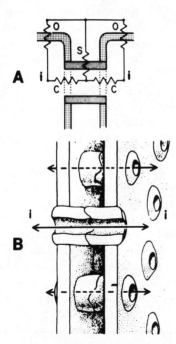

Figure 1. Cell-to-cell channel. Concept as formulated in 1965 (A) and vested in protein in 1974 (B). C-C, junctional aqueous membrane channel; S, its insulation from exterior. [From Loewenstein (36, 41).]

[307]

went onto a sand dune near the Montemar Marine Station in Chile where I had spent that month. The second went on the blackboard of the Columbia University lab where, on my return, I immediately thrashed the idea about with Sidney Socolar. Sidney had joined the lab in 1959 and eventually moved with me to Miami. Over all these years (until his recent retirement) it became second nature to us to toss and turn our ideas and results. I owe much to his ironclad reasoning; more than once he kept me in rein.

I based the cell-to-cell–channel concept solely on the permeability properties of the junction, and this is why the concept has endured. In 1965 there was not much to guide us as to the structure and chemical nature of the channel. There were still no clear ideas even about the nature of the much-longer-known inorganic ion channels, and many workers in the field of ion transport still thought in terms of channels made of membrane lipid. It was not until 1972 that Jon Singer and Garth Nicholson developed the general rationale of channels as transmembrane proteins (65). The cell-to-cell channel could then be readily envisioned as a pair of tightly joined proteins spanning the intercellular gap, and this is how I dressed the unit in 1974 (41). The basic concept remained unchanged, but the unit was now endowed with somewhat more specific attributes; the conductive channel core became a space inside the protein lined by hydrophilic amino acid residues and the channel insulation, hydrophobic residues tightly abutting in the intermembrane gap.

By then also the first pertinent details of structure had been found under the electron microscope. In 1967 Jean-Paul Revel and Manfred Karnovsky resolved the "gap junction" in the junctional membrane complex (56), and the membrane particles clustered on both sides of that structure became obvious candidates for the channel halves. In 1977 the particles were analyzed by Don Caspar, Dan Goodenough, and their colleagues (11a), and in 1980 the channel was resolved by Nigel Unwin and Guido Zampighi to an 18-Å resolution (73).

What are the functions of this channel? This question came in sync with the finding of communication in the electrically silent epithelium. From the beginning I had little doubt that such a large, ubiquitous, and (as I suspected and found out later) ancient channel between cells had adapted to many different functions. A basic function, tissue homeostasis, came immediately to mind. Only a few years before, Magnus Gregersen at Columbia had given me Claude Bernard's great work *Leçons sur les Phénomènes de la Vie*, and so I could hardly miss seeing in the cell-to-cell channel an instrument for extending Bernard's classic homeostatic concept to the intracellular milieu. Where it existed in sufficient number, the channel would provide a means for evening out differences in concentration of inorganic ions or other permeant species in the cells of the community. This was another proposal I brought to the 1965 New York

Academy of Sciences conference. In my talk I had the cheek then to call the cell community the "*most interior, interior milieu.*"

I was just as intrigued, however, by the possibility of cell coordinations of a more lively sort. The roots here reach back to my childhood, to my father's aphoristic repertoire. One of his favorite sayings I heard again and again was, "*It has been seen to it that the trees don't grow into the sky.*" He said it in German ("*Es ist dafür gesorgt, dass die Bäume nicht in den Himmel wachsen*") in which it has something of the flavor of a biblical imperative. I did not exactly lose sleep over it as a boy, but the sheer repetition stamped it in. I thought then that it was just one of those folkloric fleas-in-the-ear that grownups use to instill compliance. Only in my late teens did I learn that it was a quote from Goethe's "Poetry and Truth," and I pricked up my ears: What was the poet really after here? Surely the deep Goethe was connoting all kinds of organismic growth. I began to read his works avidly, hoping to learn *how* "*it has been seen to,*" but neither Goethe-the-writer nor Goethe-the-naturalist had anything to offer beyond the say-so; nor, for that matter, had anyone else a century later.

Thus I became fixed early on the problem of cellular growth control. In the 1950s I had learned some information theory; not much, but enough to see, in 1963–1965, the problem in gross outline as one of gene control in which the regulating information must be contained and looped within the cell population whose growth is controlled. By then results were on hand from the laboratories of Michael Abercrombie, Renato Dulbecco, Harry Rubin, Leo Sachs, and Michael Stoker, showing the first attempts at analyzing the problem in tissue culture. The "*has been seen to*" then translated into "*self-containment of information circuitry*": the senders of the information signals and the relays for the control were all within the cell population. Everybody, of course, was then thinking in terms of humoral communication—the only form of intercellular communication known—in which the signals are extracellular and diffuse through the spaces between the cells. However, there was something very unsatisfactory about this. The extracellular space is extremely irregular and often open to the space around an organ or tissue. It is hard to see how, in such a system, growth-controlling information could be cued into concentration parameters with anything near the precision required for organogenesis.

The finding of direct cell interconnection changed all that. Here we had, in the interconnected cell community, a self-contained system—a system bounded by a high diffusion barrier and of finite volume—in which information for control could be simply coded into molecular concentration of signals. I found this so unique and irresistibly alluring that in November 1963, in our first full paper reporting the finding in *Drosophila* gland, we ventured to write that the connective pathway might serve to disseminate gene-controlling factors through the cell community (44). This set us on the arduous track of the physiology of growth and differentiation.

At that time I received good advice from Heinz Kröger. Heinz, a cytogeneticist from the Eidgenössische Technische Hochschule of Zurich, had come to our lab to learn some tricks of electrophysiology. As it turned out, I learned more from him, as he became our yearly guest when he passed through New York to his lecture engagements in the United States. He taught me about gene control, especially about the exciting work on chromosome "puffing" that was then going on. He also was patient enough to listen to all my trial balloons, many of which I sent up while we were walking in the Greenbrook Sanctuary along the south palisades of the Hudson River, a wonderful woodland with brooks and a lake and spectacular river views, not far from where I lived. There we discussed what kind of molecules might take the junctional path to exert control over genes. Kröger believed that inorganic ions might do it. I favored larger molecules with more information content, something of the size of a steroid hormone. Steroids are about the size of fluorescein, which we just had shown to pass through the cell junctions, and Wolfgang Beerman had shown that the steroid ecdysterone activates genes. I was even willing to speculate on small proteins or short RNAs; we learned only in 1966 that such molecules would not pass.

By 1965 I had mulled over the potential of cell connectivity for mutual cellular control, particularly for control of growth, enough to try out an experiment. Richard Penn in our laboratory had just shown that cells in rat liver were coupled (53). Thus, on the assumption that such coupling is necessary for growth control, Kanno and I set out to test whether the coupling was altered in liver cells that had lost their control over growth. There were several experimental mouse and rat liver tumors available. In four such tumor types, we found no coupling within the limits of our electrical detection system (45). The cell-to-cell pathway was much less conductive, if not completely absent.

The results came just in time to be presented in October 1965 at the New York Academy of Sciences Membrane Conference. They encouraged me to put forward yet a third hypothesis at that conference: the cell-to-cell pathway would serve to transmit information for the control of cellular growth and differentiation (36).

I based the part of the hypothesis concerning differentiation on the fact that the pathway was present in embryonic tissue. The *Drosophila* salivary gland on which we made our initial observation was embryonic (from the larva's third instar period), and my colleague Shizuo Ito (from Kumamoto University in Japan) had found that year that the early embryo of the newt *Triturus* (up to the morula stage), a vertebrate, was interconnected throughout. Above all I was guided by a priori considerations. The junctional form of communication seemed ideally suited for the short-range cellular

[310]

interactions experimental embryologists had long known to be involved in embryonic development. The junctional pathway not only was a plausible means for conveying morphogenetic information from cell to cell, but the entire signal traffic could take place within the cell community, in isolation from the external environment. Such insulation, of course, is paramount in an embryo—the autodeterministic system par excellence—that shapes itself with information emanating from within.

In December of that same year (1965), David Potter, Edwin Furshpan, and Edwin Lennox from Harvard and the Salk Institute announced that cells in the squid embryo, even cells that had differentiated, were electrically coupled and transferred dyes (54). The developmental ship was launched. It sailed slowly, but there was never serious doubt that it would endure. Only as recently as 1985, it received a powerful push from the finding by Anne Warner, Sara Guthrie, and Norton Gilula from University College London and Baylor University that embryonic development is disrupted when the junctional pathway is blocked by antibodies to the cell-to-cell channel protein (74). If these developmental defects prove to be specific to junctional channel closure, the hypothesis and the experiments will join the roster of good deeds.

In 1967 I spent part of my sabbatical in Europe studying the literature of experimental embryology. I had never had a course in embryology. As my readings took me from Speman to Needham, Waddington, and Grobstein, I learned one of the facts that had transpired early from the work in experimental induction: the inducers did not seem to enter into the chemical machinery of differentiation. I was particularly struck by the studies by Johannes Holtfreter showing that a variety of unspecific, seemingly unrelated agents can induce neural differentiation (26). Many of these disparate agents ("unnatural inducers"), as we had just learned from our studies with the *Chironomus* gland, produced uncoupling of junctional communication. Thus, by the end of 1967, I extended my original proposition to a more general hypothesis in which both coupling and uncoupling of the junctional pathway were instrumental in differentiation. Central here was the knowledge that the cell-to-cell channel could close. The gene-controlling signal flow might thus be turned on or off in the embryo, permitting subdivision of cell communities in the embryo and, hence, territorial differentiation. In 1968 I formulated this hypothesis in a theoretical paper (38) and presented it at the 27th Symposium of the Society for Developmental Biology (39).

Meanwhile my thoughts on growth control had jelled into a model in which the cell community self-regulates its growth by simple dilution of signal molecules (38, 40). I based this model, "the asynchronous-dilution model," on the feature that the diffusion space in the interconnected cell community increases as the number of cells in the community increase. Signals are therefore diluted if their production in the population is asynchronous. Thus, with reasonably short signal-diffusion times, the community could gauge the average number of its cell members by the steady-state chemical concentration of the signals. One could hardly imagine a simpler encoding of cell-population size, the basic information for growth control.

This chemical concentration could then be looped on the DNA replication process to obtain self-regulated growth. Here the junctional communication system casts its most alluring spell: the high diffusion boundary of the community makes the system self-contained and capable, without further ado, of demarcating a feedback loop for control of cell replication. The complete model thus simply constituted a diffusion-coupled system of asynchronous signal generators with a feedback loop (40).

I felt a bit like walking through Wonderland during these years of 1965 to 1967. As often as not I was looking over fences of fields to which I was a stranger. I knew the dangers this entailed, but I remembered Lewis Carroll, and that helped:

> When Alice came to Wonderland, the King asked her:
> "What do you know about this business?"
> "Nothing whatever?" persisted the King.
> "Nothing whatever," said Alice.
> "That's very important," the King said.

Eventually I was to learn a bit more about the ins and outs of these fields through a series of conferences on growth and differentiation that were held in the Black Forest at Titisee. These small conferences, generously endowed by Thomae GmbH and splendidly run by Hasso Schroeder, brought together many leading figures in those fields. I drew much encouragement from the exchange of ideas there.

In 1971 Alan Burton from the University of Western Ontario in Canada examined the asynchronous-diffusion model analytically (11). He endowed it with a sinusoidal oscillator and so managed a detailed quantitative description of its behavior for small cell populations (2–6 cells). Burton reasoned that chemical-reaction series with feedback loops contain the necessary inertia element for sinusoidal oscillation. This analysis brought out several important points: not only excessive signal synchrony, but also an excessively low control threshold, could give rise to decontrol of growth; a trend toward synchrony in very small cell aggregates could favor growth, providing a simple explanation for why certain cells grow when they occur as groups in culture but not when they occur as singlets. Burton sent me a preprint of his paper (the title of which made me blush), and we kept in correspondence until his death. (Although a physicist, he is, no doubt, well known to many of the readers of this volume for his many contributions to physiology.) It was not until several years later that we met in person at a conference in Houston.

There we spent enjoyable hours together and discovered that, besides growth control, we had other interests in common. In my conference talk I quoted the astronomer Arthur Eddington's witty counsel *"not to put too much weight on experimental results until they are confirmed by theory."* Burton was sitting in the front row. By the twinkle in his eyes I saw he knew exactly how I felt.

The hypothesis on growth control has worn well over the years. Its heuristic value was (and still is) limited by the lack of knowledge of the identity of the growth-regulating molecules (a handicap shared by all workers in the field), but it made several predictions that could be tested even without knowing these molecules.

Two such predictions, a paired set, were immediately explored. First, dividing cell ensembles that are growth-regulation competent would be communication competent, that is, competent to make cell-to-cell channels. Second, dividing cell ensembles that are communication incompetent would be growth-regulation incompetent, that is, they would be potentially or actually cancerous. Both predictions were borne out by the results of experiments performed over the ensuing years. All normal tissues, whether of slow or fast growth, turned out to be channel competent. The only normal adult tissue cell types known to be channel incompetent are skeletal muscles and most nerve fibers of vertebrates. However, these cells (nicely fitting the hypothesis) are not dividing. When they are still dividing in the embryo, even these cells are communication competent. The second prediction touched off a search for channel-deficient cells. Asim Jamakosmanovič, who came from the University of Sarajevo, turned to thyroid tumors, and Carmia Borek, who came from the Weizmann Institute of Science in Israel, and Higashino attacked tumor cell cultures. After getting our first taste of cell culture, our efforts in the cellular growth field, particularly Roobik Azarnia's, went in that direction. Roobik had come in 1970 from the University of Miami. Little did he know then that he would soon be back there when we all moved from New York to Miami. At that time also, William Larsen came from Case Western Reserve University and brought his electron-microscope expertise to bear on the problem.

The hunt for communication-deficient cells eventually was on in several laboratories. By 1979, when I made a tally, thirty-five cultured mammalian cell strains had been found that were radically channel deficient; they were all cancerous (42). Azarnia, Larsen, and I analyzed two of these cell systems genetically by somatic cell hybridization (4, 6). This was long and tedious work, but it was the only way we knew to cut short the risk that we were chasing an epiphenomenon. In the fall of 1973 we could breathe a sigh of relief: the communication deficiency and the growth-control deficiency were genetically linked; the channel competence and the growth-control competence, on one hand, and their respective opposite traits, on the other, did not segregate.

[313]

The list of communication-deficient cell types grew longer with the recent findings of such types among cells transformed by oncogenic viruses in the laboratories of Judson Sheridan (2) and our own (7a). The deficiency is more subtle here. The cells are not entirely devoid of cell-to-cell channels (and because of that the deficiency had been overlooked before), but the degree of communication is reduced. The oncogenic viruses are powerful tools for dissecting the molecular aspects of the regulation of communication, putting us at the threshold of a new exciting phase of research.

The hypothesis has borne further fruit in recent work in which cancer cells were induced to communicate with normal cells. In these experiments we made use of the phenomenon, discovered by Michael Stoker in the mid-1960s, that the growth of certain transformed cells is inhibited when they are cultured in contact with normal cells (69). My bet had long been that the phenomenon was due to junctional communication, and so was Stoker's. However, the means for demonstrating this had only recently become available, after the finding in our laboratory (recounted later) that cell-to-cell channels can be induced in communication-incompetent cells by activating cAMP-dependent phosphorylation. Thus Parmender Mehta, who came from the Kobe University in Japan, John Bertram at the University of Hawaii, and I joined transformed and normal cells in culture and induced channels between them. We chose several types of mammalian cells transformed by virus or carcinogen, which don't make channels readily with normal cells in ordinary conditions of culture. The results were striking: the growth of the transformed cells was arrested when a demonstrable communication was established with the normal partners, and the growth resumed when that communication was blocked (50).

We were pleased as Punch, and Stoker echoed similar feelings from Cambridge as we corresponded about these results. Finally, and after so many years, we felt somewhat closer to the question of identifying the elusive signals. These findings, quite unexpectedly, spun a connecting thread with the field of somatic tumor cell hybridization in which the results of Henry Harris, George Klein, Ruth Sager, and others led one to conclude that normal cells generate a highly conserved signal capable of suppressing the transformed state. The available genetic clues from that field, together with the clues about the admissible molecular sizes from our work, will be helpful guideposts, I hope, in the current search for the signal.

I now turn back to twenty years ago to tell the story of the regulation of the pathway of communication. I pick up the threads of my account of the work in 1965 and 1966 that put us on the calcium track. This began soon after the adoption of the midge *Chironomus* from the Swiss sewers.

The adoption stemmed from one of the Greenbrook walks with Heinz Kröger during which he told me about this midge's salivary

[314]

gland and its chromosomes, which he was studying. On his return to Zurich he mailed some egg masses, and soon the first Swiss midges were flying in New York. It was love at first sight. No sooner had I pulled out a gland when I knew it was the ideal preparation for many of the things I wanted to do. Where could one ever hope to find another tissue with that simplicity: a whole organ made of only thirty cells, each cell a giant of 100–150 μm in diameter, that could be easily isolated from the animal and kept alive for hours? These were cells into which one could put four micropipettes (with room to spare) and which were nicely transparent and large for single-cell photometry.

With these cells Nakas, Higashino, Socolar, and I began at once to examine the stability of the communication. Did the cells seal off the conductive pathway when they came apart? In our first explorations of this question we used procedures that had long been in use for mechanical dissociation of cells: Ca^{2+} removal, trypsin digestion, hypertonicity, and so on (51). We quickly found out that these procedures could be used in much milder form to get the pathway to seal. It was not necessary to separate the cells. A loss in the surface insulation, either in the cell junctions or at the nonjunctional cell membrane, seemed to suffice for uncoupling the communication between cells. One factor, however, turned out to be indispensable: calcium. The uncoupling occurred as long as the external free-Ca^{2+} concentration was at least about 10^{-5} M (47).

The importance of Ca^{2+} did not dawn on us at once. In fact, we were puzzled no end by our first results with chelator-calcium combinations ($EDTA-Ca^{2+}$ and $EGTA-Ca^{2+}$) showing that uncoupling ensued when the free extracellular Ca^{2+} concentration was between 10^{-3} M and 10^{-5} M but not when it was lower and actually more effective for getting the cells apart. This baffling situation resolved itself when our later results of uncoupling with trypsin, anisotonic, and alkaline actions all showed the same requirement of about 10^{-5} M Ca^{2+}. It became clear then that the above-mentioned limits of 10^{-3} M and 10^{-5} M reflected the limits of two entirely different processes: the former was the upper limit for rendering the surface insulation leaky, whereas the latter was the lower limit for sealing the junctional pathway. This realization also was the pivot for sliding together the pieces of the concept of the pathway unit, the cell-to-cell channel. I don't remember how long this took, but it was long enough for me to be annoyed with myself. I vented my feelings on the channel unit by naming its elusive insulation "S" (see Fig. 1); for the strong word beginning with that letter in both languages I was put to school. (The other acronyms, "C" and "O," were more rationally chosen; they stood for conductive element and outside membrane.)

Things moved swiftly thereafter. We next bypassed the surface insulation and microinjected Ca^{2+} into a cell on one side of the

[315]

junction. The cell promptly uncoupled electrically from its neighbors. This is how we were taken captives by Ca^{2+}.

The psychological roots may go back earlier than that, to my first years in Woods Hole, Massachusetts, in the 1950s. I was then just starting out, working in Kuffler's laboratory at Johns Hopkins University. In the summers the whole lab would move to the Marine Biological Laboratory at Woods Hole (a habit I have kept until today). Many legendary figures in physiology were in Woods Hole in the 1950s. Otto Loewi was there. He used to sit in the yard behind the Old Main chatting with whomever came by. It was an exciting treat for us novices. I had many conversations there with him about work, young and old. Our talks would mostly end on the same note; he would say: "*Ja Calcium, das ist alles.*"

Later I learned why. Calcium has many virtues that make it quite unique among cations in its ability to complex with biological structures. Its divalency allows for a wide range of binding constants with biomolecules, its radius is compatible with peptide chelation, and its charge-to-size ratio permits it to slip into small molecular holes. Its crystal-field requirements are quite flexible, bond distances and angles are adjustable, and coordination numbers can vary from six to ten. All this gives the ion a great advantage in binding to irregular geometries of coordination sites of biological molecules that can accept the ion rapidly and sequentially and fold around it, permitting graded structural modulation. Small wonder that such an engaging character has been awarded role after role in the evolution of biological signaling!

At the time of our experiments of calcium injection, we also did experiments of a simpler sort to probe the role of Ca^{2+} in communication. We damaged individual cells of the *Chironomus* gland, cutting into the cells without touching their lateral junctional regions, and asked whether the junction between these cells and the intact neighbors would seal off. This turned out to be so when the tissue was in the physiological calcium-containing medium but not when the medium was calcium free (47). As in the case of the experiments with chelators, the junctional conductance fell several orders of magnitude in the physiological medium, but the whole system was leaky in calcium-free medium.

We knew at once that we were peeking into a mechanism of great physiological importance, a survival mechanism that allowed the cell community to seal itself off from an injured member. The result immediately suggested a simple mechanism in which all the elements of the sealing reaction are built into the normal system and are critically poised: the steep chemical and electrical gradients would drive Ca^{2+} inward, and the cell-to-cell channels could evidently close when the cytoplasmic Ca^{2+} concentration was elevated. All that seemed to be required to set the reaction into motion was a discontinuity of the surface barrier (36).

During the preceding three years, ever since we met up with junctional communication, the thought of how evolution solved the problem of the weakness to injury had gnawed on me. Multicellular compartmental units are obviously more vulnerable than systems in which each cell is the unit: one damaged cell can endanger the survival of the entire community. Here we had an elegant self-sealing mechanism in which Ca^{2+} seemed to play the starring role. CELL CONNECTIVITY

Mark Twain once remarked that when writing *The Tragedy of Pudd'nhead Wilson* he was unable to follow his original plot because several of the characters developed wills of their own and did things quite different from what he had planned for them. I found myself in the same fix with that character *Calcium*. But, although not always as docile as I would have liked him to be, he was always great fun, and even today I don't mind seeing after him.

In fact, during 1966–1967, I had no reasons to complain; calcium behaved according to script. We expected that a sufficient elevation of Ca^{2+} in the cytosol ($[Ca^{2+}]_i$), regardless of source, would cause closure of the cell-to-cell channels, so Alberto Politoff from the University of Chile, Socolar, and I poisoned *Chironomus* glands with cyanide, dinitrophenol, oligomycin, and other metabolic inhibitors (53a). This was the obvious thing to do. Although not much was yet known about Ca^{2+} pumps, it was known that the low $[Ca^{2+}]_i$ was maintained far from electrochemical equilibrium, and already in 1957 Dan Gilbert and Wallace Fenn, and Alan Hodgkin and Richard Keynes had concluded from this that Ca^{2+} must be actively extruded. Our metabolic poisoning of the gland cells or their cooling to 6°C–8°C caused the junctional conductance to fall, and this uncoupling set in with delays as if it depended on a critical, metabolically driven level of a cytoplasmic constituent.

In our next attack on this problem things came out alright, but Ca^{2+} did not follow the script, although we were to learn this only several years later. It happened as follows. In January 1968 Alan Hodgkin from Cambridge University gave the Dame Hester Adrian Lecture at the Rockefeller University in New York. He visited our laboratory later during that week, and my colleagues and I spent an exhilarating afternoon discussing our data and calcium hypothesis with him. He told us about the experiments on calcium movement in nerve that he, Peter Baker, Mordecai Blaustein, and Rick Steinhardt were doing. They had substituted Li^+ for Na^+ in the extracellular medium and found that Ca^{2+} would accumulate inside the axon as a result (9). This seemed a good strategy for uncoupling communication, and it also chimed well with the idea I had distilled through my embryology readings the year before; Li^+ was an agent known, since Herbst's work in the last century, to profoundly influence embryonic development (24). Thus Birgit Rose and I set to work at once.

Birgit was a graduate student at the University of Munich who had come to do her doctoral work in my laboratory. She began by charting the conductive pathway of the salivary gland and produced the first complete map of coupling of any organ (57). No sooner had she started when she put us all to shame. She found that the giant cells were connected not only laterally, but that there also was cross talk between them via a few flat cells of the lumen, which we all had missed. The cell system was not so simple, after all! In an all-male lab (we were five), I think she enjoyed pricking our balloon, and in her inimitable prankish-charming way she put us in our place. It was by sheer luck that our bloomer was of no consequence. We used to lay pins over the lumen to weigh the gland down in the dish and so unwittingly had produced a simple preparation.

With the pathway now under control, Birgit and I looked into the effects of substituting external Li^+ for Na^+ on *Chironomus* salivary glands (58, 59). Nicely obliging, the cells uncoupled.

The following years were devoted to tying up loose ends and sprucing up the calcium story. Gilberto Oliveira-Castro came from the Biophysics Institute of the University of Rio de Janeiro, and we followed up the uncoupling phenomenon by drilling fine precision holes into the membrane of the gland cells in medium of known divalent cation concentration. Thus we showed that effective uncoupling required about 5×10^{-5} M Ca^{2+} (other divalent cations were much less effective), and that the channel closure was confined to the junction of the cell to which exogenous Ca^{2+} had direct access via the hole (52a). We established this first by means of electrical measurements. But there is something uniquely satisfying about seeing things directly with one's eyes, so we injected fluorescein, drilled one cell of the chain, and produced the picture of Figure 2. Clearly, the self-sealing mechanism operated with single-cell precision. The connected system takes good care of itself and sacrifices no more than the injured member.

Some of this lesson trickled to a different level. Gilberto is a descendant of the Portuguese viceroys of Brazil and his complete family name occupies a fair-sized paragraph. When publication time came, I suggested it might be easier on the journal if he uncoupled the longer string from the beginning of the name. This piece of Americana went over well, and ever since his productions have gone under the Oliveira-Castro epitome.

In 1971 Jean Délèze came from the University of Lausanne in Switzerland. He had been studying the mechanism whereby damaged heart tissue heals over, a phenomenon dating back to Engelmann's work of the past century (16). It was thought to reflect a healing of ordinary cell membrane. However, Délèze found out the year before he came that it probably was a junctional uncoupling akin to the one we had found in epithelium (13); he had shown earlier that it required Ca^{2+} (12). We were not set up for work on heart, but the *Chironomus* gland seemed a good model, so we

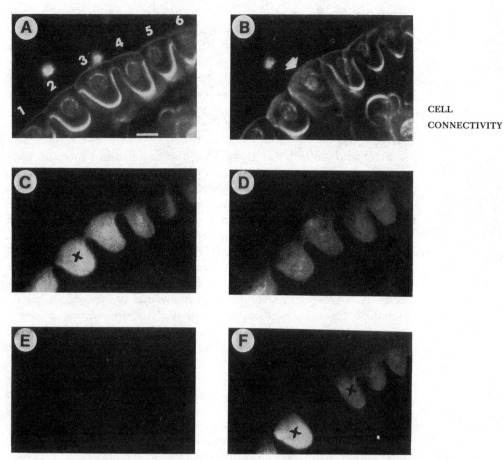

Figure 2.   Effect of Ca²⁺: cell community uncouples an injured member. A
fine hole was drilled into cell 3 (*B*) of the chain of cells of the *Chironomus*
salivary gland, in both Ca- and Mg-free medium. A half hour later cell 2 was
microinjected with fluorescein, which spread throughout cell chain (*C*).
Photographs were taken 3 min (*C*), 5 min (*D*), and 7 min (*E*) after the
injection; by the last time all fluorescein had leaked out through the perfo-
ration. Medium was then exchanged for Ca-containing medium and cells 2
and 4 were injected with fluorescein; the junctions of the injured cells have
sealed off (*F*). Fluorescein-injected cells marked *x*. Calibration 50 μm. [From
Oliveira-Castro and Loewenstein (52a).]

injected calcium, magnesium, and strontium into these cells to analyze the uncoupling process. During that year and the next at Miami,
we found that, under certain conditions calculated to give low
intracellular divalent-cation elevation, cell-to-cell passage of fluorescein was blocked by these injections, whereas electrical coupling
persisted (14). We had no reasons to doubt that fluorescein and the
small inorganic ions carrying the electric currents passed through
the same channels; we knew from the work with Oliveira-Castro and

[319]

Rose that the fluorescein passage was blocked whenever the electrical coupling was blocked. Thus it appeared that the channel closure either was graded and the reduction in the channel-open state was more limiting for the passage of the larger fluorescein molecule than for the inorganic ions—a truly selective uncoupling—or that the closure was all-or-none and involved only a fraction of the channels in the junction. [The electrical-coupling coefficient is relatively insensitive to the proportion of the channels open, except in the low range (66).]

Meanwhile Birgit was sharpening tools to demonstrate the Ca^{2+} action more directly with the aid of aequorin. A. Azzi and Britton Chance, at the University of Pennsylvania, and C. Ashley, P. Baker, A. Hodgkin, and E. Ridgway, two years before, had successfully used this Ca^{2+}-sensitive luminescent protein from the hydromedusa *Aequorea* to monitor [Ca^{2+}] in mitochondria suspensions and inside muscle and nerve cells (1, 8, 10). When we started work in January 1971, we first followed in their footsteps and used a photomultiplier to measure the light emitted by aequorin inside the *Chironomus* cells. However, within the year we knew that the technique would not do for our purposes. Our initial trials of uncoupling gave the expected Ca^{2+} signals, but they also gave Ca^{2+} signals when we least expected them: during recoupling produced by (hyperpolarizing) inward currents. We had enough confidence in the calcium hypothesis to stick with it. That was our first winter at Miami, and even the balmy Florida weather could not soften our resolve. We attributed those signals during inward current (rightly, as it turned out) to local [Ca^{2+}]$_i$ elevation around the microelectrode. It was clear that we needed a method that could give information about local Ca^{2+} concentrations inside the cells.

Fortunately just then fine-grain image intensifiers were making the scene. Astronomers were using them, and a pioneer of that development, George Reynolds from Princeton, who was our next-door neighbor in Woods Hole, put us on that scent. The US Research Office of Defense had then developed a small image intensifier to be mounted on rifles for night sniping—the "sniperscope." Rodolfo Llinás, whose lab was on the same floor in Woods Hole, had one on loan from the Army and gave it to us for a tryout. It looked promising, and that fall we obtained an improved sniperscope from the Army. It still did not quite hit the spot, but Birgit and I did manage to see with it the first aequorin fireworks inside the *Chironomus* cells—a proper backdrop for our honeymoon. We were convinced that this was the way to go.

During that year we put together a system consisting of a low-light–level television camera, a three-stage image intensifier, and a microscope, which allowed us to see and record the spatial distribution of the light emission inside the cells—the first system of this kind. In parallel we arranged a lightpipe conducting the emission to a photomultiplier. Behind these constructions, as behind those of many of our apparatus from 1970 to date, were our machinist Conrado Freites and electronics engineer Jim Gray. George Reynolds let us in on the trick of how a bit of silicone could cut down the light losses at the junctions between our preparation dish, lightpipe, and photomultiplier. He also gave us the use of the shop facilities at the Princeton physics department, where Ferdinand Schwarz knew the secret of how to bend a Lucite rod without overstressing to make an efficient lightpipe (good fiber optics were not yet on the market). Both tricks turned out to be critical for our getting good aequorin signals. Osamu Shimomura from Princeton generously supplied us with purified aequorin then and all these years.

In 1973 our apparatus finally was working with enough sensitivity, and the fireworks now were truly magnificent. Birgit and I could see local $[Ca^{2+}]_i$ elevations with a sensitivity of 5×10^{-7} M and a spatial resolution of 1–2 μm (the cell diameter was 50–100 times larger). Thus, in a series of experiments that ended in the fall of 1974, we microinjected Ca^{2+} into various regions of the cells and exposed them to calcium ionophores, metabolic poisons, or calcium-free medium (which leads to rise in cytosolic Ca^{2+}) (60). All gave the same answer: junctional conductance fell when $[Ca^{2+}]_i$ rose at the region of the cell junction; local rises of $[Ca^{2+}]_i$ elsewhere inside the cells had no effect on junctional conductance.[2]

The calcium hypothesis seemed secure. During the following years the uncoupling action of Ca^{2+} was confirmed for several tissues in other laboratories, and in 1983 P. Unwin and P. Ennis opened a new phase, demonstrating the action at the structural level in isolated-membrane preparation (72). They found by X-ray diffraction that, in the presence of 10^{-5} M Ca^{2+}, the channel subunits undergo a structural transition that displaces them radially, closing the channel bore.

In 1977 a rival of Ca^{2+} arrived on the stage: H^+. This ion had long been known to compete with Ca^{2+} for binding to cell membrane and biopolymeres. The two ions also were known to interrelate functionally in several biological mechanisms; for example, isolated mitochondria released H^+ during $[Ca^{2+}]_i$ elevation. Already in 1975 Paul De Weer, at Washington University, raised the question of whether some of the cellular phenomena ascribed to Ca^{2+} may actually be mediated by H^+ (15). This issue was stoked again two years later— but now with the urgency attached to empirical results—by the finding of R. Meech and R. Thomas (49) that Ca^{2+} injection into snail neurons causes a fall in pH_i. This bore directly on the main experiments supporting the calcium hypothesis.

Against this background Lucas Turin and Anne Warner at University College London discovered in 1977 that intracellular acidification of *Xenopus* embryo cells by exposure to CO_2 produces uncoupling (71). The effect was soon confirmed for a number of tissues. Like that of Ca^{2+}, the effect is prompt and reversible; about half of the channels in *Xenopus* junction close on reducing the cytoplasmic pH by 0.6 unit (68).

Was the action of H^+ independent of Ca^{2+}? Because acidification was well known to cause release of Ca^{2+} from intracellular stores in a variety of tissues, I naturally leaned, after so many years on the Ca^{2+} track, toward thinking that the H^+ effect was mediated by Ca^{2+}. But there was no point to shutting one's eyes to the other side of the coin: elevation of cytoplasmic Ca^{2+} might release H^+. Thus, unbiased by hypothesis, the question was: Which of the two ions mediated the channel closure, or in experimental terms: Which of the two was sufficient?

In 1978 Birgit Rose and Roger Rick got an answer to the part of the question that interested us most. They injected pH-buffered Ca^{2+} solutions into *Chironomus* gland cells while monitoring cytoplasmic Ca^{2+} and pH. The $[Ca^{2+}]_i$ elevation caused channel closure when pH_i was constant and even when it was raised to the alkaline range (62).

Clearly Ca^{2+} is sufficient for channel closure. Whether H^+ alone is sufficient is still not known. The problem is technically difficult because lowering pH_i invariably produces elevation of $[Ca^{2+}]_i$. On isolated gap-junction membrane, H^+ is not sufficient; in Unwin and Ennis's work the channel did not go into the closed configuration on exposure to pH 6 (or anywhere else in the 6.0–8.0 range), whereas it did in response to Ca^{2+} concentration comparable to that we had found to close the channels inside cells (72).

The issue is not yet closed, but pondered from the physiological point of view, there is little to sway one toward a regulatory role by H^+. The electrochemical gradients for H^+ across cells are such that, upon breaking the surface barrier, the pH_i is driven up rather than down (as it is upon any loss of membrane potential)—not precisely the sort of arrangement for getting a rapid uncoupling reaction useful for safeguarding the cell community. Weighing the possibility of a more subtle regulation of communication, such as actual intracellular signalling, H^+ comes off no better. It lacks the specificity for proteins and many of the other virtues of Ca^{2+} for information transfer. Changes in pH, in fact, commonly cause unspecific changes in protein structure through protonation—not precisely the thing intracellular signal systems are made of. On both accounts, even a priori, I find Ca^{2+} a far more attractive candidate for physiological regulation of communication and the recent experimental results speak for themselves (25a).

The molecular mechanisms by which Ca^{2+} acts on the cell-to-cell channel are still unknown. Does Ca^{2+} act on the channel protein directly or does it use an intermediary? In 1981 Michael Johnston and Fidel Ramón at Duke University took a fresh approach. They perfused both sides of the crayfish (septate) nerve junction and discovered that the junctional channels then lost both Ca^{2+} and H^+

sensitivity (31). They interpreted this loss as the result of the washout of an intermediary substance, and Ramón and Zampighi have since further substantiated this notion. Is the substance calmodulin, the ubiquitous Ca^{2+}-binding protein that mediates so many cellular actions of Ca^{2+}? Elliott Hertzberg and Norton Gilula showed that calmodulin binds to liver and lens gap-junction protein (25), and S. Girsch and Camilo Peracchia at the University of Rochester showed that calmodulin plus Ca^{2+}, but neither calmodulin alone nor Ca^{2+} alone, blocked the uptake of solute into liposomes containing lens gap-junction protein (22). Another possibility has just come to the fore: a Ca^{2+} action mediated by the diacylglycerol–protein kinase C pathway of the phosphoinositide transmembrane route. Toshihiko Yada, Birgit Rose, and I found this past year that this pathway depresses communication, and Ca^{2+} appears to be critically involved here (77). The finding spanned a bridge to the field of growth control, in which several oncogenes control communication and growth via that pathway.

Protein kinase C was the latest element joining the panel of control of the cell-to-cell channel. Two years before, Eric Wiener and I had shown that protein kinase A, the cAMP-dependent protein kinase, enhances communication (76). Thus the picture of a dual physiological control of the channel emerged in which the cAMP–protein kinase A pathway upregulates and the diacylglycerol–protein kinase C pathway downregulates communication. Each of these controls probably involves its own set of cellular phosphorylations.

The roots of the story of protein kinase A in communication go back to 1974 when W. Hax, G. Van Venrooij, and I. Vossenberg at Utrecht discovered that application of a derivative of cAMP increased junctional conductance in *Drosophila* salivary glands (23). This knowledge lay dormant until 1978 when Jean Flagg-Newton and I accidentally met with the cAMP mechanism in mammalian cell cultures. Jean had come from Harvard in 1976 to investigate the permeability properties of the mammalian cell-to-cell channels by means of the fluorescent-labeled polyaminoacids, which Ian Simpson, Birgit Rose, and I had developed for the *Chironomus* cells (64). One week Jean had forgotten to feed the cultures. We noticed then, to our surprise, that the starved cells had unusually high junctional permeabilities. It goes against one's worldly grain that something should improve by starving, but this trove soon set us on the track of cAMP (18), for it was known that serum-starved cells have increased levels of cAMP. So it came that we joined the fans of this mighty substance.

We performed three kinds of experiments: we supplied the cells with cAMP, elevated their endogenous cAMP levels by treatments with phosphodiesterase inhibitors or adenylate cyclase activators, and supplied them with the phosphorylating enzyme subunit. All had the same outcome: an increase in junctional permeability. For example, the rate of transfer of fluorescent-labeled di- or triglutamic acid through the junction of isolated pairs of rat B cells about doubled over six hours of exposure to dibutyryl cAMP. Gerhard Dahl joined us in this endeavor and showed electron-microscopically that an increase in the number of membrane particles of gap junction par-

alleled the effect (19). Increases in junctional permeability turned out to be typical of mammalian cells that make channels with one another readily, but the most dramatic effect was obtained when Azarnia, G. Dahl, and I tried the cyclic nucleotide on a cell type (C1-1D) that exhibits no junctional permeability in ordinary conditions of culture: a junctional permeability was then "induced" by the treatment (3). This knowledge stood us in good stead later on in the experiments demonstrating the junction-mediated growth control I already described.

The hallmark of all these cAMP-dependent effects—and this was so also when the effects were driven by hormones, as shown by our colleague Aurelian Radu for catecholamine and prostaglandin (55)— was a marked upregulation of communication. The cAMP pathway seemed to regulate the open state or the formation process of the cell-to-cell channels. This brought us naturally to the question of what protein kinase mediated the channel regulation. The receptor for cAMP, protein kinase A, occurs in two forms, I and II. To learn which of the two isoenzymes operates in the channel regulation, Eric Wiener and I resorted to mutant cells deficient in one or the other enzyme form, which Michael Gottesman and his colleagues at the National Institutes of Health had isolated. Mutant I⁻ turned out to be clearly communication deficient, and the correlation between this deficiency and the enzyme deficiency also held for backshifts of mutation (76). After carrying mutant I⁻ in culture for about half a year, clones appeared that had reverted to the normal communication phenotype; this reversion invariably went hand in hand with reversion to the normal enzyme phenotype. Next we put into the mutant I⁻ a purified preparation of the enzyme's catalytic subunit, a preparation from E. Krebs's laboratory in Seattle with which Glenn Kerrick provided us, and showed that the junctional permeability rose to a level comparable to that of wild-type cells. Ross Johnson and his colleagues at the University of Minnesota and Juan Saez and his colleagues at Albert Einstein, Baylor, and Rockefeller Universities have carried these studies toward the level of protein and have recently shown that MP26, the major protein of lens fiber junction, and the 27-kilodalton liver gap-junction protein are phosphoproteins (30, 62a), and Klaus Willecke and his colleagues at Essen found that the synthesis of the protein in liver gap junction is enhanced by cAMP-dependent phosphorylation (76a). The ball has now rolled to the biochemist's field.

To end my account I turn back once more to the 1960s to tell about the explorations of the channel-formation process and the eventual detection of the channels themselves. By 1965 we knew that the conductive pathway was widely present, but we had no idea how it comes about. I felt sure that the pathway is not static because the relations of cell contact in tissues are dynamic. This is certainly so in embryonic and regenerating tissues in which cells slide alongside each other; continually, old contacts are broken and new ones are made. The picture of a pathway made of unitary channels seemed best suited for such dynamism and this, as much as the knowledge concerning the pathway's insulation and permeation limit, originally encouraged me to advance the channel hypothesis. I ran the first experiments dealing with this problem in February 1966 at Montemar and in the summer of that year at Woods Hole.

The first question I addressed was: Do cell-to-cell channels form spontaneously when (separate) cells are brought into contact? I used cells from the sponges *Microciona* and *Haliclona* (39). Over many summers at Woods Hole, I had seen my neighbors Aaron Moscona and Tom Humphreys experimenting with the adhesion of these cells and learned how easy it is to isolate cells from a sponge. The time-honored way at Woods Hole (dating back to the beginning of the century) was to squeeze a piece of sponge through a fine-meshed cloth (I used a woman's nylon stocking). I then manipulated individual pairs of cells together and monitored the electric coupling across their forming junctions. Within minutes of adhesive contact, they established a communicating pathway.

Two years later Shizuo Ito and I pursued this question further with the cells from the newt embryo (28). Shizuo was an expert in amphibian embryos. He introduced me to the macromeres, the large cells in the early embryo. These cells, too, when isolated formed conductive pathways within minutes of coming into contact. With these large cells (nearly 0.5 mm in diameter), we could make conductive junctions repeatedly between a pair of cells by simply pulling the pair apart and putting it together again at a different randomly chosen spot (the pathway sealed upon disjunction). Thus it was clear that cell-to-cell channels could form in large parts of the cell membrane of these cells, including parts in which there were no open channels before. The next clue about the formation process came when we measured the cell-to-cell conductance continuously during the formation of junctions between the pairs: the conductance rose progressively to a plateau (29). This tallied nicely with the channel hypothesis; the formation of the conductive pathway could be simply explained by a progressive accretion of channel units.

The final demonstration of this accretion had to wait until later. It required a technique sensitive enough to resolve the quantum of conductance of the single channel. This quantum I had estimated from our channel probings with small peptide molecules in *Chironomus* glands to be of the order of 10^{-10} mho. Although compared to ion channels of nonjunctional cell membranes this quantum is large,

[325]

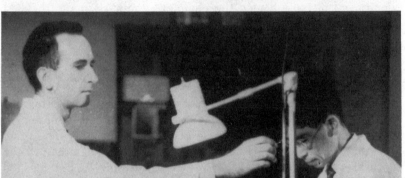

its detection is difficult; it meant to extract a transfer-voltage signal of 10^{-7} V from noise. In 1977 Socolar and I felt ready to tackle the problem with a phase-sensitive technique that gave a 10^{-8} V resolution (albeit at the expense of time resolution). Kanno, on a sabbatical leave from Hiroshima University, joined us again, and we looked for the quantal conductance step in nascent junction between pairs of amphibian embryo cells (46). At the onset of junction, against the background of low conductance, we expected to stand the best chance for detecting the single-channel event. There it was: a unitary step of 0.6071 ± 0.0072 μV (Fig. 3).

This was a jubilant moment and is a good spot to end my tale. But the best thing, perhaps, is that the story is not over. There is fresh excitement in the air just as I am writing these lines. David Paul at Harvard and Nalin Kumar and Bernie Gilula at Scripps Clinic Research Institute have cloned the DNA of the channel subunit, Gerhard Dahl and Rudolph Werner at Miami have isolated a messenger RNA that induces formation of the channel, and Roobik Azarnia, David Shalloway, and I have hit upon a regulatory gene for the channel. The work is now swiftly moving into the molecular sphere

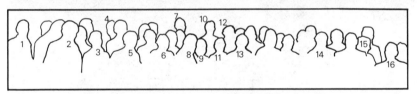

Top: At Titisee Conference on cellular growth and differentiation. *1*, R. Simantov; *2*, A. Gierer; *3*, M. Grunberg-Manago; *4*, Paul Marks; *5*, Manfred Eigen; *6*, Beatrice Mintz; *7*, Leo Sachs; *8*, David Baltimore; *9*, E. Gateff; *10*, F. Ruddle; *11*, Hasso Schroeder; *12*, H. Fischer; *13*, Werner Loewenstein; *14*, N. A. Mitchison; *15*, Peter Lawrence; *16*, Cesar Milstein.

Facing Page. *Top left*: Otto Loewi and Werner Loewenstein in Woods Hole, 1954. *Top right*: balcony of Woods Hole Old Main, 1954: Carlton Hunt, Werner Loewenstein, Steve Kuffler, Otto Hutter, Carlos Eyzaguirre. *Middle*: Sidney Socolar and Masayasu Sato at Columbia University laboratory in the 1960s. *Bottom*: Loewenstein lecturing at symposium on membrane transport in 1970. Hans Ussing and Guillermo Whittenbury at right. On the screen, the leitmotif of twenty years.

in several labs, and there is a sense of anticipation that big pieces of the communication puzzle may soon fall into place. The panorama for the travelers in the field, shown in Figure 3, will soon look different, I am sure. But will it ever be less daunting?

Looking rearward, it is plain that the unique atmosphere of Woods Hole always has loomed large in my work. It is hard to set such an atmosphere into words. Lewis Thomas captured some of it in his inimitable prose (70):

> *Biologists [in Woods Hole] seem to prefer standing on beaches, talking at each other, gesturing to indicate the way things are assembled, bending down to draw diagrams in the sand. By the end of the*

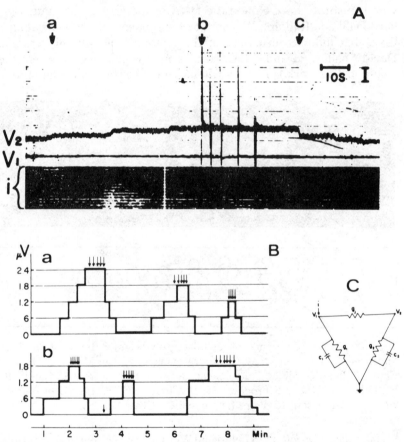

Figure 3. Opening and closing of single channels in nascent junction. Quantal steps of conductance are resolved during formation of junction between a pair of *Xenopus* embryo cells micromanipulated together. *Upsteps* correspond to spontaneous channel openings and *downsteps* to closings produced by microinjection of Ca^{2+} (Ca^{2+} injections marked by capacitative spikes in *A* and *arrows* in *B*). [From Loewenstein, Kanno, and Socolar (46).]

Adapted from cartoon by P. de Meyts.

day, the sand is crisscrossed with a mesh of ordinates, abscissas, curves to account for everything in nature.

You hear a similar sound at the close of the Friday Evening Lecture, the MBL's weekly grand occasion, when guest lecturers from around the world turn up to present their most stunning pieces of science. As the audience flows out of the auditorium, there is the same jubilant descant, the great sound of people explaining things to each other as fast as their minds will work.

Alone the change from one's habitual lair, the breathing of free air untrammeled by habitual university duties, the opportunity for quiet working and thinking, and even the knowledge during the year that, comes summer, Woods Hole would be there—all things one can't apply a yardstick to (or put in a grant application)—were immensely valuable parts of the Woods Hole mystique and, depending on one's age, romance.

Apart from that, there was for me personally the great enjoyment of seeing, year after year at Woods Hole, a lifelong friend: Steve Kuffler. I could not imagine the opening of the summer season without our bandying about experiments and jokes. Eventually, after so many years, we burst out laughing without telling the jokes. Luckily such predictability was not the rule with the experiments, but Steve always was the first to hear about the things I have chronicled here.

As this book's jib was set to give a sense of the development of ideas, I let my chronicle be steered by the hypotheses. Experiments are always rooted in theory, of course, but in biology, where serendipity has played such a dominant role, this is not always appreciated

[329]

and the parts of heralds and attestors can get mixed up. This does not happen so easily in physics. There the story lines clearly go from a few high spots of theory to the work of experimentalists. The scores in biology, rather than such grandiose fanfares, are more like strenuous minuets between theoreticians and empiricists who are not always in step and often change hats. Yet the dependency on theory is no less complete, and so even this story, which covers only two decades and is meant to give just a feel for the everyday sport of working hypothesis and experiment, has a homiletic ring to it.

NOTES

[1] The years or dates chronicled are the times of conferences, receipts of papers, or, where known, when the work was actually done.

[2] The only defector was Li^+. It produced uncoupling without detectable Ca^{2+} elevation. This posed no danger to the hypothesis, but it pointed up that, besides $[Ca^{2+}]_i$, the channel open state might be ruled by other mechanisms. We still do not know today how Li^+ works (and I guess I am not the only one writing a chapter in this book with this predicament), but the uncoupling by Li^+ has just set us on a fresh trail that, with luck, may open to the molecular mechanisms of the regulation of the channel.

BIBLIOGRAPHY

1. ASHLEY, C. C., and E. B. RIDGWAY. On the relationships between membrane potential, calcium transient and tension in single barnacle muscle fibres. *J. Physiol. Lond.* 209: 105–130, 1970.
2. ATKINSON, M. M., A. S. MENKO, R. G. JOHNSON, J. R. SHEPPARD, and J. D. SHERIDAN. Rapid and reversible reduction of junctional permeability in cells infected with a temperature-sensitive mutant of avian sarcoma virus. *J. Cell Biol.* 91: 573–578, 1981.
3. AZARNIA, R., G. DAHL, and W. R. LOEWENSTEIN. Cell junction and cyclic AMP. III. Promotion of junctional membrane permeability and junctional membrane particles in a junction-deficient cell type. *J. Membr. Biol.* 63: 133–146, 1981.
4. AZARNIA, R., W. J. LARSEN, and W. R. LOEWENSTEIN. The membrane junctions in communicating and noncommunicating cells, their hybrids, and segregants. *Proc. Natl. Acad. Sci. USA* 71: 880–884, 1974.
5. AZARNIA, R., and W. R. LOEWENSTEIN. Parallel correction of cancerous growth and of a genetic defect of cell-to-cell communication. *Nature Lond.* 241: 455–457, 1973.
6. AZARNIA, R., and W. R. LOEWENSTEIN. Intercellular communication and tissue growth. VIII. A genetic analysis of junctional communication and cancerous growth. *J. Membr. Biol.* 34: 1–37, 1977.
7. AZARNIA, R., and W. R. LOEWENSTEIN. Intercellular communication and the control of growth. X. Alteration of junctional permeability by the *src* gene. A study with temperature-sensitive mutant Rous sarcoma virus. *J. Membr. Biol.* 82: 191–205, 1984.

7a. AZARNIA, R., and W. R. LOEWENSTEIN. Polyomavirus middle T antigen downregulates junctional cell-to-cell communication. *Mol. Cell. Biol.* 7: 946–950, 1987.

8. AZZI, A., and B. CHANCE. The "energized state" of mitochondria: lifetime and ATP equivalence. *Biochim. Biophys. Acta* 189: 141–151, 1969.

9. BAKER, P. F., M. P. BLAUSTEIN, A. L. HODGKIN, and R. R. STEINHARDT. The influence of calcium on sodium efflux in squid axons. *J. Physiol. Lond.* 200: 431–450, 1969.

10. BAKER, P. F., A. L. HODGKIN, and E. B. RIDGWAY. Depolarization and calcium entry in squid giant axons. *J. Physiol. Lond.* 218: 709–755, 1971.

11. BURTON, A. C. Cellular communication, contact inhibition, cell clocks, and cancer: the impact of the work and ideas of W. R. Loewenstein. *Perspect. Biol. Med.* 14: 301–318, 1971.

11a. CASPAR, D. L. D., D. A. GOODENOUGH, L. MAKOWSKI, and W. C. PHILLIPS. Gap junction structures. I. Correlated electron microscopy and X-ray diffraction. *J. Cell Biol.* 74: 605–628, 1977.

12. DÉLÈZE, J. Effet des ions calcium sur le rétablissement du potentiel de repos après lésion des fibres cardiaques (Abstract). *Helv. Physiol. Pharmacol. Acta* 20: C47, 1962.

13. DÉLÈZE, J. The recovery of resting potential and input resistance in sheep heart injured by knife or laser. *J. Physiol. Lond.* 208: 547–564, 1970.

14. DÉLÈZE, J., and W. R. LOEWENSTEIN. Permeability of a cell junction during intracellular injection of divalent cations. *J. Membr. Biol.* 28: 71–86, 1976.

15. DE WEER, P. Aspects of the recovery process in nerve. In: *Neurophysiology*, edited by C. C. Hunt. London: Butterworths, 1975, p. 231–278. (Int. Rev. Sci. Physiol. Ser. I.)

16. ENGELMANN, T. W. Vergleichende Untersuchungen zür Lehre von der Muskel und Nervenelektricität. *Pfluegers Arch. Gesamte Physiol. Menschen Tiere* 15: 116–148, 1877.

17. FARQUHAR, M. G., and G. E. PALADE. Junctional complexes in various epithelia. *J. Cell Biol.* 17: 375–412, 1963.

18. FLAGG-NEWTON, J. L. Cyclic AMP increases junctional permeability in cultured mammalian cells (Abstract). *Biophys. J.* 25: 80a, 1979.

19. FLAGG-NEWTON, J. L., G. DAHL, and W. R. LOEWENSTEIN. Cell junction and cyclic AMP. I. Upregulation of junctional membrane permeability and junctional membrane particles by administration of cyclic nucleotide or phosphodiesterase inhibitor. *J. Membr. Biol.* 63: 105–121, 1981.

20. FURSHPAN, E. J., and D. D. POTTER. Transmission at the giant motor synapses of crayfish. *J. Physiol. Lond.* 145: 289–325, 1959.

21. GILULA, N. B., O. R. REEVES, and A. STEINBACH. Metabolic coupling, ionic coupling and cell contacts. *Nature Lond.* 235: 262–265, 1972.

22. GIRSCH, S. J., and C. PERACCHIA. Lens cell-to-cell channel protein. I. Self-assembly into liposomes and permeability regulation by calmodulin. *J. Membr. Biol.* 83: 217–225, 1985.

23. HAX, W. M. A., G. E. P. M. VAN VENROOIJ, and J. B. J. VOSSENBERG. Cell communication: a cyclic AMP mediated phenomenon. *J. Membr. Biol.* 19: 253–266, 1974.

24. HERBST, C. Experimentelle Untersuchungen über den Einfluss der veränderten chemischen Zusammensetzung des umgebenden Mediums auf

CELL
CONNECTIVITY

[331]

die Entwicklung der Thiere. II. Weiteres über die morphologische Wirkung der Lithiumsalze und ihre theoretische Bedeutung. *Mitt. Zool. Stat. Neapel* II: 136–220, 1893

25. HERTZBERG, E. L., and N. B. GILULA. Liver gap junctions and lens fiber junctions: comparative analysis and calmodulin interaction. *Cold Spring Harbor Symp. Quant. Biol.* 46: 639–645, 1982.

25a. HERTZBERG, E., and R. JOHNSON (editors). *Gap Junctions*. New York: Liss, in press. (Modern Cell Biol. Ser.)

26. HOLTFRETER, J. Neuralization and epiderminization of gastrula ectoderm. *J. Exp. Zool.* 98: 161–209, 1945.

27. ITO, S., and N. HORI. Electrical characteristics of *Triturus* egg cells during cleavage. *J. Gen. Physiol.* 49: 1019–1027, 1966.

28. ITO, S., and W. R. LOEWENSTEIN. Ionic communication between early embryonic cells. *Dev. Biol.* 19: 228–243, 1969.

29. ITO, S., E. SATO, and W. R. LOEWENSTEIN. Studies on the formation of a permeable cell membrane junction. II. Evolving junctional conductance and junctional insulation. *J. Membr. Biol.* 19: 339–355, 1974.

30. JOHNSON, K. R., P. D. LAMPE, K. C. HUR, C. F. LOUIS, and R. G. JOHNSON. A lens intercellular junction protein, MP26, is a phosphoprotein. *J. Cell Biol.* 102: 1334–1343, 1986.

31. JOHNSTON, M. F., and F. RAMÓN. Electrotonic coupling in internally perfused crayfish segmented axons. *J. Physiol. Lond.* 317: 509–518, 1981.

32. KANNO, Y., and W. R. LOEWENSTEIN. Low-resistance coupling between gland cells. Some observations on intercellular contact membranes and intercellular space. *Nature Lond.* 201: 194–195, 1964.

33. KANNO, Y., and W. R. LOEWENSTEIN. Cell-to-cell passage of large molecules. *Nature Lond.* 212: 629, 1966.

34. KANNO, Y., and W. R. LOEWENSTEIN. Intercellular diffusion. *Science Wash. DC* 143: 959–960, 1964.

35. KUFFLER, S. W., and D. D. POTTER. Glia in the leech central nervous system: physiological properties and neuron-glia relationships. *J. Neurophysiol.* 27: 290–330, 1964.

36. LOEWENSTEIN, W. R. Permeability of membrane junctions. *Ann. NY Acad. Sci.* 137: 441–472, 1966. (Conf. Biol. Membr. Recent Prog.)

37. LOEWENSTEIN, W. R. On the genesis of cellular communication. *Dev. Biol.* 15: 503–520, 1967.

38. LOEWENSTEIN, W. R. Some reflections on growth and differentiation. *Perspect. Biol. Med.* 11: 260–272, 1968.

39. LOEWENSTEIN, W. R. Communication through cell junctions. Implications in growth control and differentiation. *Dev. Biol. Suppl.* 2: 151–183, 1968.

40. LOEWENSTEIN, W. R. Transfer of information through cell junctions and growth control. In: *Proc. Eighth Canadian Cancer Res. Conf.*, 1968, edited by J. F. Morgan. New York: Pergamon, 1969, p. 162–170.

41. LOEWENSTEIN, W. R. Cellular communication by permeable junctions. In: *Cell Membranes: Biochemistry, Cell Biology and Pathology*, edited by G. Weissmann and R. Claiborne. New York: Hosp. Practice, 1974, p. 105–114.

42. LOEWENSTEIN, W. R. Junctional intercellular communication and the control of growth. *Biochim. Biophys. Acta* 560: 1–65, 1979.

43. LOEWENSTEIN, W. R. Regulation of cell-to-cell communication by phosphorylation. *Biochem. Soc. Symp.* 50: 43–58, 1985.
44. LOEWENSTEIN, W. R., and Y. KANNO. Studies on an epithelial (gland) cell junction. I. Modifications of surface membrane permeability. *J. Cell Biol.* 22: 565–586, 1964.
45. LOEWENSTEIN, W. R., and Y. KANNO. Intercellular communication and the control of tissue growth. Lack of communication between cancer cells. *Nature Lond.* 209: 1248–1249, 1966.
46. LOEWENSTEIN, W. R., Y. KANNO, and S. J. SOCOLAR. Quantum jumps of conductance during formation of membrane channels at cell-cell junction. *Nature Lond.* 274: 133–136, 1978.
47. LOEWENSTEIN, W. R., M. NAKAS, and S. J. SOCOLAR. Junctional membrane uncoupling. Permeability transformations at a cell membrane junction. *J. Gen. Physiol.* 50: 1865–1891, 1967.
48. LOEWENSTEIN, W. R., S. J. SOCOLAR, S. HIGASHINO, Y. KANNO, and N. DAVIDSON. Intercellular communication: renal, urinary bladder, sensory, and salivary gland cells. *Science Wash. DC* 149: 295–298, 1965.
49. MEECH, R. W., and R. C. THOMAS. The effect of calcium injection on the intracellular sodium and pH of snail neurones. *J. Physiol. Lond.* 265: 267–283, 1977.
50. MEHTA, P. P., J. S. BERTRAM, and W. R. LOEWENSTEIN. Growth inhibition of transformed cells correlates with their junctional communication with normal cells. *Cell* 44: 187–196, 1986.
51. NAKAS, M., S. HIGASHINO, and W. R. LOEWENSTEIN. Uncoupling of an epithelial cell membrane junction by calcium-ion removal. *Science Wash. DC* 151: 89–91, 1966.
52. NEYTON, J., and A. TRAUTMANN. Single-channel currents of an intercellular junction. *Nature Lond.* 317: 331–335, 1985.
52a. OLIVEIRA-CASTRO, G. M., and W. R. LOEWENSTEIN. Junctional membrane permeability: effects of divalent cations. *J. Membr. Biol.* 5: 51–77, 1971.
53. PENN, R. D. Ionic communication between liver cells. *J. Cell Biol.* 29: 171–173, 1966.
53a. POLITOFF, A. L., S. J. SOCOLAR, and W. R. LOEWENSTEIN. Permeability of a cell membrane junction; dependence on energy metabolism. *J. Gen. Physiol.* 53: 498–515, 1969.
54. POTTER, D. D., E. J. FURSHPAN, and E. S. LENNOX. Connections between cells of the developing squid as revealed by electrophysiological methods. *Proc. Natl. Acad. Sci. USA* 55: 328–339, 1966.
55. RADU, A., G. DAHL, and W. R. LOEWENSTEIN. Hormonal regulation of cell junction permeability: upregulation by catecholamine and prostaglandin E_1. *J. Membr. Biol.* 70: 239–251, 1982.
56. REVEL, J. P., and M. J. KARNOVSKY. Hexagonal array of subunits in intercellular junctions of the mouse heart and liver. *J. Cell Biol.* 33: C7–C12, 1967.
57. ROSE, B. Intercellular communication and some structural aspects of membrane junctions in a simple cell system. *J. Membr. Biol.* 5: 1–19, 1971.
58. ROSE, B., and W. R. LOEWENSTEIN. Depression of junctional membrane permeability by substitution of lithium for extracellular sodium. *Biochim. Biophys. Acta* 173: 146–148, 1969.
59. ROSE, B., and W. R. LOEWENSTEIN. Junctional membrane permeability.

CELL
CONNECTIVITY

Depression by substitution of Li for extracellular Na, and by long-term lack of Ca and Mg; restoration by cell repolarization. *J. Membr. Biol.* 5: 20–50, 1971.

60. Rose, B., and W. R. Loewenstein. Permeability of cell junction depends on local cytoplasmic calcium activity. *Nature Lond.* 254: 250–252, 1975.

61. Rose, B., and W. R. Loewenstein. Permeability of cell junction and the local cytoplasmic free ionized calcium concentration. A study with aequorin. *J. Membr. Biol.* 28: 87–119, 1976.

62. Rose, B., and R. Rick. Intracellular pH, intracellular free Ca, and junctional cell-cell coupling. *J. Membr. Biol.* 44: 377–415, 1978.

62a. Saez, J. C., D. C. Spray, A. C. Nairn, E. Hertzberg, P. Greengard, and M. V. L. Bennett. cAMP increases junctional conductance and stimulates phosphorylation of the 27-kDa principal gap junction polypeptide. *Proc. Natl. Acad. Sci. USA* 55: 328–339, 1966.

63. Schleiden, M. J. Beitrage zür Phytogenesis. *Müllers Arch. Anat. Physiol. Wiss. Med.* 1838: 137–176, 1838.

64. Simpson, I., B. Rose, and W. R. Loewenstein. Size limit of molecules permeating the junctional membrane channels. *Science Wash. DC* 195: 294–296, 1977.

65. Singer, S. J., and G. L. Nicholson. The fluid mosaic model of the structure of cell membranes. *Science Wash. DC* 17: 720–729, 1972.

66. Socolar, S. J. The coupling coefficient as an index of junctional conductance. Appendix. *J. Membr. Biol.* 34: 29–37, 1977.

67. Socolar, S. J., and A. L. Politoff. Uncoupling cell junctions of a glandular epithelium by depolarizing current. *Science Wash. DC* 172: 492–494, 1971.

68. Spray, D. C., A. L. Harris, and M. V. L. Bennett. Gap junctional conductance is a simple and sensitive function of intracellular pH. *Science Wash. DC* 211: 712–715, 1981.

69. Stoker, M. G. P. Regulation of growth and orientation in hamster cells transformed by polyoma virus. *Virology* 24: 165–174, 1964.

70. Thomas, L. *The Lives of a Cell. Notes of a Biology Watcher.* New York: Viking, 1974, p. 58.

71. Turin, L., and A. Warner. Carbon dioxide reversibly abolishes ionic communication between cells of early amphibian embryo. *Nature Lond.* 270: 56–69, 1977.

72. Unwin, P. N. T., and P. D. Ennis. Calcium-mediated changes in gap junction structure: evidence from the low angle X-ray pattern. *J. Cell Biol.* 97: 1459–1466, 1983.

73. Unwin, P. N. T., and G. Zampighi. Structure of the junction between communicating cells. *Nature Lond.* 283: 545–549, 1980.

74. Warner, A. E., S. C. Guthrie, and N. B. Gilula. Antibodies to gap-junctional protein selectively disrupt junctional communication in the early amphibian embryo. *Nature Lond.* 311: 127–131, 1984.

75. Weidmann, S. The electrical constants of Purkinje fibres. *J. Physiol. Lond.* 118: 348–360, 1952.

76. Wiener, E. C., and W. R. Loewenstein. Correction of cell-cell communication defect by introduction of a protein kinase into mutant cells. *Nature Lond.* 305: 433–435, 1983.

76a. Traub, O., J. Look, D. Paul, and K. Willecke. Cyclic adenosine monophosphate stimulates biosynthesis and phosphorylation of the 26 kDa gap junction protein in cultured mouse hepatocytes. *Eur. J. Cell Biol.* 43: 48–54, 1987.

77. YADA, T., B. ROSE, and W. R. LOEWENSTEIN. Diacylglycerol downregulates junctional membrane permeability. TMB-8 blocks this effect. *J. Membr. Biol.* 88: 217–232, 1985.

XIV

Epithelial Transport: Frog Skin as a Model System

HANS H. USSING

EPITHELIAL
TRANSPORT

FOR nearly four decades the isolated frog skin has played an important role as a model system for studies of epithelial electrolyte transport. Actually, the preparation had provided valuable observations much earlier. Galeotti (8) had demonstrated that the potential difference (inside positive) across the skin depended on the presence of Na^+ (or Li^+) in the outside medium. Huf (14) showed by direct chemical analysis that the frog skin could bring about a net Cl^- transport from the outside in when bathed on both sides with Ringer's solution. A few years later Krogh (26, 27) showed that salt-depleted frogs could take up both Cl^- and Na^+ from exceedingly dilute solutions. It was, however, the advent of suitable isotopic tracers for Na^+, Cl^-, and K^+ that made possible a detailed kinetic analysis of the electrolyte transport processes, and this period started just after the Second World War.

The development of the current frog skin model went through several more or less well-defined phases. The first was the black-box phase. It was necessary to develop methods that allowed one to distinguish between active and passive transport, even if the object was a multilayered structure whose detailed properties along the path were unknown. The main tools in this period became the flux-ratio analysis (46) and the short-circuiting technique (56). The flux-ratio analysis led to the conclusion that under most conditions Cl^- transport across the skin was passive, being determined by the difference between its electrochemical potentials on the two sides of the preparation. However, Na^+ was subject to active transport.

The short-circuiting method demonstrated that the Na^+ transport was solely responsible for the electric asymmetry across the skin and that the short-circuit current was exactly equal to the net transport of Na$^+$. This direct demonstration of active Na$^+$ transport acquired interest, even outside the epithelial field, because at that time there was a lively discussion of the reason for the low Na$^+$ concentration in the cytoplasm of most cells. Whereas many assumed it was due to active Na$^+$ extrusion, there were powerful schools that maintained that the cytoplasm simply had a low affinity for Na$^+$ as opposed to K$^+$. The identity between active Na$^+$ transport and short-circuit current of the frog skin also made it relatively easy to study the relation between active Na$^+$ transport and O$_2$ consumption. Zerahn (58) and Leaf and Renshaw (32) independently demonstrated that there was a stoichiometric relationship between the two quantities, so that for 18 Na$^+$ transported there was an additional consumption of one molecule of O$_2$. This again meant that 3 Na$^+$ were transported for each molecule of ATP consumed in the process. Later the same relationship was found for Na$^+$ transport in other tissues and cells.

The second phase in the development of the frog skin model was the proposal of the two-membrane theory (19, 54). The postulates were that the transporting cell must have an outward-facing membrane that is selectively but passively permeable to Na$^+$ and tight to K$^+$. The inward-facing membrane must be selectively but passively permeable to K$^+$; furthermore, it must possess an active transport mechanism for Na$^+$ (which might be an Na$^+$-K$^+$ exchange pump or a "pure" Na$^+$ pump). The model arose as an attempt to solve the paradox that the isotope experiments had ascribed the skin potential to the effect of active Na$^+$ transport, whereas ion-substitution experiments indicated that the skin potential was the sum of an outer Na$^+$ diffusion potential and an inner K$^+$ diffusion potential. The model turned out to be generally applicable to a multitude of epithelia, either in its original form or with cotransport or exchange mechanisms replacing the specific Na$^+$ channel in the outside membrane.

In the beginning, frog skin as a model system suffered from the drawback that the anatomical localization of the two ion-selective membranes was uncertain because the frog skin has a multilayered epithelium. Then came the surprising solution to the problem when Ussing and Windhager (55) and Farquhar and Palade (5) independently proposed that the vast majority of the epithelial cells formed a three-dimensional functional syncytium, so that the outward-facing membrane of the outermost living cell layer was the Na$^+$-selective membrane of the model, whereas all the basolateral membranes had the K$^+$ channels and the Na$^+$ pumps. In a way, then, the whole epithelium acted as one all.

For a long time the exact nature of the Na$^+$ pump of epithelia was a matter of dispute. In our original model we assumed a Na$^+$-K$^+$ exchange pump. Several workers in the field were unable to find any coupling between Na$^+$ transport and K$^+$ exchange and therefore assumed a pure electrogenic Na$^+$ pump. Finally, Robert Nielsen (37) from our institute showed that the pump was the type that transports

3 Na$^+$ out of the cell and returns 2 K$^+$ to the cell for each molecule of ATP broken down. This type of pump was already described in several cell types and fits the properties of the Na$^+$-K$^+$-ATPase. Thus one may say that the last-mentioned finding of the coupling ratio between active Na$^+$ and K$^+$ transport completes the two-membrane theory. The limited number of pages at my disposal does not allow even cursory coverage of the many recent studies of the mechanisms behind the individual steps of the model. Also I must refrain from discussing in detail what one might call the third wave or the Cl$^-$ wave of the model emerging right now. Until recently it was assumed that the passive inward transport of Cl$^-$ took place partly via leaks in the tight seals between the cells and partly through the ordinary epithelial cells. It now appears that the two pathways mentioned are insignificant and that the so-called mitochondria-rich cells (MR cells) provide the all-important specific Cl$^-$ pathway. This was first suggested by my former associate Cornelis Voûte (57) and has now been substantiated in various ways.

The principal cells (i.e., cells of the syncytium), on the other hand, are tight to Cl$^-$, both apically and basolaterally, but maintain a high cytoplasmic Cl$^-$ concentration, apparently via a basolateral Na$^+$-K$^+$-2Cl$^-$ cotransport (see ref. 50). The new version of the frog skin model evidently gives a greatly improved description of the handling of electrolytes in the frog skin, but we still do not know whether it represents a pattern common to other epithelia or whether it is specific to this one tissue.

BEGINNING OF ISOTOPE AGE

During the last years of the Second World War, August Krogh, the head of the Zoophysiological Laboratory, University of Copenhagen, had initiated a systematic study of cellular membrane transport using isotope tracers. Because of his close contact with Niels Bohr and The Niels Bohr Institute, cyclotron-produced isotopes were easily available.

Krogh hoped to expand these studies as soon as the war was over and to get support from the Rockefeller Foundation for the project. I have in my files a copy of the memorandum that was the basis for his grant application. The memorandum was very detailed and was based on his own experiments as well as a literature study, which was published in the form of his Croonian Lecture (28). I find it appropriate to cite the first page of the memorandum, which shows Krogh's clear appreciation of the problem to be tackled.

Memorandum, concerning the use of isotopes for determinations of ion permeabilities of cell surfaces and living membranes generally.

As a result of studies made mainly during the last 10 years or so and to a considerable extent by means of isotopes it is now recognized

[339]

1. that cell surfaces and membranes considered impermeable to certain ions are in fact permeable although the permeability is often of a low order,

2. that in view of this permeability the conceptions regarding the mode of preservation of the large differences in ionic composition between cells and their surroundings and between aquatic animals and the water have to be thoroughly revised and

3. that such a revision has in a number of cases demonstrated the existence of an active transport, requiring energy, of certain ions through cell surfaces and living membranes against concentration gradients.

When a membrane, say the surface of a red blood corpuscle, is passively permeable to an ion, e.g., potassium, and a many times higher potassium concentration is nevertheless maintained inside the cell than outside, a quantitative determination of the potassium permeability can not be made by studying the effect of varying the outside concentration or by any other means which may stimulate or inhibit the active transport processes, but only by means of a potassium isotope, and even utilizing the isotope only under certain specified conditions.

We must look upon the cell as being normally in a steady state in which the passive loss of potassium to the surrounding medium is exactly made good by the active potassium uptake.

Krogh got the grant, but in the meantime he had retired from his chair, and he asked me to head the project in the initial phase. At the time I was involved in studies of protein turnover, using deuterium-labeled amino acids, and I was reluctant to leave that promising field, even for a short period. However, I consented to give the electrolyte project a try.

One of the experimental objects with a high priority on Krogh's list was striated muscle, and we started experiments with frog sartorius. The muscle was loaded with radioactive Na^+ (^{24}Na), and we followed the washout of the radioactivity into Ringer's solution. We could resolve the washout curve into a very fast component and a slower one. The fast component was identified as the washout of the interspaces and the slower one as the exchange of Na^+ between muscle fibers and medium. Parenthetically such a simple treatment of the data is only possible for an ion that, like Na^+, is present in high concentration in the interspaces and low in the fibers, so that recycling is insignificant. In the case of K^+, recycling has to be taken into account. That problem was solved by Harris and Burn (11). For Na^+, however, our method should give reasonably correct values.

The slow component of the washout curve obviously gives exactly the figure Krogh had desired, namely, the expulsion of Na^+ from the fiber. Remembering that the muscle fiber interior is $\sim$80–90 mV negative relative to the medium and that the Na^+ concentration in the fiber is much lower than that of Ringer's solution, it is clear that the exit of Na^+ from the fiber must be against the electrical and the chemical gradients and thus must be due to active transport. The entry could be a passive process.

With respect to the energy consumed by the exit process, I was in the lucky position that I could calculate its rate from the isotope experiment. At the same time I could calculate the electric potential and the concentration gradient against which the transport had to take place. Thus I could calculate the amount of energy consumed per unit time by the Na^+ transport process.

Then I got a shock. It turned out that the amount of energy required was more than the amount of energy made available by the metabolism of the muscle. How was this possible? Then it dawned on me that one possibility had been overlooked. If the membrane had a mechanism that coupled the passage of a given ion in one direction to the movement in the opposite direction of another ion of the same species, then the exchange might not cost any metabolic energy. The phenomenon of a one-to-one exchange despite an electrochemical gradient for the species was named *exchange diffusion* (33, 44).

The lesson learned from the experiment mentioned is that it is not fully apparent from the isotope-exchange experiment whether a true measure of the energy-consuming active transport process can be obtained.

On the other hand, if the possibility of metabolic energy consumption can be ruled out, an entirely new way of crossing a membrane has been found. The exchange-diffusion phenomenon, which was first seen in muscle, later turned out to occur rather frequently in living membranes. In the particular case of the red cell, the exchange system is shared by Cl^-, HCO_3^-, and other small anions, and it serves a very important physiological purpose, namely, the Cl^--HCO_3^- shift in lungs and in tissues.

Since we could not test the theory of a close correlation between metabolism and transport in the steady-state situation (because of the possibility of exchange diffusion), we felt that we had to work with systems where there is a manifest net transport of ions, that is, epithelia. Our first choice was the axolotl, which takes up NaCl through the skin, even from rather dilute solutions. We measured the uptake both by direct chemical analysis for Na^+ and Cl^- and by measuring the disappearance of the appropriate isotopes from the bathing solution. These measurements seemed to agree very well.

Our prime object with the experiments was to find out how this NaCl uptake was regulated. Very early in the game we observed that the Na^+ uptake was strongly stimulated when antidiuretic hormone was injected into the animals (16).

One day a disaster occurred. One animal died during the experiment because the O_2 supply had been clogged. We found that this dead animal had taken up radioactive Na^+ almost as fast as a living animal—not quite, but still beyond any doubt. Thus we had to think about an explanation. Chemical analysis showed that, as expected, the dead animal had suffered a net loss of Na^+, but there was another possibility that so far had been disregarded. Was it conceivable that the dead animal developed a potential of the right magnitude and direction to account for the isotope uptake? The answer actually is no, but the mere possibility of such an explanation showed us that it would be rash to interpret movements of isotopes of electrically charged substances without knowing the electric potential difference across the membrane across which the transport is taking place and without knowledge of the way such a potential difference ought to influence the movement of ions. Curiously the effects of potential differences on the movements of ions had been largely overlooked in tracer work until then.

Enter Frog Skin as Test Object

We turned to the isolated, surviving frog skin as a test object. The frog skin has been a favorite object of electrophysiologists for more than a hundred years ever since du Bois-Reymond, the founder of electrophysiology, observed an electric potential difference between the inside and the outside of the frog skin (2). As already mentioned, Huf (14) found that isolated frog skin would transport Cl^- from the outside to the inside if the skin was in contact with Ringer's solution on both sides, and shortly afterward Krogh (26, 27) demonstrated that frogs in need of salt would take up both Na^+ and Cl^- even from very dilute outside solutions. Clearly the frog skin (both before and after isolation) could transport NaCl inward and produce an electric potential. Thus the frog skin seemed well suited as a test object for experiments concerning the relationship between ion fluxes and transmembrane potentials.

The first series of experiments (45) had already given a clear result: net Na^+ influx could proceed against the combined effects of the electric field and a concentration gradient. In other words, the transport was active. On the other hand, Cl^- uptake took place only if the electrochemical potential was higher in the outside than in the inside medium.

Thus the net uptake might be due to electrodiffusion, but was that really the case? We were facing not only the problem of a possible exchange diffusion but also the question of how the electric field influences the movement of the tracer. I looked in the textbooks and handbooks but did not find any useful treatment of the subject. I thought that there must be some kind of simple solution to the problem. I discussed it with my good friend Kaj Linderstrøm-Lang of the Carlsberg Laboratory.

[342]

Linderstrøm-Lang and I decided to tackle the problem during the weekend. So we met and each of us presented a solution to the problem. They looked similar but they were not identical. After a long discussion it dawned on us that for solving the differential equation Linderstrøm-Lang had used the constant–concentration gradient assumption, whereas I had used the constant-field assumption. Incidentally the solution I had produced already existed in the literature since it was identical to that developed by Goldman (10). His paper had not reached Denmark because of the war conditions. After having studied the two solutions, two things became clear to me. *1*) There is no reason that the potential gradient across an epithelium should be linear; on the other hand, it is equally unreasonable that the concentration gradient for the isotope should be linear. In fact, both gradients are likely to be very unlinear, and their shapes are not accessible to measurement. Thus apparently the problem had no general solution. *2*) If, however, I took either one of the solutions and calculated the flux in one direction and divided it by the flux in the opposite direction the result, the flux ratio, was the same. So I was led to the working hypothesis that the flux ratio does not depend on the assumptions chosen for the potential and concentration profiles. The flux ratio might be the same whether the membrane is homogeneous or a composite structure like an epithelium. In a way, then, the flux ratio would be a kind of state function, depending on the conditions in the bathing solutions and the electric potential difference between them but independent of the conditions along the transport path. This hunch turned out to be correct (46).

EPITHELIAL
TRANSPORT

Before proceeding any further, remember that there are two ways to interpret the behavior of an isotopic tracer. It can be seen as a substance that happens to have practically the same properties as the macrospecies for which it is tracing or it can be seen as a representative for the species it is tracing. Both concepts have advantages and disadvantages from a theoretical point of view. In a way they are complementary, and it is very important not to mix concepts developed for one of the alternatives with those belonging to the other point of view. In the following we stick to the definition that the tracer represents the ion it is tracing. This leads to the concept of the unidirectional flux: if a certain percentage of the tracer in solution one passes unit area of the membrane to solution two in unit time, then the same percentage of the macrospecies must have passed from solution one to solution two.

The ratio between the two opposite unidirectional fluxes of a passively transported ion varies with the mucosal and serosal activities (a_m and a_s) of the ion under investigation on the two sides of the membrane and with the transmembrane serosal and mucosal electric potential difference ($V_z - V_m$) according to the following equation

$$\frac{J_{ms}}{J_{sm}} = \frac{a_m}{a_s} \exp\left(- \frac{zF(V_s - V_m)}{RT}\right)$$

where J_{ms} is mucosal to serosal flux, J_{sm} is serosal to mucosal flux, z is valence, F is Faraday constant, and RT is gas constant $\times$ absolute temperature. In the derivation of this equation, no assumptions have been made with respect to homogeneity of the membrane or the shapes of the gradients of electrical potential and chemical activity. It is assumed that the ions under study do not associate with other moving particles during their passage through the membrane.

The equation shows that for an ion moving under the influence of an electrochemical potential gradient without interacting with other moving particles, the flux ratio is determined solely by the drop in electrochemical potential across the membrane, being independent of changes in potential, resistance, activity coefficient, and available diffusion area along the path. The membrane may consist of any number of different layers in series (and in parallel), and the equation is also valid if there is electric current flow through the membrane. It is only required that a continuous transport pathway exists.

The influx and outflux in principle can be determined simultaneously with two different isotopes of the same element (^{22}Na and ^{26}Na, ^{36}Cl and ^{38}Cl, etc.), or one can do parallel determinations of the two fluxes. Thus left and right side of the belly skin of a frog may yield flux ratios that are virtually as good as ratios obtained by double labeling with two tracers.

The usefulness of the equation lies in the fact that it singles out the cases where the behavior of an ion justifies more careful scrutiny. We have already met one type of deviation from ideal behavior, namely, exchange diffusion, characterized by the fact that the flux from left to right over that from right to left is one, independent of concentration and potential difference.

Another type of deviation from ideal behavior is shown by Na$^+$ in the isolated frog skin. Only a minute fraction of the Na$^+$ influx observed can be accounted for by passive movement. Nearly the total influx is an uphill transport against the combined effects of concentration and potential gradients.

On the other hand, if the skin has been poisoned with dinitrophenol, the flux ratios found are very nearly those expected from passive diffusion. Thus, if the ionic species studied is subject to active transport, it is reflected in a deviation from the expected flux ratio.

A flux-ratio analysis for Cl$^-$ in frog skin indicates that the movement is passive, even in cases where a manifest net inward transport is observed (17). This is so under the most varied conditions for potential and concentration difference across the skin. In every experiment the flux ratios found and those calculated agree completely. In *Rana temporaria*, at least, the Cl$^-$ behaves exactly as one would expect a passive ion to do.

[344]

The Na$^+$, on the other hand, is clearly subject to active transport. It was tempting then to propose that the potential difference across the skin was due to the active Na$^+$ transport. At that time the ideas concerning the origin of bioelectric potentials were rather conflicting. It was generally believed that resting potentials of nerve and muscle fibers were essentially K$^+$ diffusion potentials. Respiratory carbonic acid was assumed to contribute to epithelial potentials by splitting into H$^+$ that could pass one boundary of the cell, whereas HCO$_3^-$ might pass faster through the other.

Some workers believed in so-called adsorption potentials based on the (erroneous) assumption that asymmetric adsorption of ions to cell membranes could give rise to measurable potential differences between the bathing solutions. The idea of a role for Na$^+$ in the action potential of nerves had just begun to take shape.

Another commonly held belief at the time was that bioelectric potentials were oxidation reduction potentials. Thus it was very important to be able to prove that active Na$^+$ transport was solely responsible for the frog skin potential.

One point of theoretical nature also haunted me at the time. Although the Debuye-Huckle theory allowed the calculation of single-ion activities, it was a dogma among many physicochemists that the activity coefficient for an individual ion species could not be determined. The activity coefficient was a property of a salt. Nevertheless, in the flux-ratio equation it is necessary to ascribe an activity coefficient to each individual ion species. The justification of this procedure is now generally accepted, but it was clear to me that worry about the uncertainty of the activity coefficients for Na$^+$ in the inside and outside media might reduce the impact of our data in favor of active Na$^+$ transport. Another problem was the uncertainty arising from the diffusion potentials, since KCl bridges were used to connect the bathing solutions with the reversible calomel electrodes. In fact, some physicochemists considered the potential difference between solutions of different compositions to be undefined.

SHORT-CIRCUITED FROG SKIN

It finally occurred to me that all the problems just mentioned could be circumvented in one stroke if it was possible to short-circuit the skin potential and let it act as a battery, producing electric current with identical solutions on both sides. Under such conditions the activity coefficient must be the same on both sides and the diffusion gradients of the KCl bridges would cancel. If a total short circuit could be achieved, every single ionic species would be at the same electrochemical potential on both sides of the skin and therefore none of them would have the "right" to pass faster in one direction than in the other, unless it were subject to active transport. A partial short circuit could be achieved by connecting the inside and outside bathing solutions, using, say, Ag-AgCl electrodes.

A rough calculation showed that the Na^+ influx might be of the same order of magnitude as the short-circuit current if a full short circuit could be brought about. However, the resistance of the solutions and of the electrodes was too large for firm conclusions to be drawn. In the end we solved the problem by putting an external electromotive force in series with the skin to overcome external resistances so that the skin would have only to overcome its own internal resistance (56).

The skin has no way of knowing that the outer resistance is being overcome by the applied electromotive force. It only "knows" that the outer resistance has vanished so that both sides have the same electrochemical potential for all ion species.

When both influx and outflux of Na^+ were expressed in electric units so that we could compare them directly with the short-circuit current, it turned out that despite identical electrochemical potentials on both sides, the Na^+ influx was far greater than the outflux, so that nearly all Na^+ movement must have been due to active transport. Furthermore, the net Na^+ flux (i.e., influx minus outflux) was exactly equal to the short-circuit current. Thus the total electric asymmetry of the skin can be explained as the result of active Na^+ transport from the outside to the inside of the skin. There is no other contribution to the short-circuit current. Clearly the battery overcomes the whole resistance of the outer circuit right up to the points on either side of the skin between which the potential difference is zero, but without a potential drop the battery cannot make the current pass through the rather high resistance of the skin. For this crucial distance the current is driven by the electromotive force of the Na^+ pump. Incidentally we were fortunate to have chosen the frog skin as our experimental object. In several epithelia more than one active transport process contributes to the short-circuit current, and leak paths of different kinds may complicate the picture up to the point where no safe conclusions can be drawn. However, for the frog skin the answer was so unambiguous that the existence of specific active transport of Na^+ could no longer be disputed.

As one should expect, the Cl^- exhibited completely passive behavior under the conditions of the short-circuit experiment. Influx and outflux were identical within the accuracy of the experiment. Consequently, Cl^- is not contributing to the short-circuit current.

POSSIBLE ROLE OF SOLVENT DRAG

Until now we have assumed that passively moving ions are influenced only by the concentration gradient and the electric field. However, if we want to be honest we must admit that flow of solvent through the system, whether driven by osmosis or other forces, might influence the behavior of the ions. Under the assumption that water and ions follow the same pathway, such a solvent drag might, at least in theory, be quite important. It seemed clear that solvent drag provided by the osmotically transported water is not the sole reason for the NaCl uptake. Even if the osmosis is reduced to virtually zero by addition of nonpenetrating osmotically active substances to the outside solution, the NaCl uptake continues. However, a role for solvent flow could not be dismissed as easily as that. A few years earlier, Maurice Visscher and his group had advanced his ingenious fluid-circuit hypothesis (15). In essence this hypothesis proposes that solute-transporting epithelia of the gut have the capacity to pump solution with at least some of the solutes from lumen to blood and to pump pure water in the opposite direction, thus bringing about a net transfer of solute with relatively less transfer of solvent. Could it be possible that such a mechanism was responsible for the transport of Na$^+$ through frog skin?

We shall make the (reasonable) assumption that an interaction between solute and solvent flow is only possible to the extent solvent and solute make use of the same transport path. If, for instance, water molecules and solute molecules dissolve individually in the cell membranes and pass by diffusion, the interaction can be disregarded. The worrisome possibility is that solute and solvent pass together through narrow pores or channels. In such places the linear rate of flow might be appreciable and the interaction correspondingly great. If such a transport path was present in combination with the hypothetical return flow of water by way of a pure solvent pump, the net result might mimic specific active transport. For our present purpose we may consider the solvent drag as a kind of field acting on the diffusing ions.

From this point the mathematical procedure becomes exactly analogous to that followed for the flux ratio without solvent drag. In the presence of solvent drag, however, the equation shows that the flux ratio is not independent of the nature of the path. Besides the terms we are already familiar with, all referring to the properties of the inside and outside bathing solutions, there is a new term describing the effect of the solvent drag on the flux ratio. As one might expect, the term is proportional to the volume rate of water flow and inversely proportional to the free diffusion coefficient of the ion in question. The latter relationship occurs because the flow rate is a larger fraction of the diffusion rate for a large ion than for a small one. The third factor of the solvent-drag term is an integral describing the shape of the path the ion and the solvent have to follow (18, 46a).

$$\ln \frac{J_{ms}}{J_{sm}} = \ln \frac{a_m}{a_s} - \frac{zF(V_s - V_m)}{RT} + \frac{J_w}{D} \int_0^{x_0} \frac{1}{A} \, dx$$

[347]

where J_w is the rate of volume flow of water through unit area of membrane, D is the diffusion constant in water of the ion in question, and A is the fraction of area available for diffusion at any distance x through the membrane, and x_o is the total thickness of the membrane. The integral can be evaluated by studying the flux ratio of a substance that is not subject to active transport and diffuses through water-filled pores. Clearly the integral term cannot be evaluated directly. However, note that it is the same for all species that follow the path in question. The path can widen and contract in an unknown way, but the "shape integral" is the same for all species. This is where we have our chance. Let us assume that we want to test the hypothesis that Na^+ passes the skin via a system that takes up solution from the outside medium and carries it across the skin by osmosis. Then an uncharged, hydrophilic substance having about the same diffusion coefficient as Na^+ should follow the same route.

Determination of the flux ratio for this test substance plus knowledge of its diffusion coefficient would suffice to calculate the solvent-drag term for Na^+. We chose thiourea and acetamide. Both substances had about the right size and both could be obtained in two differently labeled forms. Thiourea labeled with ^{35}S was commercially available, whereas we ourselves synthesized ^{14}C-labeled thiourea. Acetamide could be obtained commercially labeled with ^{14}C either in position one or position two. Special chemical procedures were developed so that the different labels could be isolated and measured separately (52).

In order to obtain the largest possible solvent-drag effects, we used isolated toad skin stimulated with antidiuretic hormone, which especially in the spring produces an enormous increase in the water permeability of this experimental tissue. The inside solution was Ringer's, whereas the outside solution was either 10% Ringer's or Ringer's solution. It turned out that the flux ratio for both test substances was strongly dependent on the rate of net water flow. When the latter was close to zero, as it is with no osmotic gradient, the flux ratio does not differ significantly from one. From these experiments we can conclude two things. 1) Although osmotic water flux may create a significant solvent drag on the test substances and thus, by inference, on Na^+, this cannot explain the high flux ratios for Na^+, which may be 100 compared with 2 accounted for by solvent drag. (We now know the solvent drag in Na^+ to be virtually zero.) 2) A fluid-circuit pumping system is not likely to exist in the toad skin. If such a system had been operational, the flux ratios for the test substances could not become one when both bathing solutions were Ringer's.

RELATION BETWEEN ACTIVE SODIUM
TRANSPORT AND METABOLISM

The fact that the short-circuit current is equal to the net active Na^+ transport under nearly all circumstances makes the short-circuited frog skin an ideal preparation for studies of the relationship between active Na^+ transport and metabolism. At the time when we developed the short-circuiting technique, there were already some indications that a stoichiometric relationship might exist between the rate of active ion transport and metabolic rate. Thus Lundegårdh (34) reported that in plant roots there existed a one-to-one relationship between anions taken up and electrons passing through the respiratory chain, and Conway (1) had suggested that a similar relationship might exist between Na^+ transported and respiratory electrons in animal cells. The problem was now attacked independently by Zerahn (58) and by Leaf and Renshaw (32), both groups using the short-circuited frog skin and both measuring the respiratory rate of the preparation during short circuiting and rest. There was complete agreement between the findings: short circuiting (and thus active Na^+ transport) did increase the rate of O_2 consumption in a regular manner, so that the increase in O_2 consumption over and above the resting value was proportional to the amount of current drawn from the skin. However, the proportionality factor was far from the one electron–to–one ion relationship assumed by Conway and Lundegårdh. In fact 18 Na^+ were transported for each molecule of extra O_2 consumed, or in other words, 4.5 Na^+ per electron.

As an alternative to the direct involvement of electron flow in the active transport process, we favored a role of ATP in the process. Early experiments in our laboratory demonstrated that the short-circuit current could be completely inhibited by dinitrophenol.

Other experiments supported the idea, which in itself became even more attractive after the discovery by Skou (40) and others of the Na^+-K^+-ATPase as an essential part of the Na^+ transport system of many cells. Thus Koefoed-Johnsen (16a) found that the Na^+ transport of frog skin was completely inhibited by ouabain, which had been shown to be a specific inhibitor of the ATPase. Additional evidence came from experiments published by Kristensen and Schousboe (25). When the ATP/ADP ratio of frog skin is reduced by metabolic inhibitors like monoiodoacetate, the active sodium transport is reduced pari passu with the ATP/ADP ratio. These and many other inhibitor experiments showed rather convincingly that the active Na^+ transport depends on the consumption of ATP.

Two-Membrane Theory

So far we had been studying the extensive factor of the Na^+ transport mechanism, that is, the correlation between the amount of Na^+ transported and the amount of energy consumed in the process. However, it was also necessary to obtain information about the intensive factor, in other words, the electromotive force of the "Na^+ battery." We had noted a peculiar effect on the Na^+ transport exerted by the K^+ concentration of the inside bathing solution: both short-circuit current and skin potential decreased when the K^+ concentration increased. On the other hand, both short-circuit current and potential showed a positive correlation with the Na^+ concentration of the outside bath, where K^+ had no effect. These qualitative relationships gave us a feeling that the outward- and inward-facing membranes of the transporting cells must possess different ionic selectivities. We decided to attack the problem along two different lines. Koefoed-Johnsen and I were to study the ionic selectivities of the skin, whereas Tom Hoshiko and Lise Engbœk, using microelectrodes, would study the potential profile through the skin. Both studies seemed to substantiate our working hypothesis.

The microelectrode studies (13) indicated that usually, when the skin was pierced by a microelectrode from the outside, the potential increased in two steps that might well be located to the outer and inner membrane of a cell. The selectivity study, however, gave results that could be quantitatively explained in terms of our working hypothesis (19, 54).

Our experimental approach was based on the following consideration: to get a measure of the electromotive force of the transport mechanism, it is imperative to reduce the flow of shunting anions through the skin (cf. Fig. 1). Assuming for the sake of argument that the pump itself has no passive leak for Na^+, our main object would be to reduce the ion flow through R_{shunt}. We used two different approaches. One was to replace Cl^- by the less-permeating SO_4^{2-}, a procedure that usually does lead to high potentials. The other procedure was to add a trace of Cu^{2+} to the outside bath. Already in

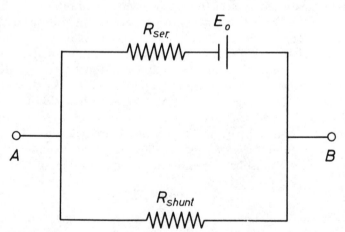

Figure 1. Black-box representation of frog skin epithelium. Sodium battery, with electromotive force (E_o), is placed in series with sodium resistance (R_{ser}) and is shunted by all passively diffusing ions, mainly chloride (R_{shunt}). If inside and outside (terminals A and B) are connected via a lead of zero effective resistance, the epithelium is short-circuited and no current is running through R_{shunt}.

1947 Koefoed-Johnsen and I had made the accidental observation that such a trace of Cu^{2+} had no poisonous effect on the frog skin but often increased the skin potential dramatically. I had used the Cu^{2+} treatment as a routine procedure to obtain high and stable potentials in a study of the flux ratio of iodide in frog skin (46), and later we had shown (56) that Cu^{2+} in the outside bath has no effect on the short-circuit current. Thus Cu^{2+} must act as an inhibitor of Cl^- flow through the shunt.

Whether one uses the sulfate procedure or the Cu^{2+} treatment, one obviously cannot reduce the shunt to zero. For one thing, the flux of cations through the shunt is not reduced by the procedures mentioned: the flow of the major anion usually is the most important contributor to the shunt current. Otherwise the whole setup could not lead to net NaCl uptake through the skin.

In experiments where the shunt current was reduced, the outward-facing side of the skin behaved like an almost ideal Na^+ electrode, whereas the inward-facing side behaved like a K^+ electrode. In the range from 1 to 100 mmol of Na^+ in the outside bath, the potential increases linearly with the logarithm of the concentration, the slope being close to the theoretical value of 59 mV. Similarly, the potential decreases with increasing K^+ concentration in the inside bath, the slope again being close to the theoretical value.

It should be emphasized that the slopes for both the Na^+ and K^+ sensitivities are only ideal for skins with high potentials. This is only natural, since, as a look at Figure 1 will show, a sizable shunt conductance must lead to an attenuation of the potential response.

We have now established that the outside of the skin behaves like an Na^+-selective electrode and the inside like a K^+ electrode. How can we reconcile this with the fact that active Na^+ transport was found to be solely responsible for the electric asymmetry of the skin? Our solution to the riddle is shown in Figure 2. We consider an idealized epithelial cell, and initially we assume that the shunt conductance to anions and other passive ions is effectively blocked. The outward-facing membrane is selectively, but passively, permeable to Na^+ and impermeable to K^+. The inward-facing membrane is permeable to K^+ but impermeable to passive permeation of Na^+. However, the latter membrane is provided with an Na^+ pump, which can pump Na^+ from the cell toward the inside bathing solution. The Na^+ pump is visualized as an Na^+-K^+ exchange pump, but a pure Na^+ pump would serve the purpose equally well. The choice of an Na^+-K^+ exchange pump was made partly because, as already mentioned, a minimal concentration of K^+ in the inside bath is necessary for the Na^+ pump. In part the choice was due to our wish to please Daniel Tosteson (who was working in our laboratory at that time) as well as other workers who favored the idea of such a pump in the red cell membrane.

The pump was placed at the inward-facing membrane in order to

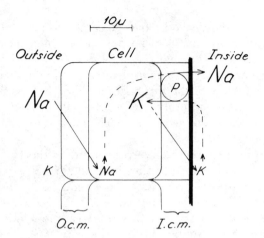

Figure 2. Simple two-membrane model of epithelium. *Oblique arrows,*
passive though highly selective diffusion; *dashed arrows,* movements of Na[+]
and K[+] through the pump.

keep the cellular Na[+] low, but many other reasons support this
localization. A look at the model suffices to show that, under open-
circuit conditions, the pump can do nothing but replace cellular Na[+]
with K[+] (coming from the inside solution). This process will increase
the potential drop at the outer as well as the inner membrane, and
the pumping must stop when the pump is working against some
maximal electrochemical potential gradient, which then would represent the maximal electromotive force of the Na[+] battery.

If, on the other hand, the whole assembly is short-circuited, Na[+]
is pumped toward the inside solution from the cell, whereas the
cellular Na[+] is replenished from the outside by passive electrodiffusion through the Na[+]-selective membrane. If the pump were a one-
to-one Na[+]-K[+] exchange pump, the pumping process itself does not
give rise to net current flow. The current is carried by K[+] diffusing
passively through the inward-facing membrane because the cellular
concentration of K[+] is higher than it is in the inside bathing solution.
Because the outward-facing membrane is tight to K[+], the K[+] pumped
into the cell from the inside solution must escape quantitatively
through the inward-facing membrane, K[+] moves in a closed circuit,
and the only net current flow through the system is exhibited by
Na[+]. If we finally consider the normal frog skin with Cl[-]-Ringer's on
both sides, the Cl[-] shunt will bring about a partial short circuit, so
that the potential drops and both Na[+] and Cl[-] undergo net inward
transport.

The model is built of a minimum number of "building stones," the
existence of which was already more or less well established in other
cell types. Although at the time the Na[+]-K[+]-ATPase was not yet
described, there was good evidence that red cells, nerve cells, and
muscle cells possessed an Na[+] pump; for red cells at least, there was

good evidence for a coupling between an active Na^+ outflux and an active K^+ influx. The Hodgkin-Huxley-Katz theory for action potentials was based on the assumption of specific passive channels for both Na^+ and K^+. The fact that they were only open for short periods at a time obviously did not preclude the possibility that similar entities could permanently exist in epithelial cells.

Because the current in the shunt branch is carried mainly by Cl^-, it is an expression of the fact that the inward-going current in the Na^+ branch is compensated by an inward-going current of negatively charged Cl^-. In other words, in the open-circuit case the net result of current flow through the equivalent circuit is a net inward transport of NaCl.

The properties of the active Na^+ transport of the frog skin are now discussed in more detail. As previously mentioned, the net active transport of Na^+ under nearly all conditions is equal to the short-circuit current. Thus we can use the short-circuit current as a convenient measure of the activity of the Na^+ pump. Let us look first at the role of the composition of the bathing solutions. If, for instance, we leave K^+ out of the inside bathing solution, we observe that the current drops to a very low value. It is immaterial whether or not K^+ is present on the outside of the skin. Thus a small amount of K^+ in the inside bath is apparently absolutely necessary for the active Na^+ transport, although no net transport of K^+ is observed in either direction. On the other hand, if the K^+ concentration of the inside bath is increased above the normal value of ~2 mM, the short-circuit current continues, although at a rate lower than normal. Also the open-circuit skin potential is reduced when the inside K^+ concentration is increased. The depressant effect of K^+ on the short-circuit current is due to an effect on the inside of the skin, but an appreciable current can still be drawn from the skin when K^+ replaces 80 percent of Na^+ on both sides of the skin, and even under such conditions all current is accounted for by net transport of Na^+.

Localization of Ion-Selective Membranes

The two-membrane theory as of 1956 formally explained many of the electrophysiological properties of the frog skin, but the anatomical localization of the membranes was uncertain. First we assumed the transporting cells to be those of the stratum germinativum. The reason was that microelectrodes, coming from the outside, had to be advanced ~25 μm after touching the corneum before the first potential jump was observed. I soon realized that, because of the tough cornified layer, the epithelium bulged appreciably under the pressure of the needle before it was suddenly penetrated to some undetermined level. It was therefore very fortunate that Erich Windhager came to work with us a few years later. He brought with him a technique that allowed precise localization of the cells in which the electrode tip was located. The trick was to transfer electrophoretically a dye from the electrode to the cell after the potential measurement.

This study (55; see Fig. 3) led to the conclusion that all cell layers of the epithelium were electrically connected so that it acted almost as a three-dimensional syncytium. Independently, Farquhar and Palade (5) proposed a similar model on the basis of an electron-microscopic study of the frog skin.

The studies mentioned made it clear that the Na^+ selectivity resided in the apical membrane of the outermost living cell layer of the skin. One consequence of total coupling between all cell layers obviously would be an enormous area of membrane facing inward, and thus the capacitance of the inward-facing membrane ought to be enormous. When Peter Smith came to visit us for a year, it was decided that he should do a low-frequency impedance study of the frog skin. Our model of the epithelium required two membranes in series. If, however, one membrane has a high conductance, its capacitance might become "invisible." Our model assumes a high K^+ conductance of the inner membrane, and our plan was to reduce that conductance. Indeed, when Smith replaced ordinary Ringer's with 50% sulfate-Ringer's made isotonic by addition of sucrose, two membranes in series were clearly observed. The experiments showed that the capacitance of the outward-facing membrane was ~ 2 $\mu F/cm^2$, whereas that of the inward-facing membrane was ~ 50 times larger. The experiment thus supported the idea of coupling between all the epithelial cells.

Convincing evidence in favor of the coupling hypothesis finally came from studies with the electron microprobe (1a, 38). They showed that in ouabain-poisoned skin the Na^+-K^+ ratio could be varied over a wide range of values depending on the Na^+ concentration in the outside bath, but the ratio between the concentrations of the two ions remains the same in all cell layers. This rule also holds true in unpoisoned skins, except when the Na^+ transport is excessively stimulated by, say, antidiuretic hormone. In such skins the Na^+ concentration is highest in the apical cells and drops gradually through the skin. This is what one would expect if the gap junctions between the cells have a definite, although low, electric resistance. It appears then that the reason for the applicability of the model shown in Figure 2 is that the epithelium does behave like a single cell.

ROLE OF MITOCHONDRIA-RICH CELLS AS SHUNT PATHWAY FOR CHLORIDE

In Figure 3 two shunt pathways are indicated, one paracellular and one via the cellular compartment. For a while it was commonly believed that Cl^- largely went through the paracellular pathway. As it turned out this is very nearly correct for skins from cold-adapted frogs. The paracellular pathway has a very low conductance and is

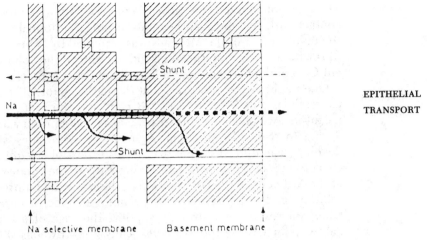

Figure 3. Diagram of "syncytium model" of frog skin epithelium showing connections between cells, Na⁺-transport paths, and shunt pathways.

quite unspecific, so that it will admit even SO_4^{2-}. If, however, the frog has been adapted to room temperature for a few days, the Cl^- conductance increases dramatically, but the excess conductance is specific for Cl^- (20). It is this excess Cl^- conductance that can be inhibited by Cu^{2+}. This behavior speaks in favor of passage through a cell membrane. Even more convincing evidence for a cellular pathway is the kinetics of the Cl^- transport. Most of the pertinent experiments have been performed on toad skin (29, 30), but the essential features of the pathway have also been found in frog skin (22).

The pathway is highly sensitive to the transepithelial potential. It can be characterized as a Cl^- channel, located in the apical cell membrane. At short circuit (i.e., when the apical membrane is polarized) the channel is closed and the Cl^- conductance is zero. As the potential difference is increased toward the normal situation (inside positive), the Cl^- conductance is activated in a manner that formally resembles the gating of cation channels in nerve. Within certain limits, the higher the clamping potential, the higher the Cl^- conductance. Note that the Cl^- fluxes, as measured with isotopes, strictly obey the flux-ratio equation, evidence for simple passive electrodiffusion. Again, the behavior of Cl^- points to a cellular pathway. Yet the cells responsible cannot be the principal cells (i.e., the syncytium), since these have been shown to be practically tight to Cl^-, both apically and basolaterally, as we shall see next.

The first clue to the riddle came when my former associate Cornelis Voûte reported (57) that there is a linear correlation between the number of MR cells per unit skin area and its Cl^- conductance. The MR cells make up only a small fraction of the epithelial volume.

[355]

They are not coupled electrically to the syncytium, but they all make contact with the apical surface of the skin. Additional evidence for the involvement of the MR cells was provided by Scheffey and Katz (39) who, using a vibrating electrode, localized the voltage-dependent Cl^- conductance of toad skin to points near MR cells.

One might argue, however, that the pathway for Cl^- could be via paracellular slits between the MR cells and the syncytium. In the meantime, we had been able to arrange a collaboration with Kenneth Spring to study the problem. As many readers know, Spring has developed an optical system that makes it possible to follow volume changes in individual cells in situ. Our study of the volume responses of the MR cells to the composition of the inside and outside bathing solutions and to applied ingoing and outgoing electric current has so far given results that strongly support the idea of a transcellular pathway for Cl^- via the MR cells. Thus, the MR cells shrink when the outside medium is made Cl^- free, and the volume is recovered when the outside medium is again made Cl^--Ringer's (42). Most of the MR cells swell when the epithelium is clamped to serosa-positive voltages, granted that the apical solution contains Cl^-. If the latter is Cl^- free, the cells stay shrunk (7). It is interesting to note that MR cells of short-circuited epithelia, with ordinary Ringer's on both sides, have a relatively large volume. However, the cells shrink by some 15%, not only if Cl^- is absent in the apical bath, but also if the latter is made Na^+ free, or if Na^+ entry is inhibited by amiloride (31). This indicates that the MR cells have apical Na^+ channels like the syncytium cells. However, it also substantiates model calculations by Kristensen (21), which showed that if a cell serves as a transport path for Cl^-, its permeability to that ion must vary in parallel with the rate of Na^+ entry. In the 1986 version of the frog skin epithelial model shown in Figure 4, our present knowledge of the MR cells is given in schematic form.

Handling of Chloride by Syncytium Cells

Some years ago (35) we observed that the frog skin epithelium loses KCl to the inside bath when the latter is diluted to half strength and that the KCl loss was recovered in normal Ringer's. Such a loss of KCl is also seen when the skin is exposed to solutions lacking K^+ (47) or Cl^- (6, 47). From the loss of volume in such experiments, we calculated the Cl^- concentration in the epithelial cells to be 40–50 mM (35), and very similar figures have been found by Dorge et al. (1a), using the electron microprobe. If we calculate the ionic product of cellular K^+ and Cl^-, it is clear that it is much higher than that for Ringer's solution. Thus either K^+ or Cl^- or both must be accumulated actively in the cells. Already the data obtained by Ussing and Windhager (55) suggested that Cl^- must be present in concentrations over and above its equilibrium concentration, and these measurements with improved technique [see, e.g., Nagel (36)] indicate that Cl^- of

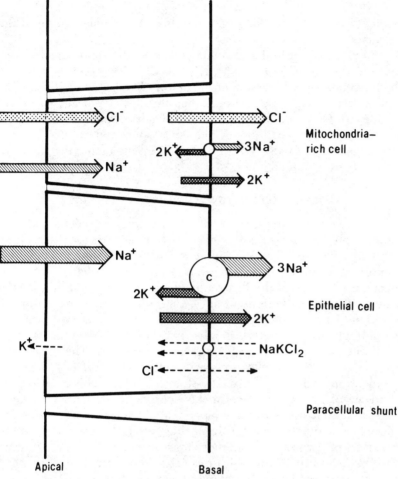

Mitochondria–
rich cell

Epithelial cell

Paracellular shunt

Apical   Basal

Figure 4. Representation of transport pathways in frog skin epithelium. Two cell types are involved. Mitochondria-rich cells (flask cells) are relatively scarce but represent a conductive pathway for Cl⁻. Ordinary epithelial cells form a three-dimensional syncytium that acts like a single cell. It is normally tight to Cl⁻ but is the main pathway for the active Na⁺ transport. The apical membrane has a selective Na⁺ channel, whereas the basolateral membrane possesses the Na⁺-K⁺ pump and a passive K⁺ leak. Moreover it is the seat of an Na⁺-K⁺-2Cl⁻ cotransport and a variable Cl⁻ leak. Normally both are virtually silent but serve the volume regulation of the cell. *Dashed arrow* (at apical membrane) indicates small K⁺ leak.

the syncytium cells must be actively accumulated. All data also showed that the permeability of the apical membrane to Cl⁻ was practically nil under all circumstances. Our recent studies have shown that even the basolateral membrane is normally rather tight to Cl⁻ but that Cl⁻ channels in that membrane can be activated either by osmotic swelling of the cells or by depolarization of the basolateral membrane (49–51).

The recovery of KCl and volume requires the simultaneous presence in the inside bath of Na^+, K^+, and Cl^- (50), and it is inhibited by bumetanide and furosemide. Thus very likely it is an Na^+-K^+-Cl^- cotransport of the type first described for Ehrlich cells by Geck and Heinz (9). Conceivably it is activated when cellular Cl^- is too low, but the question is still open. The cotransport and the Cl^- channel are indicated in Figure 4 with dashed lines, indicating that they are normally dormant.

MAKING MULTIPATHWAY MODEL OPERATIONAL

A model with more than one pathway for each ion species certainly poses problems. Thus the flux-ratio analysis and the short-circuiting method originally were developed under the assumption of only one important pathway. In recent years, however, we have run into a peculiar property of the flux ratio, which now comes in handy. Let us consider an epithelium (or any tissue) separating two well-mixed solutions. The system is assumed to be in a steady state with respect to all transport parameters. Furthermore, let us assume that there is only one type of pathway for a given ion species. Then it can be shown (43, 48) that the pre-steady-state flux ratio is time independent and equal to the steady-state flux ratio. This is true from the time of first appearance of the isotopes on the "arrival sides," and the statement is correct independent of the transport mechanism (diffusion, active transport, solvent drag, etc.); this means that if the flux ratio changes with time, there must be more than one pathway. Two pathways that differ with respect to flux ratio as well as mean passage time can be easily separated mathematically.

So far we have used the method for separating the cellular Na^+ fluxes from those via the paracellular shunt (3, 53), and quite recently we have separated the small cellular K^+ leak from the paracellular flux. Interestingly it turned out that the K^+ channels of the epithelium cells exhibit single filing, much like those of nerve fibers (12).

In addition, Larsen and Rasmussen (30) have developed a mathematical model of amphibian skin with two types of transporting cells. Such a model may serve to resolve the complex reaction of the whole tissue into the individual events in the two types.

These examples indicate that the latest version of the model is manageable and that the frog skin may continue to be useful as a test object.

BIBLIOGRAPHY

1. CONWAY, E. J. Some aspects of ion transport through membranes. *Symp. Soc. Exp. Biol.* 8: 297–324, 1954.

1a.Dörge, A., R. Rick, U. Katz, and K. Thurau. Determination of intracellular electrolyte concentration in amphibian epithelia with the electron microprobe analysis. In: *Water Transport Across Epithelia*, edited by H. H. Ussing, N. Bindslev, N. A. Lassen, and O. Sten-Knudsen. Copenhagen: Munksgaard, 1981. (Alfred Benzon Symp. 15.)

2. Du Bois-Reymond, E. *Untersuchungen über Tierische Elektrizitat*, Berlin: 1848.

3. Eskesen, K., J. J. Lim, and H. H. Ussing. Evaluation of transport pathways for Na^+ across frog skin epithelium by means of pre-steady state flux ratio. *J. Membr. Biol.* 86: 99–104, 1985.

4. Eskesen, K., and H. H. Ussing. Single-file diffusion through K^+-channels in frog skin epithelium. *J. Membr. Biol.* 91: 245–250, 1986.

5. Farquhar, M. G., and G. E. Palade. Functional organization of amphibian skin. *Proc. Natl. Acad. Sci. USA* 51: 569–577, 1964.

6. Ferreira, K. T., and H. G. Ferreira. The regulation of volume and ion composition in frog skin. *Biochim. Biophys. Acta* 646: 193–202, 1981.

7. Foskett, J. K., and H. H. Ussing. Localization of chloride conductance to mitochondria-rich cells in frog skin epithelium. *J. Membr. Biol.* 91: 251–258, 1986.

8. Galeotti, G. Concerning the E. M. F. which is generated at the surface of animal membranes on contact with different electrolytes. *Z. Phys. Chem.* 49: 542–562, 1904.

9. Geck, P., and E. Heinz. Coupling of ion flows in cell suspension systems. *Ann. NY Acad. Sci.* 341: 57–63, 1980.

10. Goldman, D. E. Potential, impedance and rectification in membranes. *J. Gen. Physiol.* 27: 37–60, 1944.

11. Harris, E. V., and G. P. Burn. The transfer of sodium and potassium ions between muscle and surrounding medium. *Trans. Faraday Soc.* 45: 508–528, 1949.

12. Hodgkin, A. L., and R. D. Keynes. The potassium permeability of a giant nerve fibre. *J. Physiol. Lond.* 128: 61–88, 1955.

13. Hoshiko, T., and L. Engbœk. Microelectrode study of the frog skin potential. In: *Abstr. Commun. 20th Int. Physiol. Congr. Brussels, 1956,* p. 443.

14. Huf, E. Versuche über den Zusammenhang zwischen Stoffwechsel, Potentialbildung und Funktion der Froschhaut. *Pfluegers Arch. Gesamte Physiol. Menschen Tiere* 235: 655–673, 1935.

15. Ingraham, R. C., H. C. Peters, and M. Visscher. On the movement of materials across living membranes against concentration gradients. *J. Phys. Chem.* 42: 141–150, 1938.

16. Jørgensen, B. C., H. Levi, and H. H. Ussing. On the influence of neurohypophyseal principles on the sodium metabolism in the axolotl (*Ambystoma mexicanum*). *Acta Physiol. Scand.* 12: 350–371, 1946.

16a.Koefoed-Johnsen, V. The effect of G-strophanthin (ouabain) on the active transport of sodium through the isolated frog skin. *Acta Physiol. Scand. Suppl.* 145: 87–88, 1957.

17. Koefoed-Johnsen, V., H. Levi, and H. H. Ussing. The mode of passage of chloride ions through the isolated frog skin. *Acta Physiol. Scand.* 28: 150–163, 1952.

18. Koefoed-Johnsen, V., and H. H. Ussing. The contributions of diffusion

and flow to the passage of D_2O through living membranes. *Acta Physiol. Scand.* 28: 60–76, 1953.

19. KOEFOED-JOHNSEN, V., and H. H. USSING. The nature of the frog skin potential. *Acta Physiol. Scand.* 42: 298–308, 1958.

20. KOEFOED-JOHNSEN, V., and H. H. USSING. Transport pathways in frog skin and their modification by copper ions. In: *Secretory Mechanisms of Exocrine Glands*, edited by N. A. Thorn and O. H. Petersen. Copenhagen: Munksgaard, 1974, p. 411–422.

21. KRISTENSEN, P. Effect of amiloride on chloride transport across amphibian epithelia. *J. Membr. Biol.* 40S: 167–185, 1978.

22. KRISTENSEN, P. Is chloride transfer in frog skin localized to a special cell type? *Acta Physiol. Scand.* 113: 123–124, 1981.

23. KRISTENSEN, P. Chloride transport in frog skin. In: *Chloride Transport in Biological Membranes*, edited by J. A. Zadunaisky. New York: Academic, 1982, p. 319–332.

24. KRISTENSEN, P., and E. H. LARSEN. Relation between chloride exchange diffusion and a conductive chloride pathway across the isolated skin of the toad (*Bufo bufo*). *Acta Physiol. Scand.* 102: 22–34, 1978.

25. KRISTENSEN, P., and A. SCHOUSBOE. The influence of anaerobic condition on sodium transport and adenine nucleotide levels in the isolated skin of the frog *Rana temporaria*. *Biochim. Biophys. Acta* 173: 206–212, 1969.

26. KROGH, A. Osmotic regulation in the frog (*R. esculenta*) by active absorption of chloride ions. *Scand. Arch. Physiol.* 76: 60–74, 1937.

27. KROGH, A. The active absorption of ions in some fresh water animals. *Z. Vgl. Physiol.* 25: 335–350, 1938.

28. KROGH, A. The active and passive exchange of inorganic ions through the surface of living cells and through living membranes generally. *Proc. R. Soc. Lond. B Biol. Sci.* 131–200, 1946.

29. LARSEN, E. H., and P. KRISTENSEN. Properties of a conductive cellular chloride pathway in the skin of the toad (*Bufo bufo*). *Acta Physiol. Scand.* 102: 1–21, 1978.

30. LARSEN, E. H., and B. E. RASMUSSEN. A mathematical model of amphibian skin epithelium with two types of transporting cellular units. *Pfluegers Arch.* 405, Suppl. 1: S50–S58, 1985.

31. LARSEN, E. H., H. H. USSING, and K. R. SPRING. Volume response of mitochondria-rich cells of toad skin to amiloride and Na-free outside medium. *Federation Proc.* 45: 746, 1986.

32. LEAF, A., and A. RENSHAW. Ion transport and respiration of isolated frog skin. *Biochem. J.* 65: 82–90, 1957.

33. LEVI, H., and H. H. USSING. The exchange of sodium and chloride across the fibre membrane of the isolated frog sartorius. *Acta Physiol. Scand.* 16: 232–249, 1948.

34. LUNDEGÅRDH, H. Anion respiration. *Symp. Soc. Exp. Biol.* 8: 262–296, 1954.

35. MacROBBIE, E. A. C., and H. H. USSING. Osmotic behavior of the epithelial cells of frog skin. *Acta Physiol. Scand.* 53: 348–365, 1961.

36. NAGEL, W. The intracellular electrical potential profile of frog skin epithelium. *Pfluegers Arch.* 365: 135–143, 1976.

37. NIELSEN, R. A 3 to 2 coupling of the Na-K pump in frog skin disclosed by the effect of Ba. *Acta Physiol. Scand.* 107: 189–191, 1979.

38. Rick, R., A. Dorge, E. von Arnim, and K. Thurau. Electron microprobe analysis of frog skin epithelium: evidence for a syncytial sodium transport compartment. *J. Membr. Biol.* 39: 313–331, 1978.

39. Scheffey, C., and U. Katz. Chloride conductance pathway across toad skin is located to the mitochondria rich cells of the epithelium. *Biol. Bull. Woods Hole* 167: S528, 1984.

40. Skou, J. C. The influence of some cations on adenosine-triphosphatase from peripheral nerves. *Biochim. Biophys. Acta* 23: 394–401, 1957.

41. Smith, P. G. The low-frequency electrical impedance of the isolated frog skin. *Acta Physiol. Scand.* 81: 355–366, 1971.

42. Spring, K. R., and H. H. Ussing. The volume of mitochondria-rich cells of frog skin epithelium. *J. Membr. Biol.* In press.

43. Sten-Knudsen, O., and H. H. Ussing. The flux ratio equation under non-stationary conditions. *J. Membr. Biol.* 63: 233–242, 1981.

44. Ussing, H. H. Interpretation of the exchange of radiosodium in isolated muscle. *Nature Lond.* 160: 262, 1947.

45. Ussing, H. H. The active ion transport through the isolated frog skin in the light of tracer studies. *Acta Physiol. Scand.* 17: 1–37, 1949.

46. Ussing, H. H. The distinction by means of tracers between active transport and diffusion. *Acta Physiol. Scand.* 19: 43–56, 1949.

46a. Ussing, H. H. Some aspects of the application of tracers in permeability studies. *Adv. Enzymol.* 13: 21–65, 1952.

47. Ussing, H. H. Relationship between osmotic reactions and active sodium transport in the frog skin epithelium. *Acta Physiol. Scand.* 63: 141–155, 1965.

48. Ussing, H. H. Interpretation of tracer fluxes. In: *Membrane Transport in Biology*, edited by G. Giebisch, D. C. Tosteson, and H. H. Ussing. Berlin: Springer-Verlag, 1978, vol. 1, p. 115–140.

49. Ussing, H. H. Volume regulation of frog skin epithelium. *Acta Physiol. Scand.* 114: 363–369, 1982.

50. Ussing, H. H. Volume regulation and basolateral co-transport of sodium potassium and chloride in frog skin epithelium. *Pfluegers Arch.* 405, Suppl. 1: S2–S7, 1985.

51. Ussing, H. H. Epithelial cell volume regulation illustrated by experiments in frog skin. *Renal Physiol.* 9: 38–46, 1986.

52. Ussing, H. H., and B. Andersen. The relation between solvent drag and active transport of ions. In: *Proc. Int. Congr. Biochem., 3rd, Brussels.* New York: Academic, 1955.

53. Ussing, H. H., K. Eskesen, and J. Lim. The flux-ratio transient as a tool for separating pathways in epithelia. In: *Epithelial Ion and Water Transport*, edited by A. D. C. Macknight and J. B. Leaser. New York: Raven, 1981, p. 257–264.

54. Ussing, H. H., and V. Koefoed-Johnsen. Nature of the frog skin potential. In: *Abstr. Commun. 20th Int. Physiol. Congr. Brussels, 1956,* vol. 2, p. 511.

55. Ussing, H. H., and E. E. Windhager. Nature of shunt path and active sodium transport path through frog skin epithelium. *Acta Physiol. Scand.* 61: 484–504, 1964.

56. Ussing, H. H., and K. Zerahn. Active transport of sodium as the source of electric current in the short-circuited isolated frog skin. *Acta Physiol. Scand.* 23: 110–127, 1951.

57. Voûte, C. L., and W. Meier. The mitochondria-rich cell of frog skin as hormone-sensitive "shunt path." *J. Membr. Biol.* S40: 151–165, 1978.
58. Zerahn, K. Oxygen consumption and active sodium transport in the isolated and short-circuited frog skin. *Acta Physiol. Scand.* 36: 300–318, 1956.

MEMBRANE
TRANSPORT

XV

Flow and Diffusion Through Biological Membranes

JOHN R. PAPPENHEIMER

THE centennial of the American Physiological Society coincides with the 100th anniversary of the publication of Jacobus van't Hoff's theorem relating the osmotic pressure of solutions to the gas laws (69). It is fitting to call attention to this coincidence of anniversaries in a volume devoted to the people and ideas of membrane-transport research because van't Hoff's brilliant generalization provided the theoretical basis for innumerable subsequent papers on the osmotic behavior and permeability of biological membranes. This chapter is no exception: it is concerned largely with the application of van't Hoff's theorem to fluid movement and effective osmotic pressures across organized biological membranes such as capillary endothelium, glomerular membranes, or leaky epithelia. It is also a personal account of my contributions to this field from 1946 to 1954, including the people who influenced my thinking and the twists of fortune that shaped the course of my research. Ideas generated during this period have since been modified, expanded, and supplemented by new methods and new concepts, including the development of irreversible thermodynamics as applied to membrane transport (23). The recent American Physiological Society *Handbook of Physiology* volume *Microcirculation* (54b) is gratifying testimony to the advances that have been made during the last thirty years. Nevertheless many features of the 1946–1954 era remain unchanged, and my purpose, indeed my charge, in this chapter is to provide a personal account of how these earlier ideas came about. I mention only a few of the subsequent advances that pertain most directly to the original concepts.

In 1946 I started to investigate edema formation in isolated perfused mammalian tissue. The reasons for doing so were not especially commendable; they were based more on what I was equipped to do at the time than on natural curiosity or on any clearly defined goal. The National Institutes of Health (NIH) had not yet started large-scale support for university-based medical research, but even if it had I don't think I could have written an acceptable request for a grant. Certainly I had no inkling that the project would lead rather swiftly to a logical sequence of advances in the physiology of the microcirculation and the biophysics of membrane transport. In retrospect it seems easy to trace the flow of ideas and cite the people who directly or indirectly contributed to the advances. Nevertheless I am quite aware that personal recollections of this kind are often biased and that the sharp edges of memory may become smoothly rounded in the course of fifty years.

Setting the Stage: 1935–1945

Fifty years ago the permeability of peripheral capillaries was defined in terms of net transcapillary filtration or absorption. The elegant micropuncture methods developed by E. M. Landis (29) in 1927 for single capillaries had essentially proved Starling's "hypothesis" of 1896 (63), namely, that the rate of net fluid movement is proportional to the transmembrane differences between hydrostatic and osmotic forces. The "constant" of proportionality, expressed as volume flow rate per unit area of membrane per unit pressure difference, has the dimensions of a flux per unit driving force, and it was accepted as a definitive measure of permeability both in peripheral capillaries (30) and in renal glomerular membranes (26). This measure of capillary permeability was considered analogous to the membrane permeability of single cells as estimated from rates of swelling or shrinkage under the influence of an osmotic gradient (34). The concept was extended to human extremities during the 1930s, especially by Krogh et al. (27) and by Landis and Gibbon (31), who devised the pressure plethysmograph to distinguish transcapillary filtration from venous filling during changes of venous pressure in the human forearm. In this case the filtration constant, or measure of capillary permeability, was expressed as milliliters per minute of swelling in the arm per unit increase of venous pressure per 100 ml tissue. It was assumed that increases of venous pressure produced by a pneumatic cuff would produce proportionate changes of mean intracapillary hydrostatic force for filtration, but the absolute value of the capillary pressure, and hence the driving force, remained indeterminate.

The relation of the filtration constant, determined by plethysmography, to the permeability coefficients of solutes was poorly understood. The use of isotopes as tracers in studies of permeability was in its infancy. Pioneer investigations by Krogh, Hevesey, and colleagues (19–21) in 1935 and in 1940 and by Visscher et al. (70) in1944 had shown large discrepancies between absorption or filtration of fluid on the one hand and diffusive fluxes of D_2O on the other. The significance of these discrepancies for the fundamental understanding of membrane permeability in general, and of capillary permeability in particular, was not recognized. Ionic pumps were unknown, and it was thought that cell membranes were essentially impermeable to sodium. With regard to capillary permeability to solutes, the most that could be said with assurance was that peripheral capillaries were "almost" impermeable to the large plasma proteins and were "freely" permeable to all smaller solutes. In the words of August Krogh (26):

When I review all the facts that have come to my notice I have no hesitation in saying that there is no trustworthy evidence of the capillaries having any power of hindering or favoring the passage by diffusion of all kinds of crystalloids through the endothelium.

These were the main concepts and technical approaches to capillary permeability when I moved to Harvard from the University of Pennsylvania in December of 1945 to become right-hand man to E. M. Landis, who had himself recently moved to Harvard as chairman of the Department of Physiology at the Medical School. I had had several years experience with isolated, blood-perfused organs (12, 43) including studies of renal glomerular capillary pressure (13), and it seemed to me that one might study transcapillary exchange in perfused organs in which independent variables such as arterial and venous pressures, protein osmotic pressure, and temperature could be adjusted over a wide range of values while dependent variables such as blood flow, transcapillary fluid movement, and diffusion of tracers could be measured continuously with a precision not possible in the intact organism. Landis was the world-wide leading authority in this field, and I felt very fortunate to be able to start this work in his department (and only two doors down the hall). Although my ideas for investigating edema formation were lamentably vague, I nevertheless had a scientific background that was especially appropriate for development of the field. As an undergraduate at Harvard from 1932 to 1936, I had concentrated in biophysical chemistry and biology; my tutor was Jeffries Wyman, Jr., who introduced me to quantitative biology in general and physical chemistry of proteins and amino acids in particular. In the parlance of the 1980s, Wyman would be considered a molecular biologist, and indeed his contributions to our understanding of the mechanism of the Bohr effect and the general theory of allosteric reactions of proteins are cornerstones in the modern theory of protein function (39, 75). Our work on the surface tension of solutions of dipolar ions was published in the *Journal of the American Chemical Society* in 1936 (47). (Not long ago I was flattered on two counts when a well-known membrane biophysicist asked me if my father had written this enduring paper.)

In the summer of 1935 I took the physiology course at Woods

Hole Biological Laboratories in Massachusetts, where my instructors included such distinguished biophysical chemists as Leonor Michaelis and Rudolf Höber. The latter taught me how to perfuse frog kidneys via the portal system and how to cannulate frog ureters. Later the 1945 edition of Höber's *Physical Chemistry of Cells and Tissues* became one of the most well-worn books in my scientific library. At the end of the course I was awarded the Collecting Net Prize of fifty dollars (a lot in those days) to come back the next summer to work on the kinetics of CO_2 transport and carbamino formation in fish blood with J. K. W. Ferguson. We could not find any carbamino transport in fish blood, but we did find an anomalous distribution of bicarbonate and chloride in elasmobranch red cells and duly published our findings in the *Biological Bulletin* (15).

In the autumn of 1936, I sailed for England (there were no transatlantic airlines then) with a letter of introduction to Sir Joseph Barcroft from Jeffries Wyman. I hoped to study fetal physiology, a field that was developing rapidly in Barcroft's laboratory at Cambridge (1). Barcroft gave me a small research project on fetal red cells and arranged for me to take the Part II Honors course in physiology, an experience for which I am endlessly grateful. Our lecturers included E. D. Adrian, B. H. C. Matthews, F. R. Winton, E. B. Verney, W. A. H. Rushton, and F. J. W. Roughton; I developed life-long associations with all of them. Alan Hodgkin was a teaching assistant in the practical class; his job was to come in early in the morning to decerebrate the cats for our class experiments. During the spring term I was introduced to Winton's experiments on isolated perfused mammalian kidneys, and this opened the door to a new world for me.

Winton was then Reader in Physiology at Cambridge and the author, with L. E. Bayliss, of a widely used elementary textbook of physiology. He trained at University College in London under Starling, Verney, Evans, and Hill; in his hands the original Starling-Verney heart-lung-kidney preparation (66) developed into an extraordinarily sophisticated technique for perfusing isolated organs from a pump-lung circulation. (In the 1980s it would be called a life-support system.) I was spellbound by the sight of an isolated dog kidney sitting on a glass plate producing clear golden urine from thick red blood. The perfusion apparatus was itself a complex inorganic organism—a maze of motors, plumbing, and electrical devices for measuring, recording, or controlling pressures, flows, temperatures, ion concentrations, and blood oxygen saturation. Winton was an expert with instruments, and he designed most of his own transducers, amplifiers, recording oscillographs (there were no commercial oscillographs in the 1930s) and so forth. Two foreign postdoctoral fellows were working with Winton at this time: James A. Shannon (59) (later director of the NIH) was there to compare inulin with creatinine clearance (60), and Kurt Kramer from Göttingen was

there to develop, with Winton and G. A. Millikan, a spectrophotometric device for continuous recording of arteriovenous oxygen
differences in flowing blood, using a prototype selenium barrier cell
made in Germany (25). All of this appealed enormously to my
scientific senses; it seemed to be the ideal combination of physics,
chemistry, and physiology—to be the ideal compromise between in
vitro research on the one hand and the unsatisfying complexities of
whole-animal research on the other. I was also swayed by the fact
that Winton was a fine cellist and his wife, Bessie Rawlins, was one
of the leading concert violinists in England. They invited me to play
second cello in the great Schubert C-major quintet, and that settled
matters for me.

Winton, together with Grace Eggleton, was about to start an
investigation of renal oxygen consumption as a function of osmotic
work of urine formation, and he welcomed me as a graduate student,
possibly because of my previous training in physical chemistry and
blood gas transport. From Winton I learned the techniques and art
of perfusing mammalian organs, including the design of transducers,
amplifiers, and instrumentation in general. At the same time I learned
about physical factors affecting the formation of urine, the topic of
Winton's *Physiological Review* article in 1937 (74). Few modern
students of renal physiology are familiar with Winton's ingenious
method for estimating glomerular capillary pressure, first published
in 1931 (72) and refined in 1940 (13). Essentially Winton was able
to calculate mean glomerular capillary pressure at any given arterial
pressure from the change in ureter pressure required to maintain
constant filtration rate (creatinine clearance) following a step change
of venous pressure. The value so obtained was 60–70 mmHg at
normal arterial pressures, values which are in accord with modern
studies of glomerular dynamics in the canine kidney (40, 42). It is
regrettable, though perhaps understandable, that Winton's striking
work has been largely forgotten: the reasoning was complex, the
specialized and elaborate techniques did not invite repetition in
other laboratories, and some of the properties of the isolated perfused kidney were too abnormal to attract the lasting interest and
confidence of other prominent renal physiologists of the day. Nevertheless the reasoning was as sound as it was clever, and in simplified
form it provided the basis for my contributions to capillary physiology a decade later.

In 1938 Winton moved to the Chair of Pharmacology at University
College, and he took me with him as research student and teaching
assistant. Hermann Rein, one of the most prominent physiologists in
Germany, had just published a paper alleging that metabolism of
resting mammalian muscle is regulated by the sympathetic nervous
system, and Winton suggested that I look into this in perfused
hindlimb preparations, utilizing the methods for recording blood
oxygen he had developed with Kramer and Millikan and that we had

used for the studies of renal metabolism (12). This project turned out to be interesting and productive (43), but its only relevance to this chapter is that it gave me experience with the perfused hindlimb preparation, which I eventually used for studying capillary physiology.

MEMBRANE
TRANSPORT
War came on 2 September 1939. I was on vacation at a music house party in Harvard, Massachusetts; passports were canceled along with my job at University College. Winton became Dean of the Medical School, which was removed to Surrey before its buildings were destroyed during the blitz of 1940. I sought help from A. N. Richards at the University of Pennsylvania. He had just given a Royal Society Croonian Lecture entitled "Processes of Urine Formation" (55), and of course he knew Winton; he was also an old friend of my father's. Richards set me up with a temporary appointment in Bazett's Department of Physiology and with an "emergency" research grant of $250. At the same time, my friend Glenn Millikan, who was also cast adrift from his job in England, came to Philadelphia to join Detlev Bronk's group at the Johnson Foundation for Medical Physics. Both Richards and Bronk anticipated the future and beginning in 1939 were devoted almost full time to war-related activities. Richards became chairman of the National Research Council (NRC) Committee on Medical Research, and was charged (among other responsibilities) with mass production of penicillin. Bronk was concerned with aviation medicine, including the training of thousands of air force personnel for operations in nonpressurized aircraft at altitudes above 20,000 feet. Bronk worked directly out of the Air Surgeon's office in Washington, but he retained a small research group at the Johnson Foundation to carry on applied research on oxygen equipment and on visual problems. I soon found myself working in the Johnson Foundation, helping Millikan develop the ear oximeter (37), drawing on my experience with spectrophotometric determination of oxyhemoglobin in perfusion systems. For the next six years, Glenn and I worked closely on oxygen-demand valves, chemical oxygen generators (39), carbon monoxide poisoning in tanks and in military aircraft, positive-pressure breathing, and other problems in applied physiology for the military. Some of our work was carried out in actual flight tests and in military installations. Although this war-related work bore no direct relation to membrane transport, it added greatly to my own expertise in research. Members of Bronk's group included Keffer Hartline, Frank Brink, Martin Larrabee, Millikan, John Lilly and John Hervey; they were all experts in instrumentation, especially electronics. The working hours at the Johnson Foundation were from noon to midnight, and we dined together almost every evening. One would have to be very impermeable indeed not to learn by diffusion and osmosis from this close association with such alert and knowledgeable minds. By 1945 the war's end was in sight and all of us were eager to return to academic

life. The project I was then working on, namely, the storage of oxygen in the form of perchlorates, was already in pilot production at a chemical factory in Pittsburgh (57), but it seemed unlikely to me that it would ever be applied to its intended use in the field as a source of emergency oxygen in aircraft or for portable welding apparatus.

Soon after VJ day, I applied for a position with Landis. I arrived in his department in January 1946, but it was not until June that the perfusion apparatus was ready for action. From then on matters moved with unexpected rapidity, and the vague plan for studying edema formation was replaced by the specific objective of characterizing the permeability characteristics of mammalian capillaries. After only a few experiments, I saw how to set the mean capillary pressure in perfused muscle to known values, how to measure the effective osmotic pressure exerted by the plasma proteins across the capillary walls, and how to relate these to net filtration or absorption. On 26 July I left for a skiing vacation on Mt. Lassen, Oregon, with Lilly. Landis was away, but I left him a four-page memo dated 23 July 1946 entitled "A General Equation Relating Arterial, Venous, and Osmotic Pressures to the Rate of Passage of Fluid in and out of Capillaries," which included the first measurements of the filtration coefficient in mammalian capillaries. Perhaps it was on the strength of this that he arranged for support from the Life Insurance Medical Research Fund; he also suggested that I take on a postdoctoral fellow, Armando Soto-Rivera, from Venezuela.

Isogravimetric Capillary Pressure and Pre- and Postcapillary Resistances to Blood Flow 1946–1948

The pump-lung perfusion system I constructed at Harvard is illustrated diagrammatically in Figure 1. I soon found that at any given arterial perfusion pressure (P_a) there was a unique value of venous outflow pressure (P_v) at which the perfused limb remained at constant weight. I called this the *isogravimetric* state, an impure amalgam of Greek and Latin that reflects my poor training in the classics. Nevertheless the term became part of the jargon in the field. In the isogravimetric state, the mean capillary pressure ($\overline{P}_c$) exactly balanced the mean transcapillary osmotic forces so that no net filtration or absorption occurred. Any increase of either arterial or venous pressures above isogravimetric values caused net filtration of fluid from blood to tissues, as reflected by a steady gain of weight; conversely, any decrement of arterial or venous pressures below their isogravimetric values resulted in net absorption. The absolute value of the isogravimetric capillary pressure could be determined by reducing the arterial pressure in steps and at each step raising venous pressure to maintain capillary pressure at its isogravimetric value. Clearly, if this procedure were continued until arterial and

[369]

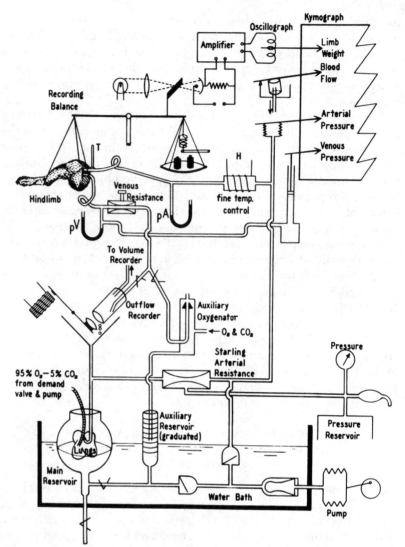

Figure 1. Perfusion circuit and weight-recording systems for measuring capillary hydrostatic pressure, effective transcapillary osmotic pressure, resistances to blood flow, and capillary filtration coefficient. Plastics were not yet available and perfusionware was made from glass and latex tubing. Flowmeters, manometers, water baths, amplifiers, recording equipment, and so forth were constructed in the lab. The annual budget for equipment and supplies, including animals, was less than $1,000. [From Pappenheimer and Soto-Rivera (49).]

venous pressures became equal, the isogravimetric capillary pressures would be defined. In practice this technique could not be followed to the point at which arterial and venous pressures were equalized because at this point there would be no blood flow and the normal capillary permeability would be compromised by anoxia.

This problem was solved (as shown in Figure 3) with reference to the schema of Figure 2. Isogravimetric blood flow (Fig. 2) is a linear function of isogravimetric venous pressure (Fig. 3); the intercept on the y-axis is the isogravimetric capillary pressure, and the slope is postcapillary resistance to blood flow as defined in Figure 2. Once the postcapillary resistance is known from the slope, then the mean capillary pressure ($\overline{P}_c$) can be calculated during any nonisogravimetric state of net filtration of absorption from the relation $\overline{P}_c = \dot{Q}R_v + P_v$.

The schema shown in Figure 2 has been widely used in subsequent investigations of the microcirculation and in textbooks of circulatory physiology. I think that Bayliss and Starling in 1894 were the first to recognize that capillary pressure might be regulated by resistances to blood flow on the two sides of the circulation (3), although they did not provide an explicit equation defining capillary pressure, as in Figure 2. Richards and Plant in 1922 (56) and Winton in 1931 (73) recognized that glomerular capillary pressure must depend upon relative afferent and efferent resistances, a concept that was

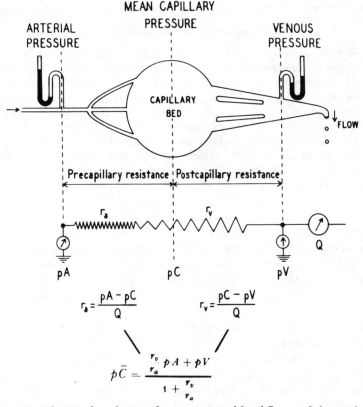

Figure 2. Electrical analogue of resistances to blood flow and their relation to mean capillary pressure (P_c), blood flow ($\dot{Q}$), and perfusion pressures (P_a, P_v). The values of R_v or R_a are determined as in Figure 3.

[371]

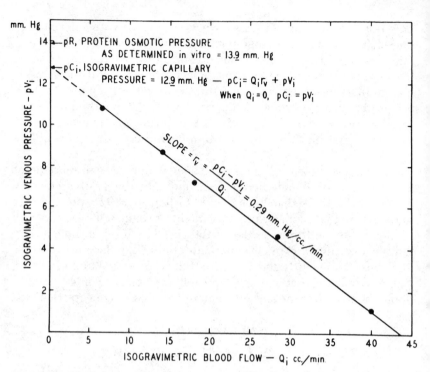

Figure 3.   Evaluation of isogravimetric capillary pressure (pC$_i$) and postcapillary resistance to blood flow (R$_v$) in isolated hindlimb of cat. The venous
pressure required to maintain the isogravimetric state is plotted as a function
of blood flow (Q̇), as this is altered by changing arterial perfusion pressure.
The slope is r$_v$, and extrapolation to zero flow yields pC$_i$. Once r$_v$ and pC$_i$
are determined, the mean capillary pressure during filtration or absorption
is given by p$\overline{C}$ = Q̇r$_v$ + pV, and the transcapillary force for filtration is given
by p$\overline{C}$ − pC$_i$. Sensitivity of the system was such that a 0.3-mmHg change in
pV produced an easily detectable continuous change in limb weight. [From
Pappenheimer and Soto-Rivera (49).]

taken up by Homer Smith in 1943 (61). My use of the isogravimetric
state to evaluate peripheral capillary pressure from changes in arterial and venous pressures is closely analogous to Winton's use of
constant urine flow (72) or glomerular filtration rate (14) to evaluate
glomerular capillary pressure from changes in ureter and venous
pressures.

There is one more analogy to the isogravimetric technique that is
of interest in terms of ideas, although it has no obvious physiological
significance. The solution for capillary pressure as a function of
arterial pressure, venous pressure and the ratio of post- to precapil-
lary resistances (R$_a$/R$_v$) is exactly analogous to the solution for membrane potential (E$_m$) as a function of Na$^+$ and K$^+$ equilibrium potentials (E$_{Na^+}$/E$_{K^+}$) and the ratio of membrane conductances (reciprocal
resistances) to Na$^+$ and K$^+$ (G$_{Na^+}$/G$_{K^+}$). Thus

$$P_c = \frac{(R_v/R_a)P_a + P_v}{1 + R_v/R_a} \qquad (1)$$

and

$$E_m = \frac{(G_{Na^+}/G_{K^+})E_{Na^+} + E_{K^+}}{1 + G_{Na^+}/G_{K^+}} \qquad (2)$$

Values for R_v and for R_a are obtained by clamping P_c (isogravimetric state) while P_a and P_v are varied experimentally. Similarly, values for G_{Na^+} and G_{K^+} are obtained by clamping E_m (voltage clamp) as E_{Na^+} and E_{K^+} are varied by changing extracellular concentrations of Na^+ and K^+. The voltage-clamp technique has had far more wide-ranging consequences for physiology and biophysics than the isogravimetric technique, but both were developed at about the same time, and the analogy is formally exact.

The isogravimetric capillary pressure, defined and measured as above, turned out to be $93 \pm 1.5\%$ of the in vitro osmotic pressure of the plasma proteins at protein concentrations ranging from ~3%–9% corresponding to in vitro protein pressures of ~8–38 mmHg. Filtration or absorption rates were exactly proportional to the difference between mean capillary pressure and isogravimetric capillary pressure (effective osmotic pressure) over a wide range of capillary or protein pressures. The filtration coefficient averaged 0.011 ml·min^{-1}·mmHg^{-1} per 100 g muscle at 37°C. The coefficient varied inversely with the viscosity of water as the temperature of perfusing blood was varied from 10° to 40°C as would be expected of viscous flow through porous media. The total capillary surface area in 100 g of tissue (mostly muscle) was estimated to be 7,000 cm^2, leading to a filtration coefficient of about 2×10^{-9} cm^3·s^{-1}·dyne^{-1} per cm^2 of capillary surface. These results (49) proved Starling's hypothesis for mammalian capillaries, thereby supplementing Landis's results on single capillaries of frog mesentery, but they did not by themselves add anything new conceptually to the field of capillary permeability. Nevertheless an important new deduction was implicit in the data, a deduction that came to light when I started to compare the filtration coefficients of living capillary membranes with those of artificial porous membranes.

As early as 1872, Guérot (18) had suggested that the porosity of an artificial membrane might be equated with that of an ideal membrane of equivalent filtration coefficient (K_f) and traversed by rectilinear, cylindrical pores of radius r and length Δx (membrane thickness) in which flow obeyed Poiseuille's law. In such an ideal membrane, the equivalent pore radius (r_e) would be defined

$$r_e = \sqrt{\frac{8\eta K_f}{A_p/\Delta x}} \qquad (3)$$

In collodion membranes of known thickness (Δx), a rough measure of pore area per unit path length $(A_p/\Delta x)$ can be made from the ratio of dry to wet weight, and this was a widely used method of calibrating collodion membranes in terms of equivalent pore radius (16). For membranes having rectangular slit pores of width (w) instead of cylindrical pores, the corresponding expression is

$$w_e = 2 \sqrt{\frac{3\eta K_f}{A_p/\Delta x}} \qquad (4)$$

Collodion membranes of equivalent pore radius 50 Å or equivalent slit width 62 Å retain 99% of serum albumin during ultrafiltration and contain 64% water (14). The filtration coefficient of such a membrane, adjusted to a membrane thickness comparable to that of a capillary wall $(0.5 \ \mu)$ is ~1,500-fold greater than what we found in muscle capillaries. It follows that if transcapillary filtration or absorption occurs by Poiseuille (viscous) flow through channels penetrating the capillary wall, then the fractional surface area occupied by such channels is ~0.64/1,500 or 0.04% of the capillary surface. This small proportion of the capillary wall might easily be confined to interendothelial channels. This conclusion from measurements of filtration coefficient was strongly supported by subsequent experiments on transcapillary diffusion and molecular sieving of lipid-insoluble molecules, as described in the next section. Recent measurements by Olesen and Crone (41) of the electrical resistance of endothelium in single muscle capillaries also indicate that only ~0.02% of the endothelial surface is available for conductance of ionic currents.

In 1947 the American Physiological Society awarded me a travel grant to present these results at the Seventeenth International Congress of Physiology at Oxford. It was the first congress after the war, and physiologists were happily reunited after many years of forced separation. It was a moving contrast to the grim congress in Zurich that I had attended in 1938, just prior to the invasion of Czechoslovakia. August Krogh listened to my paper at Oxford; afterward he came "backstage" to say nice things about it, and of course I was thrilled because his work in respiratory, comparative, and capillary physiology was (and still is) such a large part of our heritage. At the closing plenary session in the beautiful Sheldonian Theatre, Albert Szent-Györgyi gave a speech of thanks in nine different languages.

OSMOTIC TRANSIENTS AND RESTRICTED DIFFUSION
1948–1953

Pore Area per Unit Path Length

In the preceding section, it was explained how the permeability of porous artificial membranes can be evaluated in terms of equivalent pore radius or slit width. The principles and methods were well

[374]

established in the 1930s (14, 16), and I wondered whether it might be possible to characterize the permeability of living capillaries with this approach. We had already solved the problem of measuring the filtration coefficient and had shown that this coefficient varied inversely with the viscosity of water over a temperature range of 10°–40°C, suggesting that viscous flow does in fact occur. It remained to solve for $A_p/\Delta x$ in order to evaluate effective pore dimensions by Equation 2.

Three methods of estimating $A_p/\Delta x$ had been used for artificial membranes. The first involved the ratio of wet to dry weight of membrane (18, 14), the second involved electrical conductance (22, 36), and the third involved diffusion of small molecules (35). In 1929 Manegold (35) was the first to recognize that $A_p/\Delta x$ could be evaluated by measurements of diffusion. Thus rearrangement of Fick's law yields

$$A_p/\Delta x = J_s/D\Delta C \tag{5}$$

where J_s is the diffusive flux, D is the diffusion coefficient, and ΔC is the transmembrane concentration difference. Given two well-stirred compartments separated by an artificial membrane, it is easy to measure J_s and ΔC for test molecules. The value of $A_p/\Delta x$ obtained from Equation 5 can then be inserted in Equations 3 and 4 to solve for equivalent pore radius (r_e) or slit width (w_e). Manegold used urea, sugar, and HCl as test molecules to evaluate $A_p/\Delta x$, r_e, and w_e in collodion membranes. Isotopic water was not available to him as it was to Gene Renkin a generation later (51). In the perfused hindlimb it was easy to measure the J_s of any test molecule from the product of blood flow and arteriovenous concentration difference, and indeed Isidore Edelman and I had already done this using D_2O as a tracer. The only remaining unknown was the effective mean concentration difference ($\Delta \overline{C}$) across the capillary walls during the diffusion process. I emphasize "only" because the definition and the measurement of $\Delta \overline{C}$ during transcapillary passage of small molecules remain controversial to the present day. The problem is extremely complex, especially for rapidly diffusing molecules such as isotopic water or ions. There is not only a concentration gradient across the capillary membranes but also along the length of the capillary, both inside and outside, and all three vary with time and rate of blood flow. I spent many hours trying unsuccessfully to solve the problem mathematically, using the analogy of heat loss from liquids flowing through pipes with leaky insulation. The resulting Bessel functions were too much for my weak mathematical skills, and besides, I did not think that anyone, including myself, would really believe results based on an elaborate mathematical model. Several years later Renkin (52), and independently Crone (5), offered a simplified mathematical solution based on assumptions that the longitudinal concentration of a test solute within the capillary would decrease exponen-

[375]

tially and that the concentration outside the capillary would be zero early in the diffusion process. With these two assumptions the driving force for diffusion across the membrane would be the mean longitudinal concentration, which by elementary calculus is

$$\Delta\overline{C} = (C_a - C_v) \div \ln(C_a/C_v) \tag{6}$$

and, since $J_s = Q(C_a - C_v)$, one could substitute in Equation 6 to obtain

$$A_p/\Delta x = Q/D \ \ln(C_a/C_v) \tag{7}$$

A more familiar form of this same equation is

$$PS = D \frac{A}{\Delta X} p = -Q \ln(1 - E) \tag{8}$$

where E is the extraction ratio.

This treatment of the indicator dilution method has been widely used in various ways to estimate the permeability of capillaries in organs in which the circulation can be confined to a single entering artery and exiting vein. It has the obvious limitation that if the test molecule moves out of the capillaries so rapidly that the extracapillary concentration cannot be neglected or that C_v is zero, then $A_p/\Delta x$ or permeability is indeterminate. An additional complication in whole organs stems from heterogeneity of capillary permeabilities and transit times. These and other problems are discussed in recent reviews (2, 7).

My own solution to the problem in 1949 was based on van't Hoff's law and the transient osmotic pressures developed by hypertonic solutes as they diffuse across capillary walls. The fact that injections of hypertonic solutions cause osmotic withdrawal of fluid from the tissue spaces to blood was noted by Starling in 1896 (64). The amount of hemodilution was greater for Na_2SO_4 than for an equimolecular dose of NaCl, and this was interpreted by Starling (65) to mean that capillaries were less permeable to Na_2SO_4 than to NaCl.

In the perfused hindlimb, the osmotic withdrawal of fluid could be prevented by raising the mean hydrostatic pressure in the capillaries by known amounts (Eq. 1) just sufficient to maintain the isogravimetric state. In effect the hindlimb preparation was a sensitive, living osmometer with an extremely rapid response time. We could therefore obtain a continuous measure of the excess partial osmotic pressure generated by a test solute (over and above the preexisting protein pressure) as it diffused across the capillary walls.

By "we" I include Gene Renkin, who came to Harvard as a graduate student in 1948. At first Renkin was busy taking courses, and we did not start systematic investigations of osmotic transients until Luis Borrero came to join us in the autumn of 1949. Renkin came to our department by chance. As an undergraduate at Tufts, he had studied insect physiology in the Department of Biology with Ken Roeder. He applied to the Harvard graduate school in biology, but he mentioned the word "physiology" in his application; an unknown and unknowing secretary inadvertently sent him to me at the medical school instead of the biology department in Cambridge where he really intended to go. It was an error that had long-lasting and happy consequences for capillary physiology and I hope for Renkin as well. The 1984 American Physiological Society *Handbook of Physiology* volume on microcirculation (54b), of which Gene Renkin was editor, is testimony to his continued devotion and preeminence in the field.

I'm not sure when I first thought that the complex transcapillary concentration difference ($\Delta\overline{C}$) responsible for the diffusion process at any time during a transient might be linked to the observed increment of osmotic pressure via van't Hoff's law. In 1949 irreversible thermodynamics was in its infancy, and it was not until 1951 that Staverman (67) defined the osmotic reflection coefficient. It is understandable, then, that I assumed van't Hoff's law would hold without a correction factor, and $\Delta\overline{C}$ could be expressed as $RT\,\Delta\Pi$. Substitution in Equation 5 yielded

$$A_p/\Delta x = J_s/D\Delta\overline{C} = J_sRT/D\Delta\Pi = Q(C_a - C_v)RT/D\Delta\Pi \qquad (9)$$

where all quantities on the right are known or could be determined experimentally. The first experimental tests of Equation 9 were made in February 1950 using ferrocyanide or glucose as test molecules. The values obtained for $A_p/\Delta x$ were ~0.7×10^5 cm/100 g perfused tissue, corresponding to a fractional pore area of about 0.05% of the total estimated capillary surface. This was very close to the value we had inferred from filtration coefficients, as described previously. It was significant also that the values of $A_p/\Delta x$ calculated in this way were independent of time in the diffusion process and of the rate of blood flow, as befitted a geometrical property of wall structure. When this value for $A_p/\Delta x$ was substituted in Poiseuille's law (Eq. 3), it yielded an equivalent pore radius of 37–40 Å, which was close to the equivalent diffusion radius of serum albumin (36 Å) to which the capillaries were almost impermeable. Thus, from diffusion rates of small lipid-soluble molecules such as glucose, we were able to estimate the dimensions of aqueous channels penetrating the capillary walls, which accounted not only for the diffusion of glucose but also for the hydrodynamic conductance on the one hand and the permeability to serum albumin on the other. These results were first presented in 1950 at the 18th International Congress of Physiology in Copenhagen under the title "Filtration and molecular diffusion from the capillaries in muscle, with deductions concerning the number and dimensions of ultramicroscopic openings in the capillary wall."

Later we found that the diffusion area calculated from Equation 9 was less for large than for small molecules and therefore represented a "virtual" rather than an anatomical area. This led to the theory of restricted diffusion and its relation to osmotic reflection coefficients, as described in the next section.

Restricted Diffusion and Osmotic Reflection Coefficients

The relation we found experimentally between molecular size and pore area per unit path length, $A_p/\Delta x$, is shown in Figure 4. A simple explanation of Figure 4 in terms of a distribution of pore sizes was not consistent with the data nor with the observed filtration coeffi-

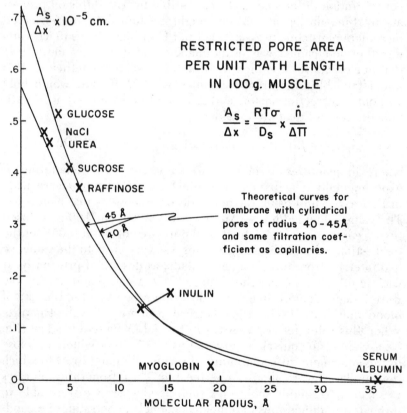

Figure 4. The decrease of transcapillary diffusion rates as a function of molecular size is far greater than can be accounted for by free diffusion, and this led to the theory of restricted diffusion. The data shown here, expressed in terms of restricted pore area per unit path length available to the solute, are based on the data of 1950–1951, after revision in 1963 to include osmotic reflection coefficients estimated as in Equation 12. Other revisions, based on the parallel pathway hypothesis of 1970 and using different estimates of osmotic reflection coefficient, lead to lower estimates of pore area and slightly higher estimates of equivalent pore radius. See Renkin and Curry (54) or Curry (8) for recent reviews. [From Landis and Pappenheimer (33).]

cient: some other explanation was needed, and this led to the theory
of restricted diffusion.

In 1929 Manegold (35) had already noted that diffusion of sucrose
relative to urea through collodion membranes was slower than could
be accounted for on the basis of free diffusion, and in 1930 Friedman
and Kraemer (17) reported a similar phenomenon for diffusion of
sugars through gelatin gels. In 1936 J. D. Ferry (16) used the term
impedance to diffusion to describe this phenomenon. In 1951 (47)
we used the terms *restricted pore area* and *restricted diffusion coefficient* (*D'*) to describe the results summarized in Figure 4, and we
proposed further that the ratio of restricted- to free-diffusion coefficients (*D'/D*) could be expressed in terms of the ratio of molecular-
diffusion radius (*a*) to equivalent pore radius (*r_e*) (48). Thus

$$D'/D = (1 - a/r_e)^2/(1 + 2.4a/r_e) \qquad (10)$$

The numerator of Equation 10 expresses steric hindrance to the
diffusing molecules at the entrance to the pore, as suggested by
Ferry (16); the denominator expresses the effects of the stationary
walls of the pore on the kinetic movements of the diffusing molecules
and is taken directly from the Ladenburg (28) correction to Stokes's
law for spherical particles sedimenting in a cylindrical vessel of
radius comparable with the spheres.

More accurate (and more complex) equations describing restricted
diffusion have since been published [for reviews see Crone and
Christenson (6) or Curry (8)], but the original form accounted fairly
well for the observed decrease in restricted pore area as a function
of molecular size. When corrected to the size of a water molecule,
the $A_p/\Delta x$ was ~1.2×10^5 cm, and combining this value with the
filtration coefficient yielded an equivalent pore radius of 30 Å or a
slit width of 36 Å.

All of these results were published in our 1951 paper in the
American Journal of Physiology (48). In the same year Staverman
(67) published a seminal paper on corrections to van't Hoff's law in
permeable-membrane systems. Staverman described an "osmotic
reflection coefficient" (*σ*) such that

$$\Delta\Pi = \sigma\Delta CRT \qquad (11)$$

In membranes having aqueous pores, the values of *σ* could vary from
unity in perfectly semipermeable membranes (as in van't Hoff's law)
to zero in membranes in which the pores were so large that diffusion
of solute was unrestricted relative to water. It followed from Staverman's paper that our estimates of $\Delta\bar{C}$ from $\Delta\Pi$ were too low, and
hence our estimate of $A_p/\Delta x$ (Eq. 9) was too high and equivalent
pore radius too low (Eq. 3)—but how large was the error?

The magnitude of the Staverman corrections to van't Hoff's law
was left unspecified, and in 1951 there were no osmometers with
response times rapid enough to measure osmotic pressures at zero

flow during diffusion of small solutes through relatively thick artificial porous membranes. It seemed reasonable to suppose, however, that the reflection coefficient would be related to restricted diffusion and hence through Equation 10 to molecular and pore dimensions. As a first hunch I proposed in 1953 (44) that

$$\sigma = 1 - D'_s/D'_w \tag{12}$$

where D'_s/D'_w is the ratio of restricted diffusion coefficients of the solute to the solvent through the membrane, both of which could be estimated in terms of molecular and pore dimensions, as in Equation 10. With this correction, the calculated $A_p/\Delta x$ for water was revised downward to $0.5 - 0.6 \times 10^5$ cm/100 g muscle, and the corresponding equivalent pore radius was revised upward to 40–45 Å (33).

During the next twenty-five years, there were several theoretical formulations of osmotic reflection coefficient as a function of molecular and pore dimensions [for reviews see Crone and Christenson (6) or Curry (8)], and all of them predicted much lower reflection coefficients than did Equation 12. Indeed, it is obvious that Equation 12 must be incorrect for the limiting case in which $D'_s/D'_w = D_s/D_w$. These theoretical predictions were fortified by measurements of hydrodynamic and osmotic flow through artificial membranes of known pore size; reflection coefficients determined experimentally in such model flow systems were considerably lower than those predicted by Equation 12. These results led to a paradox that was well expressed by Tosteson (68) at a symposium on capillary permeability in 1970. Thus, if reflection coefficients were as low as indicated in model systems, then estimates of pore size from osmotic transients in living capillaries must be far too low and this in turn would lead to further reduction of calculated reflection coefficients and so on until, by iterative corrections, the estimated pore dimensions would be infinitely large!

A possible explanation of this paradox was put forward at the 1970 symposium (46) and has since been explored experimentally (9). In artificial membranes the passage of solutes takes place through the same channels as does the flow of water. In living capillaries, however, most of the surface area (99.9%+) consists of plasma membranes that are essentially impermeable to lipid-insoluble molecules ($\sigma = 1$) but are nevertheless slightly permeable to water, which can pass by diffusion or hydrodynamic flow (or both) through very small pores that exclude ions and other lipid-insoluble solutes. This additional (parallel) pathway for bulk flow of water could have profound effects on osmotic reflection coefficients in the membrane as a whole.

Experimental evidence that water can penetrate areas in the capillary wall that exclude small solutes was first obtained by Yudilevitch and Alvarez (76) in heart muscle and subsequently by Curry et al. (9) in single capillaries. The quantitative effects of this additional pathway for water on the osmotic reflection coefficients of the whole membrane are still unclear, but certainly they operate to reconcile the general theory of restricted diffusion and osmotic reflection with the experimental data obtained from living capillaries. Estimates of $A_p/\Delta x$ and r_e in muscle capillaries, based on the parallel-pathway hypothesis, are 0.15×10^5 cm/100 g and 45–50 Å, respectively (54). This value of $A_p/\Delta x$ is close to that estimated by the indicator-dilution technique (Eq. 7) or by the electrical conductance technique (41), which are independent of assumptions about osmotic reflection coefficients.

Hydrodynamic Flow and Diffusion of Water Through Porous Membranes

It was mentioned in the introduction that pioneer experiments by Hevesy, Hofer, and Krogh (20) in 1935 and by Visscher et al. (70) in 1944 had revealed large discrepancies, as measured by diffusion of heavy water and permeability measured by osmotic flow. Thus, in frog skin or intestinal epithelia, the transport of water by osmotic flow is 10 to 100 times greater than net diffusional flux of water measured simultaneously by D_2O.

Reasons for this large and important discrepancy became obvious in 1950, after we had formulated the relations between filtration, diffusion, and equivalent pore dimensions. Thus the diffusion of water through membranes having cylindrical pores is proportional to the cross-sectional area of the pores, or $N\Pi r^2$. In contrast, hydrodynamic flow through the same pores is proportional to $N\Pi r^4$. It follows that the ratio of hydrodynamic to diffusional flow is proportional to r^2. In 1953 I showed that the constant of proportionality is $8\eta D_{H_2O}\bar{V}_{H_2O/RT} = 6.1 \times 10^{14}$ (48). In red cells having equivalent pore radii of 4–5 Å, the ratio of hydrodynamic to diffusional flow of water during osmotic shifts is about 1:1 (62); in contrast, this ratio is more than 100:1 through interendothelial slits of equivalent pore radius 45 Å. Intermediate values, such as those found experimentally in frog skin (20) or intestine (70), may be expected of membranes having intercellular leaks of equivalent pore radius 10–20 Å.

Similar reasoning was advanced independently by Hans Ussing and his colleagues (24) during 1953, and indeed we corresponded about this and about osmotic reflection coefficients during 1952.

Clarification of the differences between diffusional permeability and hydrodynamic conductance and their relation to pore dimensions was one of the main contributions of this period.

Molecular Sieving: Application to Renal Glomerular Membranes and Leaky Epithelia

In 1894 Bayliss and Starling (3) showed that the protein concentration in intestinal lymph varies inversely with the rate of lymph flow, an observation that was extended to peripheral vascular beds by Drinker et al. (71), Landis et al. (32), and others during the 1920s and 1930s. In the perfused hindlimb it was easy to show that not only proteins but also smaller molecules such as inulin or even raffinose were subject to molecular sieving during high rates of ultrafiltration through capillary walls (48). During ultrafiltration through porous membranes, the concentration of solutes in the ultrafiltrate depends on a "race" between hydrodynamic flow of solvent through the pores on the one hand and restricted diffusion of the solute on the other. In 1953 I (44) proposed that this might be expressed quantitatively as

$$C_2/C_1 = \frac{D'/D + [(D'A_p)/(J_v\Delta x)]}{1 + [(D'A_p)/(J_v\Delta x)]} \tag{13}$$

where C_1 and C_2 are concentrations in filtrand and filtrate, respectively, and J_v is the rate of filtration. When there is no restriction to diffusion $(D' = D)$ or when there is no ultrafiltration $(J_v = 0$, dialysis), then $C_2 = C_1$.

When Equation 13 is expressed in the language of contemporary irreversible thermodynamics, it becomes the general equation of flow of solute (J_s) and of solvent (J_v) through porous membranes (6, 23, 62)

$$J_s = \omega RT(C_1 - C_2) + (1 - \sigma)J_v\overline{C} \tag{14}$$

where $\omega RT = D'A_p/\Delta x$ and $(1 - \sigma) = D'/D$ and $\overline{C}$ is mean concentration of solute in the membrane. Equation 14 is identical with Equation 13 when $\overline{C} = C_1$. In practice $\overline{C}$ is always less than C_1, and various modifications of Equation 13 have been formulated to express more exactly the molecular sieving of proteins and other macromolecules in lymph or glomerular filtrate [for recent reviews see Curry (8), Renkin (53), or Deen et al. (10).] Nevertheless the 1953 equation (Eq. 13) predicted with reasonable accuracy the molecular sieving of solutes through calibrated porous membranes (51), and it seemed to me that the theory might be used to characterize the permeability of renal glomerular membranes, which normally sustain a high rate of ultrafiltration. Thus the sieve coefficient (C_2/C_1) can be evaluated experimentally by indirect clearance methods, and J_v is the clearance of inulin, and D' is a function of molecular radius and effective pore radius as defined by Equation 10. It followed that $A_p/\Delta x$ and r_e could be evaluated from standard clearance measurements of substances such as small proteins or dextran fractions, which are subject to molecular sieving at the glomerulus. Once $A_p/\Delta x$ and r_e are determined in this way, it should be possible to evaluate the filtration coefficient of glomerular membranes and the transglomerular membrane pressure head for filtration (Eq. 3). Using data for glomerular clearances of myoglobin, ovalbumin, and hemoglobin, I estimated an equivalent pore radius of 37.5 Å, an $A_p/\Delta x$ of 16×10^6 cm/100 g of kidney, a filtration coefficient of 3.1 ml·

min^{-1}·100 g^{-1}, and a filtration pressure head (hydrostatic minus osmotic pressure) of 52 mmHg (45). These values of filtration coefficient and pressure head (obtained from the theory of molecular sieving and standard clearance techniques) are reasonably close to modern estimates, based on direct measurements of hydrostatic pressures in dog kidneys (40, 42).

In 1954 I was invited by the Unitarian Service Committee to join a team of renal physiologists on a "goodwill" tour of several German universities. The tour started in Göttingen with a symposium on the kidney, and I used the occasion to launch the theory of molecular sieving for characterization of glomerular permeability. One purpose of the tour was to promote international exchange and understanding, and for this reason we gave our lectures in German and the work was published in *Klinische Wochenschrift* under the title "Über die Permeabilität der Glomerulummembranen in der Niere" (45). This paper is rarely cited in the English literature, but it did lay the groundwork for subsequent studies of glomerular permeability to macromolecules [for reviews see Renkin and Gilmore (54a) or Deen et al. (12)]. Of particular importance was the subsequent discovery that molecular sieving through glomerular membranes depends on molecular charge as well as size, a fact that has been interpreted by Deen et al. (11) in terms of negative charges in the glomerular membranes.

At the opening of the Göttingen symposium, I read a poem in "Gemixte Pickles" that may have been more effective in fulfilling the aims of the Unitarian Service Committee than all our esoteric scientific communications; at any rate it was printed in the form shown in Figure 5 and distributed widely in Germany.

The theory of molecular sieving has not been applied systematically to "leaky" epithelia such as mucosa of the small intestine or proximal renal tubules. It is likely, however, that paracellular channels contribute importantly to the permeability and transport properties of these epithelia. It has already been mentioned that observations by Visscher et al. (70) on the discrepancy between water flow and net diffusional flux of isotopic water in the absorbing intestine are most easily accounted for by paracellular channels 20 Å or more in equivalent radius. Boulpaep (4) found that renal proximal tubules allow (restricted) passage of molecules as large as raffinose (M_r 504) during rapid fluid absorption stimulated by saline loading. The low transtubular electrical resistance, combined with permeability to raffinose, suggested paracellular channels as large as 34 Å. Channels of equivalent radius 10 Å have been postulated by Schultz (58) to account for low transmucosal electrical potentials relative to ion transport in the small intestine. Recently, Karen Reiss and I have found that polar molecules ranging in size from creatinine to inulin pass through the intestinal mucosa of rats during maximal fluid absorption activated by the glucose-Na$^+$ cotransport system

"Das Kidney„

von John R. Pappenheimer

Listen, my friends, Sie werden hören
Wie Urine ist aus Blut gebören

Es ist ein Epic, gar kein Myth
Von Ludwig bis zu Homer Smith

Das Blut ist durch die beide Nieren
Von Circulation abgesteeren

Durch Bowman's Kapsul strömt das Blut
For lack of any other route

Though Bowman hat es nie geguessed
Ist Filtrat dann aus Blut gepresset

Das Filtrat rein, das Filtrat pure
Ist noch nicht Urine to be sure

Wasser und Salz, Glukose süß
Are taken from the little Fluß

Durch Tubuluszell ins Blut zurück
In spite of intrarenal Druck

Mit Heidenhain'scher Kraft, die Zelle
Break all the rules of Fick'scher Quelle

I'd be delighted to explain
Secretion à la Heidenhain

But leider ist der true Technik
Ein Patent of the Harnfabrik.

Zur Eröffnung des Symposions über Physiologie
und Patho-Physiologie der Niere in Göttingen
vom 18. — 22. Juni 1954.

Figure 5. "Das Kidney," by John R. Pappenheimer, distributed at the *Symposions über Physiologie und Patho-Physiologie der Niere*, Göttingen, 22 June 1954.

(50). In all these cases it is proposed that the apical seals (zonulae occludens) of mucosal or proximal tubule cells become leaky during osmotic flow. The theory of molecular sieving, originally developed for studies of capillary permeability, may provide an important tool for investigating paracellular pathways in epithelia and their regulation during osmotic absorption of fluid.

Parts of this essay were published in *Annu. Rev. Physiol.* 49: 1–15, 1987. I thank the editors for permission to use some of the same wording in each of these two publications.

Author's current address: Department of Biology, Concord Field Station, Harvard University, Old Causeway Road, Bedford, MA 01730.

BIBLIOGRAPHY

1. BARCROFT, J. *Researches on Pre-natal Life.* Oxford, UK: Blackwell, 1946.
2. BASSINGTHWAIGHTE, J. B., and C. A. GORESKY. Modeling in the analysis of solute and water exchange in the microvasculature. In: *Handbook of Physiology. The Cardiovascular System, Microcirculation,* edited by E. M. Renkin and C. C. Michel. Bethesda, MD: Am. Physiol. Soc., 1984, sect. 2, vol. IV, chapt. 13, p. 549–626.
3. BAYLISS, W. M., and E. H. STARLING. Observations on venous pressures and their relationship to capillary pressures. *J. Physiol. Lond.* 16: 159–202, 1984.
4. BOULPAEP, E. L. Permeability changes of the proximal tubule of *Necturus* during saline loading. *Am. J. Physiol.* 222: 517–531, 1972.
5. CRONE, C. The permeability of capillaries in various organs as determined by the indicator diffusion method. *Acta Physiol. Scand.* 58: 292–305, 1963.
6. CRONE, C., and O. CHRISTENSEN. Transcapillary transport of small solutes and water. *Int. Rev. Physiol.* 18: 149–213, 1979.
7. CRONE, C., and D. G. LEVITT. Capillary permeability to small solutes. In: *Handbook of Physiology. The Cardiovascular System. Microcirculation,* edited by E. M. Renkin and C. C. Michel. Bethesda, MD: Am. Physiol. Soc., 1984, sect. 2, vol. IV, chapt. 10, p. 411–466.
8. CURRY, F.-R. E. Mechanics and thermodynamics of transcapillary exchange. In: *Handbook of Physiology. The Cardiovascular System. Microcirculation,* edited by E. M. Renkin and C. C. Michel. Bethesda, MD: Am. Physiol. Soc., 1984, sect. 2, vol. IV, chapt. 8, p. 309–374.
9. CURRY, F.-R. E., J. C. MASON, and C. C. MICHEL. Osmotic reflection coefficients of capillary walls to low molecular weight hydrophilic solutes measured in single perfused capillaries of the frog mesentery. *J. Physiol. Lond.* 261: 319–336, 1968.
10. DEEN, W. M., C. R. BULGER, and B. M. BRENNER. Biophysical basis of glomerular permeability. *J. Membr. Biol.* 71: 1–10, 1983.
11. DEEN, W. M., B. SATVAT, and J. M. JAMIESON. Theoretical model for glomerular filtration of charged solutes. *Am. J. Physiol.* 238 (*Renal Fluid Electrolyte Physiol.* 7): F126–F139, 1980.
12. EGGLETON, M. G., J. R. PAPPENHEIMER, and F. R. WINTON. The influence of diuretics on the osmotic work done and on the efficiency of the isolated kidney of the dog. *J. Physiol. Lond.* 97: 363–382, 1940.
13. EGGLETON, M. G., J. R. PAPPENHEIMER, and F. R. WINTON. The relation between ureter, venous, and arterial pressures in the isolated kidney of the dog. *J. Physiol. Lond.* 99: 135–152, 1940.
14. ELFORD, W. J., and J. D. FERRY. The calibration of graded collodion

membranes. *Br. J. Exp. Pathol.* 16: 1–14, 1935.

15. FERGUSON, J. K. W., S. M. HORVATH, and J. R. PAPPENHEIMER. The transport of carbon dioxide by erythrocytes and plasma in dogfish blood. *Biol. Bull. Woods Hole* 75: 381–388, 1938.

16. FERRY, J. D. Ultrafilter membranes and ultrafiltration. *Chem. Rev.* 18: 373–455, 1936.

17. FRIEDMAN, L., and E. O. KRAEMER. The structure of gelatin gels from studies of diffusion. *J. Am. Chem. Soc.* 52: 1295–1304, 1930.

18. GUÉROT, A. Sur les dimensions des intervalles poreux des membranes. *C. R. Acad. Sci. Paris* 75: 1809–1812, 1872.

19. HAHN, L., and G. HEVESY. Rate of penetration of ions through the capillary wall. *Acta Physiol. Scand.* 1: 347–361, 1940.

20. HEVESY, G. E., E. HOFER, and A. KROGH. The permeability of the skin of frogs to water as determined by D_2O and H_2O. *Scand. Arch. Physiol.* 72: 199–214, 1935.

21. HEVESY, G., and C. F. JACOBSEN. Passage of water through capillary and cell walls. *Acta Physiol. Scand.* 1: 11–18, 1940.

22. HITCHCOCK, D. I. The size of pores in collodion membranes. *J. Gen. Physiol.* 9: 745–762, 1926.

23. KATCHALSKY, A., and P. F. CURRAN. *Nonequilibrium Thermodynamics in Biophysics.* Cambridge, MA: Harvard Univ. Press, 1965.

24. KOEFOED-JOHNSEN, V., and H. H. USSING. The contribution of diffusion and flow to passage of D_2O through living membranes. *Acta Physiol. Scand.* 28: 60–76, 1953.

25. KRAMER, K., and F. R. WINTON. The influence of urea and of change in arterial pressure on the oxygen consumption of the isolated kidney of the dog. *J. Physiol. Lond.* 96: 87–103, 1939.

26. KROGH, A. *Anatomy and Physiology of Capillaries.* New Haven, CT: Yale Univ. Press, 1929, p. 348–350.

27. KROGH, A., E. M. LANDIS, and A. H. TURNER. The movement of fluid through the human capillary wall in relation to venous pressure and to the colloid osmotic pressure of the blood. *J. Clin. Invest.* 11: 63–95, 1932.

28. LADENBURG, R. Über den Einfluss von Wänden auf die Bewegung einer Kugel in einer Reibenden Flussigkeit. *Ann. Physik.* 22: 287–309, 1907.

29. LANDIS, E. M. Microinjection studies of capillary permeability. The relation between capillary pressure and the rate at which fluid passes through the walls of single capillaries. *Am. J. Physiol.* 82: 217–238, 1927.

30. LANDIS, E. M. Capillary pressure and capillary permeability. *Physiol. Rev.* 14: 404–481, 1934.

31. LANDIS, E. M., and J. H. GIBBON, JR. The effects of temperature and of tissue pressure on the movement of fluid through the human capillary wall. *J. Clin. Invest.* 12: 105–138, 1933.

32. LANDIS, E. M., L. JONAS, M. ANGEVINE, and W. ERB. The passage of fluid and protein through the human capillary wall during venous congestion. *J. Clin. Invest.* 12: 105–138, 1932.

33. LANDIS, E. M., and J. R. PAPPENHEIMER. Exchange of substances through capillary walls. In: *Handbook of Physiology. Circulation,* edited by W. F. Hamilton. Washington, DC: Am. Physiol. Soc., 1963, sect. 2, vol. II, chapt. 29, p. 961–1034.

34. LUCKÉ, B., and M. McCUTCHEON. The living cell as an osmotic system

and its permeability to water. *Physiol. Rev.* 12: 68–139, 1932.

35. MANEGOLD, E. Die Dialyse durch Kollodiummembranen und der Zusammenhang zwischen Dialyse, Diffusion und Membranstruktur. *Kolloid-Z.* 49: 372–395, 1929.

36. MANEGOLD, E., and K. SOLF. Das elektroosmotische Verhalten von Kollodium membranen abgestufte Porosität. *Kolloid-Z.* 55: 273–310, 1931.

37. MILLIKAN, G. A., J. R. PAPPENHEIMER, A. J. RAWSON, and J. HERVEY. The continuous measurement of arterial saturation in man. *Am. J. Physiol.* 133: P390, 1941.

38. MILLIKAN, G. A., and J. R. PAPPENHEIMER. Development of chemical oxygen generators for use in aircraft (Abstract). *J. Aviat. Med.* 19: 118, 1947.

39. MONOD, J., J. WYMAN, JR., and P. CHANGEAUX. On the nature of allosteric transitions: a plausible model. *J. Mol. Biol.* 12: 88–118, 1965.

40. NAVAR, L. G., P. D. BELL, R. W. WHITE, R. L. WATTS, and R. H. WILLIAMS. Evaluation of single nephron glomerular coefficient in the dog. *Kidney Int.* 12: 137–149, 1977.

41. OLESEN, S., and C. CRONE. Electrical resistance of capillary endothelium. *Biophys. J.* 42: 31–41, 1983.

42. OTT, C. E., G. R. MARCHAND, J. A. DIAZ-BUXO, and F. G. KNOX. Determinants of glomerular filtration rate in the dog. *Am. J. Physiol.* 231: 235–239, 1976.

43. PAPPENHEIMER, J. R. Vasoconstrictor nerves and oxygen consumption in the isolated perfused hindlimb muscles of the dog. *J. Physiol. Lond.* 99: 182–200, 1941.

44. PAPPENHEIMER, J. R. Passage of molecules through capillary walls. *Physiol. Rev.* 33: 387–423, 1953.

45. PAPPENHEIMER, J. R. Über die Permeabilität der Glomerlulummembranen in der Niere. *Klin. Wochenschr.* 33: 362–365, 1955.

46. PAPPENHEIMER, J. R. Osmotic reflection coefficients in capillary membranes. In: *Capillary Permeability: Transfer of Molecules and Ions Between Capillary Blood and Tissue*, edited by C. Crone and N. A. Lassen. Copenhagen: Munksgaard, 1970, p. 278–286. Alfred Benzon Symp. 2.

47. PAPPENHEIMER, J. R., M. P. LEPIE, and J. WYMAN, JR. The surface tension of aqueous solutions of dipolar ions. *J. Am. Chem. Soc.* 58: 1851–1855, 1936.

48. PAPPENHEIMER, J. R., E. M. RENKIN, and L. M. BORRERO. Filtration, diffusion and molecular sieving through peripheral capillary membranes. A contribution to the pore theory of capillary permeability. *Am. J. Physiol.* 167: 13–46, 1951.

49. PAPPENHEIMER, J. R., and A. SOTO-RIVERA. Effective osmotic pressures of the plasma proteins and the quantities associated with the capillary circulation in the hindlimbs of cats and dogs. *Am. J. Physiol.* 152: 471–491, 1948.

50. PAPPENHEIMER, J. R., and K. Z. REISS. Contribution of solvent drag through intercellular junctions to absorption of nutrients by the small intestine. *J. Membr. Biol.* In press.

51. RENKIN, E. M. Filtration, diffusion and molecular sieving through cellulose membranes. *J. Gen. Physiol.* 38: 225–243, 1954.

52. RENKIN, E. M. Transport of potassium-42 from blood to tissue in isolated mammalian skeletal muscles. *Am. J. Physiol.* 197: 1205–1210, 1959.

53. RENKIN, E. M. Capillary transport of macromolecules: pores and other

endothelial pathways. *J. Appl. Physiol.* 58: 315–325, 1985.

54. RENKIN, E. M., and F. E. CURRY. Transport of water and solutes across capillary endothelium. In: *Membrane Transport in Biology,* edited by G. Giebisch, D. C. Toteson, and H. H. Ussing. New York: Springer-Verlag, 1978, chapt. I, p. 1–45.

54a. RENKIN, E. M., and J. P. GILMORE. Glomerular filtration. In: *Handbook of Physiology. Renal Physiology,* edited by J. Orloff and R. W. Berliner. Washington, DC: Am. Physiol. Soc., 1973, sect. 8, chapt. 9, p. 185–248.

54b. RENKIN, E. M., and C. C. MICHEL (editors). *Handbook of Physiology. The Cardiovascular System. Microcirculation.* Bethesda, MD: Am. Physiol. Soc., 1984, sect. 2, vol. IV.

55. RICHARDS, A. N. Processes of urine formation. *Proc. R. Soc. Lond. B. Biol. Sci.* 126: 398–432, 1938.

56. RICHARDS, A. N., and O. H. PLANT. Urine formation in the perfused kidney. The influence of adrenaline on the volume of the perfused kidney. *Am. J. Physiol.* 59: 184–190, 1922.

57. SCHECHTER, W. H., R. R. MILLER, R. M. BOVARD, C. B. JACKSON, and J. R. PAPPENHEIMER. Chlorate candles as a source of oxygen. *Ind. Eng. Chem.* 42: 2348–2353, 1950.

58. SCHULTZ, S. G. The role of paracellular pathways in isotonic fluid transport. *Yale J. Biol. Med.* 50: 99–113, 1977.

59. SHANNON, J. A., and F. R. WINTON. The renal excretion of inulin and creatinine by the anesthetized dog and the pump-lung-kidney preparation. *J. Physiol. Lond.* 96: 87–103, 1939.

60. SMITH, H. W. *Lectures on the Kidney.* Lawrence: Univ. of Kansas Extension Div., 1943.

61. SMITH, H. W. *The Kidney.* New York: Oxford Univ. Press, 1951, chapt. 14.

62. SOLOMON, A. K. Characterization of membranes by equivalent pores. *J. Gen. Physiol.* 51: 335–364, 1968.

63. STARLING, E. H. On the absorption of fluid from the connective tissue spaces. *J. Physiol. Lond.* 19: 312–326, 1896.

64. STARLING, E. H. Physiological factors involved in the causation of dropsy. *Lancet,* 1896.

65. STARLING, E. H. *The Fluids of the Body.* Chicago, IL: Univ. of Chicago Press, 1909, p. 67–68.

66. STARLING, E. H., and E. B. VERNEY. The secretion of urine as studied on the isolated kidney. *Proc. R. Soc. Lond. B. Biol. Sci.* 97: 321–363, 1925.

67. STAVERMAN, A. J. The theory of measurement of osmotic pressure. *Recl. Trav. Chim. Pays-Bas Belg.* 70: 344–352, 1951.

68. TOSTESON, D. C. In: *Capillary Permeability,* edited by C. Crone and N. A. Lassen. Copenhagen: Munksgaard, 1970, p. 658–663. Alfred Benzon Symp. II.

69. VAN'T HOFF, J. H. Die Rolle des Osmotische Druckes in der Analogie zwischen Lösungen und Gasen. *Z. Phys. Chem.* 1: 481–508, 1887.

70. VISSCHER, M. B., E. S. FETCHER, JR., C. W. CARR, H. P. GREGOR, M. S. BUSHEY, and D. E. BARKER. Isotopic tracer studies on the movement of water and ions between intestinal fluid and blood. *Am. J. Physiol.* 142: 550–575, 1944.

71. WHITE, J. C., M. E. FIELD, and C. K. DRINKER. On the protein content and normal flow of lymph from the foot of the dog. *Am. J. Physiol.* 103: 34–44, 1933.
72. WINTON, F. R. The glomerular pressure in the isolated mammalian kidney. *J. Physiol. Lond.* 72: 361–375, 1931.
73. WINTON, F. R. The control of the glomerular pressure by vascular changes within the isolated mammalian kidney demonstrated by the actions of adrenaline. *J. Physiol. Lond.* 73: 151–162, 1931.
74. WINTON, F. R. Physical factors involved in the activities of the mammalian kidney. *Physiol. Rev.* 17: 408–435, 1937.
75. WYMAN, J., JR. Linked functions and reciprocal effects in proteins. *Adv. Protein Chem.* 19: 224–286, 1964.
76. YUDILEVICH, D. L., and O. A. ALVAREZ. Water, sodium, and thiourea transcapillary diffusion in the dog heart. *Am. J. Physiol.* 213: 308–314, 1967.

Contributors

CLAY M. ARMSTRONG, *Professor of Physiology, Department of Physiology, University of Pennsylvania School of Medicine, Philadelphia, Pennsylvania*

HUGH DAVSON, *Professor of Physiology, Departments of Physiology, St. Thomas's Hospital Medical School, King's College, London, and Southampton University, Southampton, United Kingdom*

DAVID E. GOLDMAN, *Guest Researcher, Laboratory of Biophysics, National Institutes of Health, Bethesda, Maryland*

WILHELM HASSELBACH, *Professor Doctor of Medicine; Director, Department of Physiology, Max-Planck-Institut Für Medizinische Forschung, Heidelberg, Federal Republic of Germany*

**MEMBRANE
TRANSPORT**

ERICH HEINZ, *Professor of Biochemistry, Emeritus, J. W. Goethe University, Frankfurt, Federal Republic of Germany; Guest Professor, Max Planck Institute, Dortmund, Federal Republic of Germany; Adjunct Professor of Physiology and Biophysics, Cornell University Medical College, New York, New York*

WERNER R. LOEWENSTEIN, *Professor and Chairman, Department of Physiology and Biophysics, University of Miami School of Medicine, Miami, Florida*

LORIN J. MULLINS, *Professor of Biophysics and Chairman, Department of Biophysics, University of Maryland School of Medicine, Baltimore, Maryland*

JOHN R. PAPPENHEIMER, *George Higginson Professor of Physiology Emeritus, Department of Physiology, Harvard Medical School, Boston, Massachusetts*

RICHARD J. PODOLSKY, *Chief, Laboratory of Physical Biology, National Institute of Arthritis and Musculoskeletal and Skin Diseases, National Institutes of Health, Bethesda, Maryland*

J. DAVID ROBERTSON, *James B. Duke Professor of Neurobiology; Chairman Emeritus, Department of Anatomy, Duke University Medical Center, Durham, North Carolina*

ASER ROTHSTEIN, *Director Emeritus, Research Institute; Professor, Medical Biophysics, Hospital for Sick Children, University of Toronto, Toronto, Ontario, Canada*

JENS C. SKOU, *Professor, Doctor of Medicine, Institute of Biophysics, University of Aarhus, Aarhus, Denmark*

ARTHUR K. SOLOMON, *Professor of Biophysics, Emeritus, Harvard Medical School Department of Physiology and Biophysics, Boston, Massachusetts*

DANIEL C. TOSTESON, *Dean, Harvard Medical School, Boston, Massachusetts*

[393]

MEMBRANE
TRANSPORT

HANS H. USSING, *Professor Emeritus, Institute of Biological Chemistry, Copenhagen University, Copenhagen, Denmark*

Index

Abdominal ganglia, synapses in, 56, 58

"Absolute impermeability" concept, 37–38

Acetamide and solvent drag, 348

Acetylcholine, release during muscle twitch, 187–188

Acetylcholinesterase (AChe)
anesthesia penetration and, 157
monolayer experiments, 158

Actin, myosin, ATP and, 190–191

Action potential
Hodgkin-Huxley-Katz theory, 353
plant cells, 284–285
two-factor theory of excitation, 276

Active transport
electrodiffusion and, 257–259
epithelial transport, 341–342
history of research on, 20
organic substrates, 240–242
pump mediation and ATP hydrolysis, 5
secondary, 237–259
sodium-potassium pump (Na$^+$-K$^+$-ATPase), 177–179
sodium pump, 155
solvent flow and, 347–348

Actomyosin system, calcium binding, 294

Adam, N. K., 29, 156

Adenylate kinase, 191–192

ADP
sodium-potassium pump, 168–169
See also ATP/ADP ratios

ADP-dependent Na$^+$-Na$^+$ exchange, 171–172

ADP-hydrolysis-dependent Na$^+$-Na$^+$ exchange, 171–172

Adsorption potentials, 345

Aequorin, cell-to-cell communication, 320–321

Affinity probe, band 3 binding, 214–215

Affinity ratios, 269–270

Affinity sequences,
field-strength theory, 265–266
crystal radii and hydration energy, 268–269

Affinity-type mechanism, 245–248

A-I junction
in muscle, 63–65
paired membrane structure, 66–67
Z line structure, 68

Albers-Post scheme, 170

Alcohols
cell permeability and, 37–38
Na$^+$ channel inhibition, 283–284

Aldosterone, sodium-potassium pump and, 163–164

Alkali metal cations
calculated "hydrated" radii, 276–277
field-strength theory of selectivity, 264–272
membrane interaction with, 261–262
selectivity ratios, 265–266

Alpha-chains
amino acid sequence, 165–166
molecular weight, 164–165

Alternating gate carriers, 239–240

Alveolar system in membranes, 2

American Association of Anatomists (AAA), 60

Amino acid sequence, sodium-potassium pump, 165–166

Amino acid transport, 241–248
active transport, 240–241
energetic adequacy, 243–245
gradient hypothesis, 242–243

Amino acid transport—*continued*
  molecular mechanism, 245–248
Amino groups, sodium-potassium
  pump, 174–175
Amphipathic channel, 114–115
Amphotericin B pores, 136–138
Amyl alcohol, 37
Analytical leptoscope, 86
Anesthetics (local)
  effect on lipid and protein in
    monolayer, 160
  ion selectivity and, 281–284
  sodium-potassium pump and,
    156–162
Anfinsen, Chris, 128
Animal cells
  kinetic and static studies, 24
  osmotic equilibria, 24–25
  structural studies, 22–23
Anion exchange
  band 3 protein, 203–228
  chemical probing agents for,
    206–210
  equilibrium theory and, 27–28
  historical background of,
    203–205
  permeability, 212
  transport kinetics, 212–214
Anion transport
  background on, 212–214
  carrier-type model, 213–214
  identification techniques,
    210–212
Anlyan, Bill, 104
Antidiuretic hormone (ADH)
  epithelial transport, 341–342
  solvent drag, 348
Antiport, 248
  coupling by, 241–242
APMB (2-aminophenyl-6-
    methylbenzenethiazole-3-7'-
    disulfonic acid), 218
Araldite medium, 70
Arrhenius-Kohlrausch relation,
  electrodiffusion, 252
Artifact production, 106–107
A_s/A_w ratio, 138–139
Asynchronous-dilution model,
  311–313
Atomic motion, membrane
  transport and, 11

ATP
  Ca^{2+} movement in nerve,
    286–288
  coupling for cation affinity and
    occlusion-deocclusion,
    174–175
  energetic adequacy, 245
  energy transduction and, 6
  hydrolysis, 161–162, 190–197,
    245
  intracellular Ca^{2+}, 295–296
  isolated muscle proteins,
    188–189
  sodium-potassium pump, 155,
    166–167
  sodium transport and, 3, 349
ATP/ADP ratio, sodium transport
    and, 349
ATPase
  Mg^{2+}-activated, 157–158
  Na^+ and K^+ impact on, 158–159
  nerve membrane, 158
  sodium pump activity and, 155
ATP-induced viscosity, 190
Avogadro's number, 136
Axolotl, epithelial transport,
  341–342
Axon, node of Ranvier, 89–90

Bailey, Kenneth, 190
Band 3
  anion exchange, 203–228
  characteristics of, 211–212
  conductive anion fluxes, 218–219
  conformational changes in,
    223–226
  dimer structure, 222–223
  future research trends, 226–227
  helical structure, 222–223
  models of, 215–217
  multiple bilayer crossings,
    221–223
  protein, 10
  research retrospective, 227–228
  tertiary structure, 219–223
  topology alteration, 224–226
  as transporter, 223–226
  transport models, 214–219
Barcroft, Sir Joseph, 366
Barr, Lloyd, 103
Barton, Ted, 132

MEMBRANE
TRANSPORT

Bayliss, L. E., 366
Beckman flame photometer, 146
Bendall, J. R., 190
Benzer, Seymour, 306
Berger, Robert, 297
Bernard, Claude, 308–309
Bernstein, J., 255–256
Bertram, John, 314
Beta-chains
    amino acid sequence, 165–166
    molecular weight, 164–165
Bethe, Albrecht, 188
Bethe, Hans, 188
Bilayer membrane structure,
        104–105
    electrodiffusion and, 254–255
Bimodal probes, 214–219
Bimolecular leaflet structure in
        membranes, 32–33
    in myelin, 37
Binding
    amino acid transport, 241–242
    band 3 as transporter, 223–226
    calcium, 190–191
    conservative transport and, 9
    sodium-potassium pump,
        178–179
Binding energy curve
    field-strength theory, alterations
        in, 269
    Na^+ and K^+ with water molecules,
        278
    selectivity–field-strength theory,
        264–265, 268
Bioelectric potentials, 345
Biological membranes, 1–4
    flow and diffusion, 363–384
    historical background on
        research, 19–20
    as molecule diffusion barriers,
        15–46
Biophysical Laboratory, history of,
        127–128
Black-box phase, epithelial
        transport, 337–338
    two-membrane theory, 350–351
Black lipid membrane research,
        134–136
Blood flow, pre- and postcapillary
        resistances, 369–374
Bodenheimer, Tom, 98

Bohr effect, 365
Bolis, Liana, 109
Boltzmann distribution law, 252
Borrero, Luis, 376–377
Bott, Phyllis, 143
Branton, Daniel, 106–107
Brinley, F. J., 258
Bromine isotope (^{82}Br) water
        transport, 126
Bronk, Detlev, 368
Brooks, S. C., 275
Buchanan, Jack, 126
"Building stones" concept,
        352–353
Bulk water, water transport
        measurements, 135–136
Burton, Alan, 312

Cabantchik, Ioav, 207–208
$CaEGTA^{2-}$, 297–299
Calcium
    contractile protein, 191–192
    factor variability and, 191
    radioactive, 192–193
Ca^{2+} channels
    amino acid transport, 247–248
    cell connectivity and, 315–320
    cell-to-cell communication,
        323–330
    crystal-field requirements, 316
    EGTA kinetics, 297–298
    excitation-contraction coupling,
        292–293
    internal membrane system,
        294–295
    intracellular, 295–296
    kinetic measurements, 196–197
    membrane transport and, 9
    molecular mechanisms, 322–323
    movement in nerve, 286–288
    muscular contraction messenger,
        6
    perfused skinned fibers, 296–297
    permeability and, 29–30
    pump design and function, 158,
        187–199
    reverse operation, 198–199
    selectivity, close-fit hypothesis,
        262–263
    skinned muscle fibers and,
        293–294

Ca^{2+} channels—*continued*
  uncoupling action, 321–322
  uptake by sarcoplasmic
    reticulum, 299–300
Calcium-induced calcium release,
    298–299
Calcium loading, 193–194
Caldwell, P. C., 159–160
Calmodulin, 323
cAMP
  cell-to-cell channel
    communication and,
    323–324
  -dependent phosphorylation, 314
Cancerous cells, channel-deficient
    cells, 313–314
Capacitance values, 86–87
Capillary permeability, 364–365
Capillary pressure (P$_c$),
    isogravimetric, 369–374
Capnophorin. *See* Band 3 protein
Carbamates, facilitated transport,
    41–42
Carbamino transport, 366
Carbohydrates, presence on
    membrane surfaces, 76–78
Carbon-11, 126
Carbon-14, 126
^{14}C-inulin, 144
Carbon dioxide
  band 3-anion exchange, 205
  uncoupling and, 321–322
Cardiac glycoside
  sodium-potassium pump,
    175–179
  enzyme inhibition of, 161
Carrier model
  alternating gate model, 239–240
  band 3 anion, 216–217
  bimodal affinity, 215–216
Catalyzed permeability, 42–44
Catecholamines
  cell-to-cell channel
    communication, 324–325
  sodium-potassium pump
    inhibition, 176
Cat erythrocyte, swelling of, 27
Cation exchange
  coupling for ATP and occlusion-
    deocclusion, 174–175
  permeation, 212

sodium-potassium pump and, 161
  water transport and, 129–130
Cation-site interaction, selectivity–
    field-strength theory,
    264–267
cDNA
  band 3 research, 226–227
  cell-to-cell channels, 327–328
  sodium-potassium pump,
    165–166, 179
Cell communication channels,
    111–115
Cell connectivity, 303–330
  electrical transmissions, 305
  growth regulation and, 311–314
  mutual cellular control, 310
Cell economy, membrane transport
    and, 7–8
Cell loading, 41
Cells
  conductivity measurements,
    32–33
  electrical properties of, 252–259
  membrane structure, historical
    background on, 238–240
Cell-to-cell channel concept,
    305–309
  Ca^{2+} and, 307, 315–320
  single-channel events, 326–330
Cellular functions of membrane
    transport, 10
Cellular growth
  communication competence,
    311–315
  DNA replication and, 312
  growth control, 309–311
C fiber research, 61–63
cGMP (cyclic guanine
    mononucleotide), vision
    membrane transport, 6–7
Chambers, Robert, 17, 22–23
Channels
  transport proteins and, 3–4
  water-permeable, 134–135
Chelator-calcium combinations. *See*
    EDTA and EGTA
Chemical groupings principle,
    kinetic studies and, 24
Chemical reactions, sodium-
    potassium pump, 169–173
Chemical signaling, 5–6

Chironomus cell research, 306–307
  communication pathway
     regulation, 314–315
Chloride
  shift, 204–205
  shunt pathways, 354–356
  syncytium cells and, 356–358
Cholesterol dibromide, 126
Christensen, H. N., 237, 239
Cl^-
  action potentials in plant cells,
    284–285
  epithelial transport, 339
  flux-ratio equation, 344,
    355–356
  intestinal absorption and,
    141–142
  mitochondria-rich cells, 355
  muscle contraction, 291–292
  short-circuit technique, 346
  two-membrane theory, 350–351
Clarkson, Tom, 205–206
Clathrin cages, 89
Clausen, T., 162
Close-fit hypothesis of selectivity,
  262–264
"Club" nerve endings, 98–103
  oblique section, 102–103
Cole, K. S., 256, 261, 276
Collodion membranes, flow and
    diffusion measurements,
    373–374
Colloid osmotic hemolysis, 38–39
Communication channels
  amphipathic channel, 114
  regulation of, 314–315
Communication-deficient (channel-
    deficient) cells, 313–314
Conant, James Bryant, 126–127
Conductance factor, 251
Conductive anion fluxes, 218–219
Conformational transitions
  coupling with transport, 169–173
  sodium-potassium pump,
    166–167
  terminology of, 228
Conservative transport, 9
Constant-concentration concept,
  343–344
Constant-field assumption,
  257–259, 343–344

Contiguous cell transport, 8
Continuous unit membrane phase,
  87
Contractile protein
  calcium activity and, 191–192
  calcium sensitivity, 197–198
  enzymatic properties, 190–191
Cooperativity effects, amino acid
    transport, 247
Cornelius, F., 162
Cotransport (symport)
  models of, 246
  Na^+-linked, 242–243
  of organic substrates, 237–259
  transport proteins, 3
Countertransport (antiport),
  241–242, 248
  electrolyte ions, 243
  transport proteins, 3
Coupling
  cell-to-cell channel concept,
    317–318
  cellular growth and
    differentiation, 310–311
  excitation-contraction, 291–300
  sodium-potassium pump,
    169–173
  solute/solvent fluxes, 138–141
  tracer or antiport, 241–242
Covalent bonding, band 3–anion
    exchange, 207–208
Crab nerve enzyme, research on,
  158–161
Crayfish giant fiber synapses,
  53–59, 95–97
Creatine phosphokinase, 191–192
Creatinine clearance, 367
Criegee reaction, 84–85
Crystallization, sodium-potassium
    pump, 165
Cs^+
  affinity sequences for, 266–267
  excitable membrane selectivity,
    276–277
  permeability measurements, 278
Cu^{2+}, two-membrane theory,
  350–351
Curran, Peter, 141–142
Cyclotron, water transport
    research, 125–127
Cytoskeleton and membrane
    structure, 89

Czerlinski, George, 297

DAB (diaminobutyrate), amino acid
transport, 244
DADS (4,4'-diamino-2,2'-
stilbenedisulfonic acid), 208
Dahl, Gerhard, 324
Dainty, Jack, 134
Danielli-Davson cell membrane
model, 33–34, 86, 105–106,
238
further development of, 34–35
sodium-potassium pump,
156–157
Danielli, James, 29, 45–46, 85
Davson, Hugh, 85, 292
Dean, R. B., 155, 256–257
Debye, P., 254–255
Debye-Huckle theory, 345
Decane, anesthetic action of,
281–282
Deleze, Jean, 318
Deocclusion
coupling for cation affinity,
174–175
sodium-potassium pump (Na$^+$-K$^+$-
ATPase), 167–168
Deoxycholate (DOC), 163
Dephosphorylation, Na$^+$-K$^+$
exchange, 171
Deplasmolysis, 17
selective permeability and, 19
Descriptive morphology, 96–97
Detergents, sodium-potassium
pump and, 163–164
Deuterium-labeled amino acids,
340
D$_2$O, water transport, 125
de Vries, 18
Dewey, Maynard, 103
DIDS (4,4'-diisothiocyanostilbene-
2,2'-disulfonic acid),
207–210
anion identification, 210–212
band 3 binding, 218, 220
bimodal probes, 214–215
conductive anion flux exchanges,
219
ligand binding, 223–226
Diethyl pyrocarbonate, sodium-
potassium pump, 174–175

Difference factor, 76–77
Diffusion
channel size and, 134–135
membrane structure and, 2
restricted, 374–384
transcapillary concentration
difference, 377–379
water through porous
membranes, 381
Diffusional permeability (P$_d$)
continuous flow apparatus,
131–132
water transport in red cells,
131–133
Dihydro-DIDS, 208–210
Dimensionless ratio (P$_f$/P$_d$), 135
Diosmosis, 20
Direct cell interconnection,
309–310
Discovery, 127
Discretion about research, 69–70
Dissipative transport, 9
5,5'-Dithiobis(2-nitrobenzoic acid),
148
DNDS (4,4'-dinitro-2-2'-stilbene-
disulfonic acid, 223–224
Dobzhansky, Theodosius, 304
Dog erythrocyte, swelling of, 27
Doggenweiler, Carlos, 103
Donnan equilibrium, 27
Donnan potentials
electrodiffusion and, 258–259
ion asymmetry and, 255
Double-labeling experiments, 108
Drummond, Jack, 29
Durbin, R. P., 237

Ear oximeter, 368
Ecdysterone, gene activation with,
310
Eddington, Arthur, 313
Edema formation, 363–364
EDTA-Ca^{2+}, communication
pathways, 315
EDTA (ethylenediaminetetraacetic
acid)
gap junction structure, 111
perfused skinned fibers, 296–297
Eggleton, Grace, 367
EGTA-Ca^{2+}, communication
pathways, 315

EGTA (ethyleneglycol-
bis(2-aminoethylether)-
tetraacetic acid)
Ca^{2+} buffer system, 297
Ca^{2+}-dependent Na$^+$ efflux, 287
Ca^{2+} uptake by sarcoplasmic
reticulum, 300
kinetics, 297–298
perfused skinned fibers, 296–297
Ehrlich cells, amino acid transport,
240–242
Einstein relation, electrodiffusion,
252
Eisenmann, George, 261, 264,
279–280
Electrical conductivity of cells, 25
measurements of, 32–33
Electrical impulses, propagation of,
275–288
Electrical signaling, 5
Electrodiffusion in membranes,
251–259
Electroneutrality, 212–213
Electron microscopy
membrane characterization and,
1
role in membrane transport
research, 53
synapses, 53
Electron Microscopy Society of
America (EMSA), 60
Embryonic development, of myelin,
36–37
Endocytosis, 5
membrane fusion, 88–89
Endomatrix phase for general cell,
87–88
clathrin cages of, 89
Endosmosis terminology, 20
Energetic adequacy and amino acid
transport, 243–245
Energy barriers, band 3 anions, 216
Energy transduction, 6
Engleman, Don, 105
Enzymatic properties, contractile
proteins, 190–191
Enzyme-catalyzed permeability, 24,
43–44
Enzymes
alpha-beta structure, 165
band 3 tertiary structure,
220–221

E$_1$Na$_3$ affinity for ATP, 174–175
function, 159
sodium-potassium design,
158–159
transport activity and, 161–162
Epithelial cells, connectivity
research, 305–306
Epithelial tissue, molecular sieving
and, 381–384
Epithelial transport
bioelectric potentials, 345
black-box phase, 337–338
flux-ratio equation, 343–344
frog skin model, 337–358
ion-selective membrane
localization, 353–354
isotope tracers, 339–342
multipathway model, 358
solvent drag, 346–348
two-membrane theory, 349–3553
"Equilibrium selectivity," 267–268
Equivalent pore radius, water
transport across membranes,
135
Erythrocytes
aqueous channel locus, 147–149
bilayer structure in, 85–87
catalyzed permeability and,
42–44
channel size for water transport,
134
colloidal osmotic hemolysis,
37–38
conductive anion flux exchanges,
219
dimensions of, 32
hemolysis from narcotics and
heavy metals, 37–38
hemolysis reversal in, 39–40
osmotic exchanges, 25–29
sialic acid and, 224
swelling of, 26–27
triple-layered pattern, 70–71
water fluxes in, 128–138
Esmann, M., 162
Ethyl alcohol, facilitated transport,
41
Exchange diffusion, 242, 341–342
Excitable membrane, ion
selectivity, 275–277
Excitation-contraction coupling,
291–300

Exercise and sodium-potassium
       pump, 178
Exocytosis, 5
  membrane fusion, 88–89
Exomatrix phase of cell, 87
Exosmosis. *See* Plasmolysis
Experimental induction, 311

Facilitated diffusion, 24
Facilitated transport, 41–42
Faraday constant and
       electrodiffusion, 252
Farquhar, Marilyn, 306
FDNB (fluoro-2,4-dinitrobenzene),
       206–207
Fenn, Wallace, 275
Ferguson, J. K. W., 282–283, 366
Ferritin, band 3 binding and, 224
Fick's law of diffusion, 252
  diffusional flow, 134
  osmotic transient, 375–376
Field, Michael, 142
Field-strength theory of selectivity,
       264–272
  affinity sequences, 265–266
  crystal radius and hydration
       energy, 268–269
  modification with water molecule
       removal, 269–272
  site-cation interactions and,
       264–265
Filtration coefficient, isogravimetric
       capillary pressure, 373–374
Filtration constant
  cation fluxes, 131
  permeability coefficients, 364
Finean, Brian, 74–77
Fine-grain image intensifiers,
       320–321
Fischer, Ernst, 190
Fixed-charge diffusion concept,
       257–258
  anion exchange, 206–210
Flagg-Newton, Jean, 323
Flavorepleta research, 304
Fluid circuit theory of intestinal
       absorption, 141, 347
Fluid mosaic model, 107–108
Fluorescein, cell-to-cell channels
       and, 318–320

Fluorescence quenching, 134
Fluorescent-labeled
       polyaminoacids, 323
Flux measurement, transport
       reactions, 214
Flux-ratio analysis, 343–344
  epithelial transport, 337–338
  solvent drag and, 347–348
Folch-Lees proteolipid protein, 78
Force development records, 297
Fourier analysis, sodium-potassium
       pump, 164
Freeze-fracture technique
  membrane inhomogeneity and,
       44–45
  shadow-casting technique,
       106–107
Freites, Conrado, 321
Freksa, Friedrich, 189
Frenk, Samy, 103
Fricke, H., 32–33, 254
Friedmann, E., 126
Frog lung, calcium pump, 187–199
Frog skin model of epithelial
       transport, 337–358
  mitochondria-rich cells, 356–357
  short-circuit theory, 345–346
  suitability of, 342
  "syncytium model," 354–356
Furshpan, Edwin, 98, 305

Gaffey, Cornelius, 284
Gap junction, 98–103
  cell-to-cell channel concept,
       308–309, 327–330
  evolution of terminology, 104
  function of, 8
  model of, 45
  protein structure, 113–115
  structure of, 111–115
Gardos, G., 159
Gasser, Herbert, 61
Gauss's law of electrostatics,
       251–252
General cell structure, 87–89
Genes
  channel and pump synthesis, 3–4
  peptide control, 227
  translation and transcription of,
       12–13
Geren, Betty, 61–63, 65

German terminology for plant cell structure, 16–17

Ghosts
  calcium pump, 193–194
  cation permeability of, 40
  cell loading and, 41
  hemolysis reversal and, 39–40
  leaky, band 3–anion exchange, 211–212
  See also White ghosts

Giant fibers
  cells of origin, 55–56
  transection of, 56, 58
  synapse, postsynaptic process, 95–97

Gibbs-Donnan equilibrium, 38–39
Giebisch, Gerhard, 143
Glauret, Audrey, 70
Globulin X, 188–189
Glucose, diffusion of, 238
Glucose transport, glucose consumption and, 6
Glutaraldehyde carbohydrazide (GACH), 70–71
  gap junction structure, 114–115
Glycophorin, band 3 structure, 221–222
Glynn, Ian Michael, 177
Goldschmidt radii, 269
Golgi-Rezonnico spiral apparatus, 62–63
Gorman, A. L. F., 258
Gorter, E., 31–32, 86–87, 254
Gottesman, Michael, 324
Gouy, A., 254
Gradient hypothesis
  amino acid transport, 242–243
  secondary active transport, 237–238
Gramicidin
  anesthetic action, 283
  anion permeability, 212
Gramicidin A pores, selectivity, close-fit hypothesis, 263–264
Gray, Jim, 321
Gregersen, Magnus, 308–309
Gregg, Jimmy, 52
Gregg, John, 52
Grendel, F., 31–32, 86–87, 254
Grignard reaction, 131
Grundfest, Harry, 157

Gryns, 24

H[+]
  cell-to-cell channel communication, 321–322
  sodium-potassium pump (Na^+-K^+-ATPase), 175
[3]H radiation, 128
Haldane ratio, amino acid transport, 247–248
Hamburger, 18–19, 22
Hamburger shift, 22
Handler, Phil, 105
Hanenson, Irwin, 144
Hansen, O., 162
Hanson, Jean, 189
Harris, E. J., 158
Hasselbach-Schneider solutions, 189
Hasselbach, Wilhelm, career highlights, 187–188
Hastings, A. Baird, 126–127, 142–143
Heavy metal
  gap junction structure, 113
  membrane structure and, 37–38
Hedin, 24
Heilbrunn, 292–293
Heinz, E., 240
Helmholtz, H., 254
Helmreich, Ernst, 109
Hemoglobin cation fluxes, 130–131
Hemolysis
  in animal cells, 22
  colloid osmotic, 38–39
  osmotic exchange and, 26–27
  reversal of, 39–40
Heppel, L. A., 155
Hildebrand theory of solutions, 282–283
Hill, A. V., 69, 291, 293
Hille, 280
Hober, Rudolph, 25, 366
Hodgkin, A. L., 68, 129, 159–160, 261, 284, 366
Hodgkin-Huxley-Katz theory for action potential, 353
Hofer, 125
Hoffman-Berling, H., 192
Hurst, Donald, 126

Huxley, Andrew, 68–69, 129, 261, 284
Huxley, Hugh E., 189
Hydration
    alkali metal cations and, 261–262
    partial ion, 279–280
    selectivity–field-strength theory, 264–265
Hydraulic conductivity of red cell, 133–134
Hydrodynamic flow, 381
Hydrophobic barrier model of unit membrane, 115
Hypertension, sodium-potassium pump, 176

Ideas
    origins of and sharing of, 69–70
    See also Research methodology; Publication; Science
Impedance to diffusion concept, 379
Indicator dilution technique, 375–376
Individuality principle, cell connectivity and, 303
Influenza virus, 85–86
Information theory, cellular growth control, 309–310
Inhomogeneity of membranes, 44–45
Inotropic effect, sodium-potassium pump inhibition, 176
Integral protein concept, 78
Interionic electrostatic attraction, 255
Internal dialysis technique, 285–286
Internal membrane system, 294–295
Internal perfusion technique, 285–286
Intestinal absorption
    fluid circuit theory, 141–142
    water permeability and, 141–147
Intracellular calcium, 295–296
Intracellular membranes, structure and function, 7–8
"Intrinsic efficiency" concept, 244
Inward-facing membrane, 350–353
Ionic equilibrium theory, 27

Ionizing radiation, electrodiffusion in membranes, 256–257
Ionophores, 216
Ions
    membrane penetration by, 278–281
    movement through voltage, 6
Ion selectivity. See Selectivity
Isogravimetric capillary pressure, 369–374
Isogravimetric venous pressure, 371–372
Isotopic tracers
    epithelial transport, 339–342
    flux ratio and, 344
    intestinal absorption, 141
    membrane characterization and, 1
    water transport and, 125

Jacobs, Merkel, 25–27, 125
Jensen, J., 162
Jorgensen, P. L., 162–163
Journal of Biophysical and Biochemical Cytology, 66–67
Junctional pathway research, 309–312
    cell growth and, 314
Juxtaterminal myelin region, node of Ranvier, 91–93

K+ channels
    action potentials in plant cells, 284–285
    affinity ratios, field-strength theory, 270–272
    affinity sequences for, 266–267
    ATP support of erythrocyte accumulation, 159–160
    catalyzed permeability, 42–44
    cation fluxes, 130
    cell conductivity, 25
    colloidal osmotic hemolysis, 38–39
    epithelial transport, 338
    exchange with Na+, 170–171
    excitable membrane selectivity, 276–277
    gradient hypothesis, 242–243
    isotopic tracers for, 128

kidney absorption and, 146–147
leakage due to narcotics and
heavy metals, 37–38
permeability, 280–281
recovery processes in nerve,
285–286
red cell permeability and, 29–30
selectivity, close-fit hypothesis,
262–263
selectivity, field-strength theory,
264–272
sodium-potassium pump
transport, 168–169
two-membrane theory, 349–353
K$^+$-K$^+$ exchange, 173
KCl
erythrocyte swelling and, 27
syncytium cells, 356–358
KMnO$_4$ (permanganate) fixation
technique, 70–71
myelin research, 36–37
Kanno, Yoshinobu, 303–304
Kao, Eric, 157
Karsh, F., 189–190
Katchalsky, Aharo, 139, 306
Katz, Bernard, 59, 67, 285
Kedem-Katchalsky equations,
139–140
Kempton, Albert, 126
Kerick, Glenn, 324
Keynes, R. D., 158–160
Kidney, water permeability in,
141–147
Kinetic studies
affinity and velocity mechanisms,
245–246
band 3–anion exchange, 206
cell membrane, 24
of membrane structure, 37
Kistiakowsky, George B., 125–126
Klemperer, Friedrich, 126
Klodos, I., 162
"KM 21" (Morgenstern poem), 307
Knauf, Phillip, 204–206
Knoll's formula, erythrocyte
measurements, 31–32
Kolloidchemische standpunkt, 25,
27–29
Korn, Ed, 105
Kramer, Richard, 126
Kroger, Heinz, 310, 314–315

Krogh, August, 125, 339–342, 374
Kuffler, Stephen, 304–305,
329–330
Kuhn, Thomas, 107

Landis, E. M., 141, 364–365, 369
Lane, Alan, 127
Langmuir, I., 156, 256
Langmuir trough, 31–32
Lateral dendrite, in Mauthner cell,
98–99
Latham, Robert, 126
Lathanum tracer technique,
103–104
Lead, K$^+$ leakage and, 37
"Leaky" epithelia, 381–384
Leaky ghosts, band 3 tertiary
structure, 220–221
Leçons sur les Phénomènes de la Vie,
308
Lettvin, Jerry, 103–104, 307
Li$^+$
affinity ratios, field-strength
theory, 271–272
cell-to-cell channels and,
317–318
epithelial transport, 337–358
hydration rate, 262
selectivity–field-strength theory,
265–272
sodium pump functions and, 160
Ligand binding
DIDS and, 223–226
sodium-potassium pump, 164
Lipids
anesthetic solubility in, 156
band 3 conformational change,
225–226
molecular structure, 73–75
kinetic studies, 24
role in bilayer structure, 85
surface film studies, 31–32
Lipmann, Fritz, 197
Loewi, Otto, 316
Lullies, Hans, 187–188
Lundsgaard, E., 157

Macromeres, cell-to-cell channels,
325
Macromolecules
glomerular permeability,
382–383

Macromolecules—*continued*
  membrane and rod shapes, 4–5
  quaternary structure, 4
  tertiary structure, 4
"Macula communicans," 104
Maizels, M., 158
Makinose, Madoka, 192
Mallory, Kenneth, 53
Mallory, Tracy, 53
Mannitol, leakage in intestinal
  absorption, 142–143
Marmont, G., 276
Marmor, M. F., 258
Mass spectrometer, 128
Maunsbach, A., 162
Mauthner cell, electrical synapse,
  98–103
Maxwell, Clark, 32–33
McFarlane, Margaret, 79
Mean concentration difference,
  osmotic transient, 375–376
Mehta, Parmender, 314
Membrane-bound calcium transport
  systems, 197–199
Membrane fusion, 88–89
Membranes
  bilayer permeability, 2–3
  bimolecular leaflet structure,
    32–33
  cell structure and, 1–13
  chemical reactions in, 2
  composition and thickness, 2
  damage by hemolysis, 39–40
  electrodiffusion, 251–259
  excitable, ion selectivity,
    275–277
  fluidity concepts, 93
  foreign molecule occupancy,
    281–284
  heavy metals and narcotics,
    37–38
  inhomogeneity, 28–29, 44–45
  ionic permeability of, 22
  ion penetration of, 278–281
  models of, 280
  molecular individuality of, 88–89
  outward- and inward-facing,
    349–353
  renal glomerular, 381–384
  research trends in, 30–31
  thickness of, 31–32

  structure, 29–30
  universality of, 20–21
  unstirred layer as impedance
    barrier, 134
Membrane transport
  biology and, 1–4
  different locations for, 7–8
  different substances, 8–9
  future research trends in
    physiology, 10–13
  historical background, 238–240
  mechanisms for, 8
  membrane proteins, 9–10
  physiological role of, 4–7
  role in physiology, 1–13
Mersalyl, 148–149
Mesaxon, 61–63
  at juxtaterminal myelin region,
    node of Ranvier, 91–93
  triple-layered pattern, 71–74
Metabolic inhibition
  active transport, 240–241
  cell-to-cell channels and, 317
Metabolic pathways, substrate
  penetration, 238
Metabolism
  regulation of, 6
  sodium transport and, 348–349
  transport and, 341–342
Meyer-Overton hypothesis, 156
  ion selectivity, 281–284
Mg^{2+}
  ATPase activity, 157–158
  erythrocyte "ghosts," 40
  excitation-contraction coupling,
    300–301
Michaelis constant, 171
  amino acid transport, 247–248
Michaelis, Leonor, 366
Microcirculation, 363
Microcirculation research, 370–374
Microelectrode studies, two-
  membrane theory, 350–351
Microinjection techniques,
  excitation-contraction
  coupling, 292–294
Micropuncture technique
  capillary filtration, 364
  kidney absorption and, 143–144
Microsomes
  contractile protein, 191–192

structure of, 194–195
Microtubules and membrane fusion, 89
Millikan, G. A., 367–368
Millipore filtration techniques, 198–199, 239
Mitochondria, exomatrix and, 87–88
Mitochondria-rich cells
  chloride shunt pathway, 354–356
  epithelial transport, 339
Molecular biology, 96–97
Molecular mechanism of transport, 10–11
Molecular sieving, 381–384
Molecular structure, triple-layer structure, 10–11, 73–74
Molecular weight, sodium-potassium pump, 164–165
Molecule diffusion, biological membrane barriers, 15–46
Molecules, lipid and protein, 73–75
Mond, R., 206
Monotonicity
  affinity sequences and, 266
  selectivity and, 267–268
Moore, Francis, 128
Morales, Manuel, 190
Morgenstern, 307
Mosaic membranes, 31
Motor end plate research, 60
Motor giant fiber synapses, 55–59
Mott theory of copper–copper oxide rectifier, 256–257
MP26 protein, 324
mRNA, cell-to-cell channels, 327–328
Mullins, L. J., 258, 261–264
Multipathway models for epithelial transport, 358
Mulvany, M., 162
Murexide, intracellular calcium location, 295–296
Muscle, transverse tubule system, 63–70
Muscle contraction
  excitation-contraction coupling, 291–300
  sliding filament model, 292
Mutant I⁻, 324
Myelin

embryonic development of, 36–37
protofibers, 71–72
radially repeating unit, 73–75
spiral, 61–63
triple-layered pattern, 71–74
Myelin sheath
  molecular organization, 73, 75
  structure of, 34–36, 62–63
Myofibrils
  actomyosin ATPase, 192
  A-I junctions, 63–65
Myosin, actin, ATP and, 190–191

Na⁺ channels
  affinity ratios, field-strength theory, 270–272
  affinity sequence for, 266–267
  anesthetic action, 283
  bioelectric potentials, 345
  catalyzed permeability and, 42–44
  cation fluxes, 130
  channel permeability, 280–281
  energetic adequacy, 243–245
  epithelial transport, 337–358
  exchange with K⁺, 170–171
  excitable membrane selectivity, 276–277
  flux ratio analysis, 343–344
  gradient hypothesis, 242–243
  intestinal absorption and, 141–147
  isotopic tracers for, 128
  kidney absorption and, 146–147
  membrane transport and, 9
  metabolism and, 348–349
  modification of field-strength theory, 269–270
  muscle contraction, 291–292
  pore selectivity filter model, 280
  recovery processes in nerve, 285–286
  selectivity, close-fit hypothesis, 262–263
  selectivity–field-strength theory, 264–272
  short-circuit technique, 345–346
  sodium-potassium pump transport, 168–169
  two-membrane theory, 349–353

Na$^+$ channels—*continued*
   See also Sodium potassium pump
Na$^+$/Ca^{2+} ratio, 286–287
NaCl (salt) uptake
   animal cell structure and
      function, 22
   epithelial transport, 341–342
   solvent drag, 347
Na$^+$-K$^+$-2Cl$^-$ cotransport, 243–245
Nachmansohn, David, 157, 190
Nagai, T., 192
Nageli, on plasmolysis, 17
NAP-taurine, 208–209
   band 3–anion binding, 218
   as bimodal affinity probe, 215
   DIDS and, 223–224
Narcotics
   facilitated transport, 41–42
   membrane structure and, 37–38
Nature, 299
Nernst-Planck equation, 245
   electrodiffusion, 251–255,
      258–259
Nerve
   ionic calcium movement in,
      286–288
   recovery processes in, 285–286
Net flux
   NaCl concentrations and,
      144–145
   intestinal absorption and,
      142–143
   water transport and, 125
N-Ethylmaleimide (NEM), 148–149
Neural differentiation, 311
Neurotransmitters, vision
      membrane transport, 7
"Nexus" contact, 103–104
NH$_4$, channel permeability,
      280–281
Node of Ranvier, 88–93
   juxtaterminal myelin region,
      91–93
   longitudinal section, 93
*Nonequilibrium Thermodynamics in
      Biophysics*, 139
Nonpolarized synapses, 55, 57
Norby, J. G., 162
Nuclear magnetic resonance
      technique, 215
Nuclear membrane, 78–79

Occluding junction, 104
Occlusion
   of cations, 169–170
   coupling for cation affinity,
      174–175
   sodium-potassium pump (Na$^+$-K$^+$-
      ATPase), 167–168
Octanol, action of, 283
Ohm's law, electrodiffusion, 252
Oliveira-Castro, Gilberto, 318
Oncogenic viruses, channel-
      deficient cells and, 313–314
Onsager, L., 255
Organic chemical molecules, 281
Osmosis in plant cell, 18–19
Osmotic equilibria, 24–25
Osmotic exchange in erythrocyte,
      25–29
Osmotic flow measurements,
      diffusion and, 134–135
Osmotic pressure
   isogravimetric capillary pressure,
      373–374
   membrane flow and diffusion,
      363–384
   vs. van't Hoff pressure, 138
Osmotic reflection coefficients,
      378–381
Osmotic transients, 374–384
   pore area per unit path length,
      374–378
OsO$_3$, triple-layer molecular
      structure, 84–85
OsO$_4$, triple-layer molecular
      structure, 84–85
Osterhout, W. J. V., 285
Ottolenghi, P., 162
Ouabain
   cation fluxes and, 130
   sodium-potassium pump,
      160–161, 175–176
Outward-facing membrane,
      350–353
Oxalate
   calcium storage and, 194–195
   intracellular calcium, 295–296
Oxidation-reduction potentials, 345
Oxygen consumption, sodium
      transport and, 349

Paganelli, Charles, 130

Paired membrane structure, 65–67
Palade, George, 20, 60, 104, 306
Papain, band 3 structure, 221
Parker, Fred, 53
Parpart, A. K., 27
Parrafinic molecules, membrane
    occupancy, 281–284
Partial ionic hydration, 279–281
Partition coefficient, kinetic studies,
    24
Passow, Hermann, 205–206
Pauci-molecular theory, 78–79,
    85–86
Pauling crystal radii, 268–269
Pauling, Linus, 189
pCMB (p-chloromercuribenzoate),
    148
pCMBS (p-chloromercuri-
    benzenesulfonate),
    206–207
  inhibition of urea flux, 146–149
Penicillin production, 368
Peptides
  band 3, 221–222
  genetic control of, 227
Perfluoropentane, anesthetic action,
    281–282
Perfused hindlimb preparation
  capillary pressure research,
    367–368, 370–372
Perfused skinned fibers, 296–297
Perfusion fluid analysis, 142–143
Peripheral nerve research, 61
Permeability
  capillary, 364
  catalyzed, 42–44
  cell-to-cell channel, 308
  filtration constant, 364–365
  kolloidchemische Standpunkt and,
    27–29
  membrane structure and, 29–30
  of red cells to water, 130–131
  of water in intestine and kidney,
    141–147
Permeability coefficients, 43–44
  erythrocyte membranes vs. black
    lipid bilayers, 136–138
Permease-governed transport, 24
Permeases, catalyzed permeability
    and, 43–44
Pfeffer, 18–22

Pflanzenphysiologie, 18–19
pH
  anesthetic potency, 156–157
  cell-to-cell channels, 322
  Na+ permeability and, 43–44
  positively charged ligands, 224
Phage F-2 cell research, 306
Phosphate esters, sodium transport,
    159
Phosphoenzymes
  ATP-hydrolysis-dependent
    Na+-Na+ exchange, 171–172
  sodium-potassium pump
    (Na+-K+-ATPase), 166–168
Phospholipid layer, anesthetic
    action and, 282–283
Phosphorylation
  actomyosin and, 196–197
  cAMP-dependent, 314
  Na+-K+ exchange, 170–171
Phosphorylation-
    dephosphorylation, 166–168
Photomultipliers, cell-to-cell
    communication, 320–321
Photosynthesis and ATP synthesis,
    6
Physical Chemistry of Cells and
    Tissues, The, 107–108.
Physics and Chemistry of Surfaces,
    156
Physikalische Chemie der Zelle und
    der Gewebe, 25
Physiology
  future research trends in
    membrane transport, 10–13
  membrane transport and, 1–13
Ping-Pong mechanism, 213–214
  band 3 anion, 216–217
  kinetic experiments, 218
Pinocytotic transport, 20
Planck, Max, 252, 257
Plant cells, 284–285
  membrane permeability, 134
  osmosis, 18–19
  plasmolysis, 17
  structure and functions, 15–21
Plant slime. See Protoplasmic layer
Plasma membrane
  in plant cell, 17
  protein in, 33–34
  structure and function, 7–8

Plasma membrane—*continued*
water movement across, 125–149
Plasmolysis
penetration rate and, 254
plant cell, 17
MEMBRANE   terminology, 20
TRANSPORT   Platzman, Robert, 278
Plesner, L., 162
Poiseuille's law
channel flow measurements,
134–135
transcapillary concentration
difference, 377–379
Poisson equation, 251–252,
257–258
Polarization optics, 34–36
triple-layer myelin structure,
73–74
Polarized synapse, 55, 57
Polar substance, rate of transport,
2–3
Politoff, Alberto, 317
Pomerat, Charles, 51–52
Ponder, Eric, 39
Pore area per unit path length,
374–384
Pore size, reflection coefficients
and, 380–381
Porous membrane diffusion, 381
Porter, Keith, 51, 68–69
Post, Robert Licheley, 157, 177
Postsynaptic process, 95–97
Potential difference (PD), amino
acid transport, 244–245
Potter, David, 98, 304–305
Precipitation membrane, osmosis
and, 17
Pressure plethysmograph, 364
Primordial sac. *See* Protoplasmic
layer
Pronase, anion transport, 219–220
Prostaglandin, cell-to-cell channel
communication, 324–325
Protein kinase C, 323–324
Protein membrane, 21–22
Proteins
allosteric reactions, 365
cell-to-cell channels, 324–325
labeling experiments, 107
membrane transport through,
9–10

molecular structure, 73–78
role in cell structure, 33–34
transmembrane receptor, 89
Proteolytic enzymes, band 3–anion
exchange, 211–212,
219–221
Protonation, cell-to-cell channels,
322
Protoplasmic layer (plant cell),
15–16
Proximal tubule, kidney absorption
mechanism and, 143–145
Publication, hazards and obstacles
of, 69–70, 130, 285–285
Ca^{2+}-induced Ca^{2+} release, 299
importance of, 65–67
Pump-lung perfusion system,
369–370
Pumps
transport proteins and, 3–4
See also specific types of pumps
Pyridoxal 5-phosphate (PDP)
band 3 binding, 217–218
sodium-potassium pump
(Na^+-K^+-ATPase), 174–175

Radioactive calcium, 192–193
Radioactive decay, water transport
research, 125–127
Ramon y Cajal doctrine
cell connectivity, 305
electron microscopy confirmation
of, 59
Rapid reaction method, cation
fluxes, 130–131
Rawlins, Bessie, 367
Rb^+
affinity sequences for, 266–267
excitable membrane selectivity,
276–277
permeability measurements, 278
Recovery processes in nerve,
285–286
Reflection coefficients
osmotic, 378–381
solute/solvent flux coupling,
138–141
Reflection coefficient for urea,
140–141
Reger, James, 60

Regulatory genes, cell-to-cell channels, 327–328
Rein, Hermann, 367–368
Relaxing-factor activity, 191–192
  calcium-sensitive, 197–198
Renal metabolism, 367–368
Renkin, Gene, 140, 376–377
Replication and growth regulation, 312
Research, directions and obstacles in, 63–67, 69–70, 84–85, 237, 203–204
  on biological membranes, 19–20
  discretion about sharing, 69–70
  cell-connectivity, 303–305, 307–308
  excitation-contraction coupling, 292–293
  future trends in band 3 research, 226–227
  "heroic era," 261
  isogravimetric capillary pressure, 373–374
  retrospective on band 3 research, 227–228
  sodium-potassium pump research, 162–163, 177–179
Respiration, ATP synthesis, 6
Respiratory electrons, sodium transport and, 349
Restricted diffusion, 374–384
  osmotic reflection coefficients, 378–381
Restricted pore area, 379
Reynolds, George, 320, 321
Rhodopsin, membrane transport and, 6
Richards, A. N., 142–143, 368
Ringer's solution
  ion behavior in muscle contraction, 291–292
  limits of, 40
  solvent drag, 348
RNA
  membrane molecular structure, 110–111
  See also mRNA
Robertson, J. D.
  background and training, 51–53
  Harvard period, 98–103
  London period, 67–115

Robinson, Charles, 131
Rose, Birgit, 317–320
Rothschild, Victor, 86–87
Rutherford (Lord), 125–126

Sandow, Alexander, 197
Sarcoplasmic reticulum
  calcium pump and, 197–199
  calcium uptake, 299–300
  internal membrane system, 295
  microsome structure, 194–195
Schatzmann, Hans, 144, 160–161, 177
Schiff-base reaction, band 3 binding, 218
Schmidt, C. L. A., 155
Schmidt-Lanterman cleft, 94–96
Schou, Mogens, 160
Schramm, Gerhard, 189
Schulmann, J. H., 156
Schwann cells
  C fibers in, 61–62
  in crayfish, 54
  myelin sheath and, 36
  node of Ranvier and, 89–91
  Schmidt-Lanterman cleft, 94–96
Schwannian nodal collar, 90–91
Schwartz, Arnold, 161
Schwarz, Ferdinand, 321
Science, 299
Scientific ideas
  evolution of, 69–70
  obstacles to, 129–130, 203–204
  sodium-potassium pump research and, 162–163
  S-shaped curve analogy, 129
Scolar, Sidney, 308
SDS–acrylamide-gel electrophoresis, 210–212
Secondary active transport, 237–259
Secretion and membrane transport, 4–5
Selectivity, 261–288
  action potentials in plant cells, 284–285
  anesthetic action, 281–284
  close-fit hypothesis, 262–264
  electrical impulse propagation, 275–288

Selectivity—*continued*
  epithelial transport, 353–354
  equilibrium, 267–268
  excitable membrane, 275–277
  field-strength theory, 264–272
  foreign molecule occupancy of
    membranes, 281–284

MEMBRANE
TRANSPORT
  localization of, 353–354
  recovery process in nerve,
    285–286
Self-diffusion of water, 135–136
Semipermeability concept,
  255–256
Septal synapse, 55–56
  transection of, 56, 58
Sequence data, band 3 research,
  226
Shadow-casting techniques,
  106–107
Shipp, Joe, 144
Short-circuiting technique
  epithelial transport, 337–338
  frog skin, 345–346
  two-membrane theory, 352–353
Sidel, Victor, 133
Siekewicz, Peter, 197
Sieve effect, membrane structure
  and, 31
Signaling, 5–6
  cell connectivity and, 314–315
Silver, cell permeability and, 37
Simultaneous mechanism, 214
Singer, John, 107–108
SITS (4-acetamido-4′-
    isothiocyanatostilbene-2,2′-
    disulfonic acid), 207
Skinned muscle fiber
  calcium activation and, 293–294
  internal membrane system,
    294–295
  perfused, 296–297
Skou, Jens Christian, 177
Skriver, E., 162
Sliding filament model of muscle
  contraction, 292
Slimy layer. *See* Protoplasmic layer
Smectic fluid crystal, 79–81
Smith, Homer, 16
Smith, Peter, 354
"Sniperscope," 320–321
Snow, C. P., 127
"Soap bubble" analogy of cell
  membranes, 85–86

Sodium borohydride, 217–218
Sodium dodecyl sulfate
    (SDS)–acrylamide gels
  band 3–anion exchange, 208–209
  sodium-potassium pump, 164
Sodium-potassium pump (Na^+-K^+-
    ATPase), 155–179, 193–194
  ADP-dependent Na^+-Na^+
    exchange, 171–172
  amino acid sequence, 165–166
  ATP and conformational
    transitions, 166–167
  ATP coupling for cation affinity
    and occlusion-deocclusion,
    174–175
  ATP-hydrolysis-dependent
    Na^+-Na^+ exchange, 171–172
  consecutive vs. simultaneous
    reaction, 173–174
  crystallization, 165
  energetic adequacy, 245
  importance of, 177–179
  inhibitors, 175–177
  mitochondria-rich cells, 357
  molecular weight, 164–165
  Na^+-K^+ exchange, 168–171, 173
  pathway to, 156–162
  purification, 163–164
  sodium transport and, 349
  two-membrane theory, 351–353
Sodium transport, solute movement
    and, 5
Sollner, Karl, 138–141
Solomon, A. K., 237
Solute
  coupling with solvent fluxes,
    138–141
  primacy of, 5
  solvent flow and, 347–348
Solvent drag, 346–348
Solvent fluxes, coupling with
    solutes, 138–141
Space constant of layer, 254
Spectrophotometric technique, 131
Spinous bracelet, 93
Spiral myelin, 61–63
Sponge cells, cell-to-cell channels,
    325
Spring, Kenneth, 356
Stage, David, 98
Starling hypothesis, 364
  extracellular fluid research,
    38–39

Starling-Verney heart-lung-kidney
preparation, 366
Staverman, 379–380
Steck, Ted, 211
Steinbach, H. B., 155
Steinberg, Daniel, 128–129
Stephenson, Elizabeth, 300–301
Steroids, gene control and, 310
Stoeckenius, Walther, 82–84, 105
Stoker, Michael, 314
Stokes-Einstein coefficient,
134–136
Stopped-flow microperfusion,
144–145
Straub's actin, 189
Striated muscle, epithelial
transport, 340–342
*Structure of Scientific Revolution,
The*, 107
Sulfate, two-membrane theory,
350–351
Sulfhydryl, anion exchange,
206–207
Surface pressure, sodium-potassium
pump and, 156–157
Symport, amino acid transport, 242
Synapse, crayfish, 95–97
Synaptic disk, 100–101
structure, 99–104
See also Gap junction
"Syncytium model"
chloride and, 356–358
frog skin epithelium and,
354–356

Temperature, catalyzed
permeability and, 42–44
Terminology
confusions about, 84–85
sodium-potassium pump,
160–161
Territorial differentiation, 311–312
Thermodynamics
anesthetic effect and, 281–284
chemical, 282–283
electrodiffusion in, 251
reflection coefficient and,
138–141
Thiocyanate, cell permeability of,
28–29
Thiourea and solvent drag, 348
Thyroid hormone, sodium-
potassium pump, 178

Tight junction concept, 104
Tl$^+$, permeability measurements,
278–281
Tobacco mosaic virus, molecular
construction, 110–111
Tosteson, Dan, 104
Tracer coupling, 241–242
Transcapillary concentration
difference, 377–379
Transinhibition, 241
Translocation, bimodal affinity, 215
Translocator species, amino acid
transport, 246–247
Transmembrane channels
gap junction proteins and,
111–112
molecular construction, 109–110
Transmembrane-protein concept, 8,
107–109
receptor proteins, 89
Transport
band 3 as, 223–226
coupling of, 169–173
enzyme activity and, 161–162
facilitated, 41–42
intracellular, 8
inward-facing, 225–226
outside-facing, 225–226
pathways, 125–149
secondary active, 237–259
Transport models
bimodal probes for band 3
binding, 214–219
topology alteration, 224–226
Transport proteins
channel-forming, 3–4
characteristics of, 2–3
distribution in membranes, 3
future research on, 11
gene transcription and
translation, 12
pump-forming, 3–4
series and parallel connections,
11–13
Transport site, asymmetry and, 215
Transstimulation, 241
Transverse tubule system in muscle,
67–70
Traube, 18
Triethyl-tin as carrier, 216
Triple-layered structure, 78–87
molecular interpretations of,
79–82

Tritium, cation fluxes, 131
Trypsinization of band 3, 220–221
Two-factor theory of excitation, 275–277
Two-membrane theory, epithelial transport, 338–339, 349–353

Uncoupling
  Ca^{2+} channels, 315
  cell-to-cell channels and, 318–320
  Na^+ efflux and K^+-K^+ exchange, 173
Unidirectional flux concept, 343–344
Unitarian Service Committee, 189–190
Unit membrane concept, 70–87, 105
  decline of concept, 108
  evolution of, 78–79
  GACH techniques for, 114–115
  general cell, 87–89
  hydrophobic barrier model, 114–115
  nonlipid monolayer, 77–78
  vs. theory, 82–83
Universality of membrane, 20–21
University of Aarhus, 156
"Unnatural inducers," 311
Unwin, Nigel, 111–112
Urea transport vs. water transport, 146–149
Urethane, action of, 283–284
Urey, Harold C., 125
Ussing, H. H., 155, 258

Vacuole, formation of, 21
Valinomycin, 216
  anion permeability, 212
  conductive anion flux exchanges, 219
Vanadate, sodium-potassium pump inhibitors, 176–177
van't Hoff, Jacobus, 363, 376–377, 379–380
Vasoconstriction, sodium-potassium pump, 176
Velocity-type mechanism, 245–248
Vennesland, Birgit, 126
Vesicle system, band 3–anion exchange, 211–212

Vision, membrane transport and, 6–7
Visscher, Maurice, 141, 347
Voltage-clamp technique, isogravimetric capillary pressure, 373–374
von Hevesy, 125
von Mohl, 15–16

Water
  capillary permeability, 380–381
  fluxes in red cells, 128–138
  gas phase of transport, 131–133
  membrane transport and, 8–9
  modification of field-strength theory with, 269–272
  movement across plasma membranes, 125–149
  permeability in intestine and kidney, 141–147
  porous membrane diffusion, 381
  self-diffusion of, 135–136
  transport vs. urea transport, 146–149
Weber, Annemarie, 191–192, 197, 294
Weber-Edsall solution, 189
Weber, Hans Hermann, 188–189
Weiner, Eric, 324
Werthessen, Nick, 126
White ghosts, 149
Whittembury, Guillermo, 144
Why Smash Atoms, 127
Wild-type cells, 324
Wilkie, D. R., 291–292
Windhager, Erich, 144, 353–354
Winton, F. R., 366–368
Woods Hole Laboratory, 316, 328–330
  scientific atmosphere at, 157
Wright, Ernie, 142
Wyman, Jeffries, 365

X-ray crystallography
  cell membrane structure, 34–36
  molecular structure and, 11

Young, J. Z., 67, 90–91

Zampighi, Guido, 111–112
Zero-time point technique, 131–133
  reflection coefficient and, 140
Z line structure in muscle fibers, 68–69

[414]